# Hydrogen and Fuel Cells:

# Emerging technologies and applications

# Hydrogen and Fuel Cells

## Emerging technologies and applications

A volume in the "Sustainable World" series

Bent Sørensen

*Roskilde University*
*Energy & Environment Group, Institute 2,*
*Universitetsvej 1, P O Box 260*
*DK-4000 Roskilde, Denmark*

ELSEVIER
ACADEMIC
PRESS

Amsterdam  Boston  Heidelberg  London  New York  Oxford
Paris  San Diego  San Francisco  Singapore  Sydney  Tokyo

# Elsevier Academic Press

30 Corporate Drive, Suite 400, Burlington, MA 01803, USA
525 B Street, Suite 1900, San Diego, California 92101-4495, USA
84 Theobald's Road, London WC1X 8RR, UK

**This book is printed on acid-free paper.** ∞

**Library of Congress Cataloging-in-Publication Data**

Application submitted

**British Library Cataloguing in Publication Data**

A catalogue record for this book is available from the British Library

ISBN: 0-12-655281-9

For all information on all Elsevier Academic Press publications
visit our Web site at www.books.elsevier.com

Layout and print-ready electronic-medium manuscript by author
Printed in the United States of America
04   05   06   07   08   09      9   8   7   6   5   4   3   2   1

# Preface

These years, many scientists and engineers move into the field of hydrogen and fuel cells because it is exciting and well funded. The aim is to transform the way energy is delivered and used over the coming years, with major changes in technologies for production, distribution and conversion, as a response to political instability of many oil-producing countries, uncertainty about resources and increasing concerns over environmental effects.

This book is written to the many newcomers to the field, to students at the increasing number of courses given in the subject and to well-established scientists and developers who already have in-depth knowledge in certain sub-fields but like to keep informed about the entire field from technology to policy considerations, economical and environmental assessment. I aim to provide an introduction to people with general science background, but no special hydrogen and fuel cell experience, and also to give an up-to-date account of the frontiers of research and development to readers who need to be able to connect the emerging terminology to the concepts of conventional fields of science.

At the end of each chapter are problems and discussion topics, several of which can be used for problem-oriented mini-projects.

In fact, the pace of knowledge development in the field of hydrogen and fuel cells is so rapid that about half the content of this book is based on material less than a year old (as of the time of writing) and often not even published or found in the "in press" basket of the scientific journal publishers. It has been a pleasure to work with this extremely fresh material from the desks of my colleagues. New techniques have enabled specific investigations to be performed along with the book writing that would have been impossible 5-10 years ago.

To achieve the aim stated above I try to avoid specialist jargon or, if it is important, to define and explain the special terminology that the reader will meet in the latest scientific journals, providing the connection back to the concepts familiar to people with a physics, chemistry or biology background. The policy planner and assessor will similarly find the newest ideas and methods in these fields linked back to conventional economy and to environmental and planning sciences.

First of all, I want to convey the richness of the hydrogen and fuel cell fields and to present the challenges calling for a dedicated effort by the cream of human ingenuity, which is you, dear reader.

*Bent Sørensen*
*Gilleleje, October 2004*
*boson@ruc.dk*

# Contents

# LIST OF CONTENTS

# LIST OF CONTENTS

# Units and conversion factors

## Powers of 10[□]

| Prefix | Symbol | Value | Prefix | Symbol | Value |
|--------|--------|-------|--------|--------|-------|
| atto | a | $10^{-18}$ | kilo | k | $10^3$ |
| femto | f | $10^{-15}$ | mega | M | $10^6$ |
| pico | p | $10^{-12}$ | giga | G | $10^9$ |
| nano | n | $10^{-9}$ | tera | T | $10^{12}$ |
| micro | μ | $10^{-6}$ | peta | P | $10^{15}$ |
| milli | m | $10^{-3}$ | exa | E | $10^{18}$ |

## SI units

| Basic unit | Name | Symbol |
|-----------|------|--------|
| length | metre | m |
| mass | kilogram | kg |
| time | second | s |
| electric current | ampere | A |
| temperature | Kelvin | K |
| luminous intensity | candela | cd |
| plane angle | radian | rad |
| solid angle | steradian | sr |
| amount[#] | mole | mol |

| Derived unit | Name | Symbol | Definition |
|-------------|------|--------|-----------|
| energy | joule | J | $kg\,m^2\,s^{-2}$ |
| power | watt | W | $J\,s^{-1}$ |
| force | newton | N | $J\,m^{-1}$ |
| electric charge | coulomb | C | $A\,s$ |
| potential difference | volt | V | $J\,A^{-1}\,s^{-1}$ |
| pressure | pascal | Pa | $N\,m^{-2}$ |
| electric resistance | ohm | Ω | $V\,A^{-1}$ |
| electric capacitance | farad | F | $A\,s\,V^{-1}$ |
| magnetic flux | weber | Wb | $V\,s$ |
| inductance | henry | H | $V\,s\,A^{-1}$ |
| magnetic flux density | tesla | T | $V\,s\,m^{-2}$ |
| luminous flux | lumen | lm | $cd\,sr$ |
| illumination | lux | lx | $cd\,sr\,m^{-2}$ |
| frequency | hertz | Hz | $cycle\,s^{-1}$ |

---

[□] G, T, P, E are called milliard, billion, billiard, trillion in Europe, but billion, trillion, quadrillion, quintillion in the USA. M as million is universal.

[#] The amount containing as many particles as there are atoms in 0.012 kg $^{12}$C.

## Conversion factors

| Type | Name | Symbol | Approximate value |
|------|------|--------|-------------------|
| energy | electon volt | eV | $1.6021 \times 10^{-19}$ J |
| energy | erg | erg | $10^{-7}$ J (exact) |
| energy | calorie (thermochemical) | cal | 4.184 J |
| energy | British thermal unit | Btu | 1055.06 J |
| energy | Q | Q | $10^{18}$ Btu (exact) |
| energy | quad | q | $10^{15}$ Btu (exact) |
| energy | tons oil equivalent | toe | $4.19 \times 10^{10}$ J |
| energy | barrels oil equivalent | bbl | $5.74 \times 10^{9}$ J |
| energy | tons coal equivalent | tce | $2.93 \times 10^{10}$ J |
| energy | $m^3$ of natural gas | | $3.4 \times 10^{7}$ J |
| energy | kg of methane | | $6.13 \times 10^{7}$ J |
| energy | $m^3$ of biogas | | $2.3 \times 10^{7}$ J |
| energy | litre of gasoline | | $3.29 \times 10^{7}$ J |
| energy | kg of gasoline | | $4.38 \times 10^{7}$ J |
| energy | litre of diesel oil | | $3.59 \times 10^{7}$ J |
| energy | kg of diesel oil/gasoil | | $4.27 \times 10^{7}$ J |
| energy | $m^3$ of hydrogen at 1 atm | | $1.0 \times 10^{7}$ J |
| energy | kg of hydrogen | | $1.2 \times 10^{8}$ J |
| energy | kilowatthour | kWh | $3.6 \times 10^{6}$ J |
| power | horsepower | hp | 745.7 W |
| power | kWh per year | kWh/y | 0.114 W |
| radioactivity | curie | Ci | $3.7 \times 10^{8}$ $s^{-1}$ |
| radioactivity | becqerel | Bq | $1\ s^{-1}$ |
| radiation dose | rad | rad | $10^{-2}$ J $kg^{-1}$ |
| radiation dose | gray | Gy | J $kg^{-1}$ |
| dose equivalent | rem | rem | $10^{-2}$ J $kg^{-1}$ |
| dose equivalent | sievert | Sv | J $kg^{-1}$ |
| temperature | degree Celsius | °C | K – 273.15 |
| temperature | degree Fahrenheit | °F | 9/5 C+ 32 |
| time | minute | min | 60 s (exact) |
| time | hour | h | 3600 s (exact) |
| time | year | y | 8760 h |

*continued next page*

# UNITS

| Type | Name | Symbol | Approximate value |
|------|------|--------|-------------------|
| pressure | atmosphere | atm | $1.013 \times 10^5$ Pa |
| pressure | bar | bar | $10^5$ Pa |
| pressure | pounds per square inch | psi | 6890 Pa |
| mass | ton (metric) | t | $10^3$ kg |
| mass | pound | lb | 0.4536 kg |
| mass | ounce | oz | 0.02835 kg |
| length | Ångström | Å | $10^{-10}$ m |
| length | inch | in | 0.0254 m |
| length | foot | ft | 0.3048 m |
| length | mile (statute) | mi | 1609 m |
| volume | litre | l | $10^{-3}$ m$^3$ |
| volume | gallon (US) | | $3.785 \times 10^{-3}$ m$^3$ |

# INTRODUCTION

## 1.1 The current relevance of fuel cells and hydrogen

The pollution from motor cars is, particularly in city areas, becoming increasingly unacceptable to people living in, visiting or working in the cities of the world. Demands for zero-emission vehicles have been voiced, and the automobile industry is facing louder and louder criticism for not addressing the problem. The simplest solution to reducing emissions is to make the vehicle more efficient. This is the route taken by several European car manufacturers, using a combination of low-weight car structure, low air resistance, high performance but low fuel consumption engines such as the common-rail diesel engine, brake energy recuperation, computer-optimised gearshift operation, engine close down as an alternative to idling, and so on. Presently, this leads to fuel consumption for a four-person standard car of about 3 litres of diesel fuel per 100 km or 4-5 litres of gasoline if the less efficient Otto engines are used. Three litres of diesel fuel per 100 km corresponds to 0.1 GJ or 1 MJ km$^{-1}$. The fuel-to-wheel efficiency is about 60% higher than that of current average passenger cars (27% rather than 17%), while the overall efficiency of fuel-to-transportation work (i.e., the transportation service delivered to the end-user, as e.g. measured by the number of passengers times the distance driven, cf. section 6.2) is as much as around 230% (2.3 times) better than that of present average cars (12-13 km per litre of gasoline, the energy content of which is about 10% lower than that of diesel fuel). The average efficiency of current-stock passenger cars is lowest in the USA and highest in Europe, a fairly direct consequence of the prevailing fuel prices seen by the customers (i.e., including subsidies and taxes).

Other options include electric and fuel cell-based vehicles. If the original fuel is of fossil origin, environmental pollution depends of the total well-to-transportation service energy use. For electric vehicles, assuming a motor efficiency of 80%, a transmission efficiency of 98% and a 40% fuel-to-electricity conversion efficiency at current state-of-the-art power plants, the well-to-wheel efficiency calculated after consideration of battery-cycle losses is 26%. For fuel cell cars, the hydrogen fuel-to-wheel efficiency is about 36% (see section 6.2), implying a fuel-to-wheel efficiency of around 25% for the chain starting from hydrogen production from natural gas, over proton exchange membrane (PEM) fuel cells and electric motors to wheels, all for a standard mixed driving cycle.

In the electric car case, pollution is moved from street level to the locations of power plants, which normally allow for better exhaust gas cleaning and in any case dispersal at higher altitudes that rarely leads to pollutant concentrations as high as for currently common vehicle exhaust. It follows from these estimates that the higher cost of electric or fuel cell vehicle solutions will be difficult to defend, relative to the simple efficiency improvements of thermodynamical engines using conventional fuels, and the real case for electric or fuel cell vehicles is therefore fully dependent on a transition to non-fossil fuels. Of course, this does not exclude that fossil fuel-based fuel cell vehicles could serve as demonstration projects during an interim period, where the infrastructure for hydrogen production based on sustainable energy sources is not yet available.

If one considers hydrogen produced from renewable energy such as wind power and takes the efficiency of state-of-the-art electrolysis plants as being near 80%, this also leads to an overall primary energy-to-wheel efficiency of about 25% for a fuel cell vehicle, but this time without emission of pollutants. For vehicles operated purely on batteries, the corresponding efficiency is around 50%, but the weight penalty of battery operation is usually higher than that of fuel cell operation, also if this includes a reformer and some battery storage. If both batteries and fuel cell technologies were economically viable, the optimisation would consist of balancing the weight and required engine rating to obtain the lowest cost. However, none of the two technologies is economical today, and a central question for the role of fuel cells is whether cost reductions will be easier to achieve than for batteries, despite a certain level of technology similarity.

While direct pollution from motor vehicles, whether involving particles, $SO_2$ or $NO_x$ emissions only partially reduced by filters and catalytic devices, is the most visible social (health) and environmental impact, emissions of greenhouse gases have increasingly become included in the list of impacts not to be tolerated in the future. The reason is the implied additional warming of the atmosphere and the negative consequences it may have, both regarding the stability of overall climate and also in particularly vulnerable regional contexts. Presently, climatic impacts seem to top the list of reasons to move away from fossil fuels. A few decades ago, supply security and resource depletion were quoted as the key reasons to develop alternatives based upon sustainable energy sources. The temporary decline in such worries was due to the halt of the exponential growth in energy use, achieved after the oil supply crises in 1973 and 1979 by concerted efforts to improve efficiencies of energy use. The most advanced industrialised nations have not increased their total energy used during the past three decades, despite continued economic growth. However, the transportation sector has in many parts of the world continued to increase its energy use, only compensated for by less usage of fuels for space heating and other sectors seemingly more amenable to efficiency improvements than cars, ships and aircraft. As the motor car example given above shows, there are no technical reasons for not improving the efficiency of vehicles for transportation, and indeed, catching up in this sector finally seems to be forthcoming.

It is clear that the resource issue is not gone permanently. Production of fossil fuels, or at least of oil, is expected to reach its maximum sometime in the next decade or two and then to decline despite enhanced recovery techniques, as the rate of discovering new wells continues to diminish. This will necessarily lead to price increases, although likely in an irregular way, due to the fact that the actual production costs are still very low in some regions, notably in the Middle East, and prices are therefore dependent on the day-to-day market situation and on political issues such as cartel ceilings on production and warfare in the regions of production. These conditions make the development of alternatives to use of fossil fuels increasingly more attractive, both for economic reasons and also for reasons of security of supply. Because renewable energy sources are more evenly available geographically (although the best mix of renewable sources may vary from one region to another), they are seen as attractive in scenarios placing emphasis on local control, often referred to as "decentralisation", but certainly do not eliminate the need for transmission and trade of electricity and (renewable-based) fuels or for developing suitable forms of energy storage (see Sørensen, 2004a).

The reasoning presented above will be qualified by the detailed discussion offered in the following chapters.

The possibility of using hydrogen as a general energy carrier has long been recognised (see, e.g., Sørensen, 1975, 1983, 1999; Sørensen et al., 2004).

The issues raised for attention include hydrogen production, storage and transmission, as well as the use of hydrogen, notably as fuel for fuel cells. The current hope is that fuel cells will experience a significant price reduction along with the development of new fields of application and that infrastructure problems will eventually be solved. This could happen through a series of steps, with hydrogen first being used in those niche areas where the required change in infrastructure is modest, such as fuel cell buses travelling by fixed routes from a single filling station. The present price of producing hydrogen fuel (whether from fossil or renewable energy) is higher than that of the fuels already in use, but is expected to decline if the market expands and production technology is refined. The cost of central hydrogen storage in underground facilities such as those already in use for natural gas will have a rather insignificant impact on the overall cost of hydrogen usage, while that of transmission is expected to be similar to or slightly higher than the cost of natural gas transmission. Local hydrogen storage costs, e.g. using pressure containers, are not negligible but still have a fairly modest influence on overall cost, while the critical cost item remains the fuel cell converter used in all cases where the end-use energy form is electricity, including traction through electric motors. Current fuel cell costs are way above direct cost viability, and progress in fuel cell manufacture, performance and durability are thus the critical development items for allowing the penetration of hydrogen as a general energy carrier.

The disposition of the book is as follows: hydrogen production by a long list of technical or biological systems, storage and transmission in Chapter 2, fuel cell basics in Chapter 3, fuel cell systems in Chapter 4, followed by implementation issues (including safety and norms) and scenarios for future use in Chapter 5, economic issues such as direct and life-cycle costs in Chapter 6 and rounding up in Chapter 7. The distinctions are not watertight, as it is often useful to mention systems options in connection with individual technologies or to mention implementation issues along with the technologies, but cross-references are provided.

**CHAPTER**

**2**

# HYDROGEN

## 2.1 Production of hydrogen

Hydrogen is the most abundant element in the universe. The main isotope consists of one proton and one electron occupying the lowest angular momentum zero atomic state (i.e., the electron ground state, denoted $1s$) at an energy of $-2.18 \times 10^{-18}$ J, relative to the electron being at infinity. Hydrogen is found as interstellar gas and as the chief constituent of main-sequence stars. On planets such as the Earth, hydrogen is found as part of the molecules of water, methane and organic material, whether fresh or fossilised. The average isotope ratio $^2$H to $^1$H in the natural abundance on Earth is $1.5 \times 10^{-5}$. The normal molecular form is $H_2$. Further properties are listed in Table 2.1.

By hydrogen production I mean extracting and isolating hydrogen in the form of independent molecules, at the level of purity required for a given application. The processes naturally depend on the starting point, and the currently dominant scheme of production from methane only makes sense if the energy is initially contained in methane or can easily be transferred to methane. Thus in the case of fossil fuels, the transformation of natural gas into hydrogen is relatively easy and that of oil is a little bit more elaborate, while transformation of coal requires an initial step of high-temperature gasification. For energy already transformed to electricity, electrolysis is presently the most common process for hydrogen production, although there are other methods that may become preferred in the future. Among the renewable energy sources, biomass sources require special attention, depending on their form. Photo-induced or direct thermal decomposition of water at high temperatures may be considered, while more complex, multi-step schemes are required at lower temperatures, such as those offered by steam from nuclear reactors.

| | | |
|---|---|---|
| Atomic number, H | 1 | |
| Electron binding (ionisation) energy in $1s$ ground state | 2.18 | aJ |
| Molar mass, $H_2$ | 2.016 | $10^{-3}$ kg mol$^{-1}$ |
| Average distance between nucleons, $H_2$ | 0.074 | nm |
| Dissociation energy, $H_2$ to 2H at infinite separation | 0.71 | aJ |
| Ionic conductance of diluted $H^+$ ions in water at 298 K | 0.035 | m$^2$ mol$^{-1}$ Ω$^{-1}$ |
| Density, $H_2$, at 101.33 kPa and 298 K | 0.084 | kg m$^{-3}$ |
| Melting point at 101.33 kPa | 13.8 | K |
| Boiling point at 101.33 kPa | 20.3 | K |
| Heat capacity at constant pressure and 298 K | 14.3 | kJ K$^{-1}$ kg$^{-1}$ |
| Solubility in water at 101.33 kPa and 298 K | 0.019 | m$^3$ m$^{-3}$ |

*Table 2.1.* Properties of hydrogen [a].

## 2.1.1 Steam reforming

Current industrial production of hydrogen starts from methane, $CH_4$, which is the main constituent of natural gas. A mixture of methane and water vapour at elevated temperature is undergoing the strongly endothermic reaction,

$$CH_4 + H_2O \rightarrow CO + 3H_2 - \Delta H^0, \tag{2.1}$$

where the enthalpy change $\Delta H^0$ equals 252.3 kJ mol$^{-1}$ at ambient pressure (0.1 MPa) and temperature (298 K), and 206.2 kJ mol$^{-1}$ if the input water is already in gas form. The carbon monoxide and hydrogen mixture on the right hand side of (2.1) is called "synthesis gas". This step requires a catalyst (nickel or more complex nickel on aluminium oxide, cobalt, alkali and rare earth mixtures) and typically is made to proceed at a temperature of about 850°C and a pressure of around $2.5 \times 10^6$ Pa.

The process is controlled by design of the reactor used for the reforming process, by input mixture (typical steam/methane ratios are 2 to 3, i.e. above the stoichiometric requirement) and, as mentioned, by reaction temperature and catalysts. Other reactions may take place in the reformer, such as the in-

---

[a] For units, prefixes and conversion factors see front matter, pp. xii-xiv.

verse of (2.1). In order to obtain a high conversion efficiency, some heat inputs are taken from cooling the reactants, and from the heat outputs derived from the subsequent water-gas "shift reaction" (WGS reaction), usually taking place in a separate reactor,

$$CO + H_2O \rightarrow CO_2 + H_2 - \Delta H^0, \tag{2.2}$$

with $\Delta H^0$ equal to $-41.1$ kJ/mol when all reactants are in gas form at ambient pressure and temperature, and $-5.0$ kJ/mol if the water is liquid. Heat is recovered and recycled back to the first reaction (2.1). This involves two heat exchangers and is the main reason for the high cost of producing hydrogen by steam reforming. The use of a steam/methane ratio above unity is necessary in order to avoid carbon (char) and excess CO formation (Oh *et al.*, 2003).

Industrial steam reformers typically use direct combustion of a fraction of the primary methane (although other heat sources could of course be used) to provide the heat required for the process (2.1),

$$CH_4 + 2O_2 \rightarrow CO_2 + 2H_2O - \Delta H^0, \tag{2.3}$$

where $\Delta H^0 = -802.4(g)$ or $-894.7(l)$ kJ/mol, (g) for gaseous end products and (l) for liquid, condensed water (the $CO_2$ is supposed to stay gaseous at the relevant process temperatures and pressures). The heat evolution with products in gas form, 802.4 kJ/mol, is called the "lower heat value" of methane, while the one including heat of condensation to liquid form is called the "upper heat value". Only a modest fraction of the heat required for (2.1) is obtained from recycling. The reactions (2.1) and (2.2) combine to

$$CH_4 + 2H_2O + \Delta H^0 \rightarrow CO_2 + 4H_2, \tag{2.4}$$

with $\Delta H^0 = 165$ kJ/mol for gaseous water input and $\Delta H^0 = 257.3$ kJ/mol for liquid water input. As a chemical energy conversion process, (2.4) has an ideal efficiency of 100%: the sum of the burning value for $CH_4$, 894.7 kJ/mol ((2.3) with gaseous $CO_2$ but all steam condensed) and the required heat input to the process (2.4), 257.3 kJ/mol, exactly equals the 1152 kJ/mol burning value of $4H_2$ (again steam condensed). The enthalpies depend on temperature and pressure, but considering the process as starting and ending at ambient temperature and pressure, then the theoretical efficiency does not change by additional heating and pressurising, as the energy spent in these processes could be regained at the end. In practice, only some 50% of the energy from combustion heat input contributes to the chemical process (2.1), and the remaining heat gets transferred to the product stream and is eventually recaptured and recycled, adding further to the capacity requirement for heat exchangers (Joensen and Rostrup-Nielsen, 2002). Kinetic effects influencing the catalyst action of the shift reaction (2.2) have been studied by Ovesen *et al.* (1996). The actual efficiency of industrial steam reforming of methane is rarely above 80%. As mentioned, the need for heat recycling is a dominating cost item reflected in the hydrogen product cost. If there is sulphur contamination, e.g. by $H_2S$, of the input methane used for (2.1) above the limit set by the effect on the integrity of the catalysts, a sulphur removal step must precede the synthesis gas step.

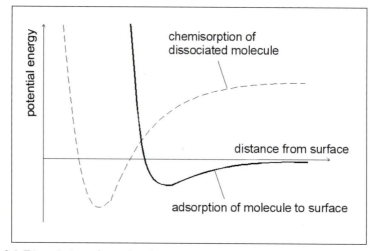

*Figure 2.1.* Dissociation of a molecule at a catalyst surface may require enough kinetic energy to overcome a potential energy barrier. Without this extra energy the molecule just gets adsorbed to the surface (Lennard-Jones, 1932).

The role of the catalyst, say a simple nickel metal surface with periodic lattice structure, is to adsorb methane molecules and step-by-step strip them of hydrogen atoms, ready for the following conversion steps. The basic mechanism of the surface dissociation was already suggested by Lennard-Jones (1932), who postulated the presence of a double-minimum potential energy as a function of the distance between surface and impinging molecule, as illustrated in Fig. 2.1. The outer minimum is due to molecular adsorption by the long-range Coulomb forces (sometimes referred to as van der Waals forces), while the inner minimum is due to dissociative chemisorption. The barrier between the two minima may have a height exceeding the potential energy of the molecule at large distances, in which case the molecule must possess enough kinetic energy to be able to penetrate the barrier (Ceyer, 1990). The thermal motion associated with elevating the process temperature achieves this. Clearly, the height of the potential barrier depends on both surface material and impinging molecule, and thus allows engineering by selection of catalyst. New catalysts with increased levels of hydrogen production have been brought into use, from the traditional Ni catalysts on $Al_2O_3$ (Yokota *et al.*, 2000) over $Ni/ZrO_2$ (Choudhary *et al.*, 2002) to a recent $Ni/Ce-ZrO_2$ catalyst, which yields 15% more hydrogen than implied by the reaction (2.1), because the excess steam used in the first reaction step contributes hydrogen production by the shift reaction (2.2), even in the first process reactor (Roh *et al.*, 2002).

The elevated temperatures used in the reactor may damage the catalysts. Of particular importance is the possible carbon formation by methane cracking:

$$CH_4 + 74.9 \text{ kJ/mol} \rightarrow C + 2H_2. \tag{2.5}$$

This carbon may appear as filaments on the catalyst surface, encapsulating catalyst atoms and largely blocking the steam reforming reaction (Clarke *et al.*, 1997). As mentioned above, excess steam is effective in preventing cracking or at least diminishing the negative effects by cleaning the catalyst surfaces.

The shift reaction (2.2) employs additional sets of catalysts, traditionally Fe or Cr oxides, but in newer plants with staged-temperature processing these would be used only at the first step at around 400°C, while new catalysts such as $Cu/ZnO/Al_2O_3$ are used for a second low-temperature step, following heat removal for the recycling mentioned above. Carbon contamination at the catalysts can proceed via the Boudouard reaction (Basile *et al.*, 2001),

$$2CO \rightarrow C + CO_2 + 172.4 \text{ kJ/mol}. \tag{2.6}$$

New catalysts are being sought, with higher stability against such effects. The final $CO_2$ removal is, on the other hand, a fairly simple process that can be achieved by water sprinkling. Current natural gas steam reforming in large industrial plants is not aimed at fuel cell use of the hydrogen produced, and the CO contamination remaining after the shift reaction stays in the range of 0.3 to 3% (Ghenciu, 2002; Ladebeck and Wagner, 2003).

For use in proton exchange membrane (PEM) fuel cells (see section 3.6), the CO contamination in the hydrogen produced must be below 50 ppm (parts per million). This is due to the poisoning limit of typical platinum catalysts used in PEM cells. The implication is the need for a final CO cleaning treatment, unless the main reaction steps (2.1) and (2.2) can be controlled so accurately that all reactants are accounted for. This CO cleaning stage may involve one of the following three techniques: preferential oxidation, methanation (the reverse of (2.1)) or membrane separation (Ghenciu, 2002). The process currently in most widespread use is partial oxidation,

$$CO + \tfrac{1}{2}O_2 \rightarrow CO_2 + 283.0 \text{ kJ/mol}, \qquad (2.7)$$

where "preferential" means that catalysts are applied, which favour this reaction over the corresponding hydrogen oxidation,

$$H_2 + \tfrac{1}{2}O_2 \rightarrow H_2O - \Delta H^0, \quad \Delta H^0 = -242(g) \text{ or } -288(l) \text{ kJ/mol}, \qquad (2.8)$$

(enthalpies at ambient temperature and pressure; in (2.7) $CO_2$ is assumed to be gaseous). Catalysts currently used are based on Pt or platinum alloys. They require high surface areas and low temperature (80-200°C), with smaller being better (Shore and Farrauto, 2003).

A way to avoid some of the problems mentioned while maintaining a high efficiency is to continually remove the hydrogen produced from the second reactor. This can be done by using a membrane allowing hydrogen molecules to pass, but none of the other reagents. A *membrane reactor* can replace both reactors of the conventional steam reforming process in achieving the overall reaction (2.4). It may consist of a tubular reactor with packed catalysts (Pd catalysts have been employed in several pilot studies) adjacent to the membrane and inlet flow channel. The removal of hydrogen produced will shift the thermodynamical equilibrium of (2.1) and (2.2) towards the product side, allowing a lowering of reaction temperature and use of less excess steam (Kikuchi *et al.*, 2000; Gallucci *et al.*, 2004). A cleaning step has, in some cases, been needed, but CO contamination can both theoretically and in practice be kept as low as 0.001%, as recently demonstrated (Yasuda *et al.*, 2004; at a total efficiency of 70%). The reactor may be small in size relative to a comparable, conventional steam reforming reactor, and the efficiency of conversion higher, at least in the laboratory experiments performed.

Reforming hydrocarbons other than methane, notably oil products, can be achieved with reactions similar to (2.1):

$$C_nH_m + n\,H_2O \rightarrow n\,CO + (n+\tfrac{1}{2}m)\,H_2 - \Delta H^0. \qquad (2.9)$$

The successive breaking of C-C terminals of higher hydrocarbons with suitable catalysts is more involved than for methane, due to differences in reaction rate and increased problems with thermal cracking (in analogy to (2.5), also called pyrolysis). To avoid this problem, the carbon stripping is usually done in a separate pre-reformer (Joensen and Rostrup-Nielsen, 2002), making this way of producing hydrogen more complex and more expensive than natural gas reforming. It is also possible to employ new types of catalysts (such as Pd/ceria) in order to reform specific higher hydrocarbons with acceptable catalyst integrity (Wang and Gorte, 2002).

## 2.1.2 Partial oxidation, autothermal and dry reforming

The moderately exothermic catalytic partial oxidation process for methane,

$$CH_4 + \tfrac{1}{2} O_2 \rightarrow CO + 2H_2 + 35.7 \text{ kJ/mol,} \qquad (2.10)$$

or more generally

$$C_nH_m + \tfrac{1}{2} n \, O_2 \rightarrow n \, CO + \tfrac{1}{2} m \, H_2 - \Delta H^0, \qquad (2.11)$$

is considerably more rapid than steam reforming. When oxygen (usually in air) and methane are passed over a suitable catalyst (e.g. $Ni/SiO_2$), the reaction (2.10) occurs, but to a degree also full oxidation (2.3) and product oxidation ((2.7) and (2.8)), opening up for further processes such as methanation (the reverse of (2.1)), the shift reaction (2.2) and dry reforming (see (2.14) below). When oxygen is supplied through air, nitrogen has to be removed from the hydrogen product. This usually takes place in a separate stage following the oxidation reactor. Partial oxidation is suitable for small-scale conversion, such as in a motor vehicle with fuel cells. The process can be stopped and started, as required for on-board reformation, and when in progress, it provides elevated temperatures that may start steam reforming along with the oxidation processes. This is called "autothermal" reforming and involves all the reactions mentioned so far, plus stoichiometric variations on (2.10) in the possible presence of water,

$$CH_4 + \tfrac{1}{2} x \, O_2 + (1 - x) \, H_2O \rightarrow CO + (3 - x) \, H_2 - \Delta H^0, \qquad (2.12)$$

written in a way suitable for $x < 1$, which need not be fulfilled, as the following special case shows:

$$CH_4 + \frac{3}{2} O_2 \rightarrow CO + 2H_2O - \Delta H^0, \quad \Delta H^0 = 611.7(l) \text{ or } 519.3(g) \text{ kJ/mol.} \quad (2.13)$$

The use of air is convenient for automotive applications, but if the autothermal scheme were to be used for large industrial plants, dedicated oxygen production for the feed would be preferable, e.g. due to concerns over the volume of $N_2$ to be handled and the associated size of heat exchangers (Rostrup-Nielsen, 2000).

Hydrogen production by partial oxidation from methane increases with process temperature, but reaches a plateau value at around 1000 K (Fukada *et al.*, 2004). The theoretical efficiency is similar to that of conventional steam reforming, but less water is required (Lutz *et al.*, 2004).

Gasoline and other higher hydrocarbons may be converted to hydrogen on board cars by the autothermal processes, using suitable catalysts (Ghenciu, 2002; Ayabe *et al.*, 2003; Semelsberger *et al.*, 2004). Partial oxidation may also be combined with the palladium-catalyst membrane reactors mentioned in section 2.1.1 (Basile *et al.*, 2001).

As an alternative to conventional steam reforming, methane could be reformed in a stream of carbon dioxide rather than steam of water,

$$CH_4 + CO_2 \rightarrow 2CO + 2H_2 + 247.3 \text{ kJ/mol}, \tag{2.14}$$

again followed by the shift reaction (2.2). Advantages of this reaction could be the disposal of $CO_2$ and the possibility of operating at fairly low temperatures, for example in combination with conventional steam reforming (Abashar, 2004).

## 2.1.3  Water electrolysis: reverse fuel cell operation

The conversion of electric energy into hydrogen (and oxygen) by water electrolysis has been known for a long time (demonstrated by Faraday in 1820 and widely used since about 1890), but if the electricity is produced by use of fossil fuels, then the cost of hydrogen obtained in this way is higher than the one associated with steam reforming of natural gas. On the other hand, high hydrogen purity needed in some applications is easier to achieve, and for this reason electrolysis hydrogen currently has a market share of about 5%. As the cost of electricity dominates the cost of hydrogen production from electrolysis, the situation is quite different, if the electricity used is surplus electricity from variable resources such as wind or solar radiation. In cases where such electricity is produced at times of no local demand and no evident option for export to other regions, the value of this electricity may be seen as zero. This makes it very favourable to store the energy for later use, and storage in the form of hydrogen is one option that may appeal, provided that uses of hydrogen or regeneration of electricity are economically attractive in a specific situation.

Conventional electrolysis uses an aqueous alkaline electrolyte, e.g. KOH or NaOH at a weight percentage of around 30, with the positive and negative electrode areas separated by a microporous diaphragm (replacing earlier asbestos diaphragms). The overall reaction at the positive electrode[a] (e.g. Ni or Fe) is

$$H_2O \rightarrow \tfrac{1}{2}O_2 + 2H^+ + 2e^-, \qquad (2.15)$$

where electrons are leaving the cell by way of the external circuit (Fig. 2.2) and where the three products may be formed by a two-step process,

$$2H_2O \rightarrow 2HO^- + 2H^+ \rightarrow H_2O + \tfrac{1}{2}O_2 + 2e^- + 2H^+. \qquad (2.16)$$

The reaction at the negative electrode is

$$2H^+ + 2e^- \rightarrow H_2, \qquad (2.17)$$

grabbing electrons from the external circuit. The hydrogen ions are to be transported through the electrolyte by the electric potential difference $V$. The role of the alkaline component is to improve on the poor ion conductivity of water, favouring use of KOH. However, this limits process temperature to values below 100°C, in order to avoid strong increases in alkali corrosion of electrodes. The overall reaction is the inverse of (2.8),

$$H_2O - \Delta H^0 \rightarrow H_2 + \tfrac{1}{2}O_2, \qquad \Delta H^0 = -242\ (g)/288\ (l)\ kJ/mol \qquad (2.18)$$

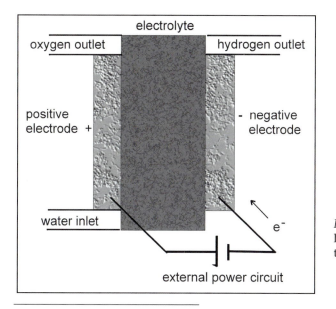

*Figure 2.2.* Schematic layout of a water electrolysis cell.

---

[a] See section 3.1 for a discussion of the names *anode* and *cathode* sometimes used in connection with the two electrodes of an electrochemical system.

$$\Delta H = \Delta G + T \Delta S. \tag{2.19}$$

At ambient pressure and temperature (298 K), the change in enthalpy and free energy for liquid water is $\Delta H = -288$ kJ mol$^{-1}$ and $\Delta G = 236$ kJ mol$^{-1}$. The electrolysis process thus requires a minimum amount of electric (high-quality) energy of 236 kJ mol$^{-1}$, while the difference between enthalpy and free energy changes, $\Delta H - \Delta G$, in theory could be heat taken from the surroundings. Since the apparent conversion efficiency would then be $\Delta H/\Delta G$, as $-\Delta H$ is also the (upper) heating value of $H_2$, it could theoretically exceed 100% by as much as 22%. However, with use of ambient heat at 25°C, the process would be very slow. Temperatures used in classical electrolysers are around 80°C, and in some cases active cooling has to be applied. The much lower conversion efficiency (50-70%) obtained with simple electrolysers in practical cases is largely a consequence of electrode "overvoltage", mainly stemming from polarisation effects. The cell potential $V$ for water electrolysis may be expressed by

$$V = V_r + V_a + V_c + Rj, \tag{2.20}$$

where $V_r$ is the reversible cell potential,

$$V_r = -\Delta G/(z\mathscr{F}) = 1.22 \text{ V}, \tag{2.21}$$

using the free energy change above, the Faraday constant $\mathscr{F} = 96\ 493$ C mol$^{-1}$, and the number $z$ of electrons involved in the reaction, here (2.15). The three next terms in (2.20), which constitute the overvoltage, have been divided into negative ("anode") and positive ("cathode") electrode parts $V_a$ and $V_c$, and the resistance contribution. The current is $j$ and $R$ is the internal resistance of the cell. The three last terms in (2.20) represent electrical losses, and the voltage efficiency $\eta_V$ of an electrolyser operating at a current $j$ is defined by

$$\eta_V = V_r / V. \tag{2.22}$$

Efforts are being made to increase the efficiency to 80% or higher. One method is to increase the operating temperature, typically to above 1500°C, and to optimise electrode design and catalyst choice.

The alkaline electrolysers described here are one type of fuel cells (described as such in section 3.4). Although a detailed discussion of the quantum chemical processes involved in such devices is the subject of Chapter 3, some explanation will be given here of the mechanisms responsible for gas dissociation and the electrode losses appearing in (2.20). How do the water molecules and ions of the electrolyte behave near the electrode surfaces, or on the surface of any catalysts placed on (or near) the electrodes in order to speed up the transfers? Experimental methods such as tunnelling electron microscopy (TEM) and a range of spectral measurements allow the establishment of visual impressions of molecular structure unthinkable to early constructors of electrochemical devices. Furthermore, recent advances in semi-classical and quantum chemistry allow insights into the mechanisms of electron transfer on a level of detail that was unheard of until the 1990s.

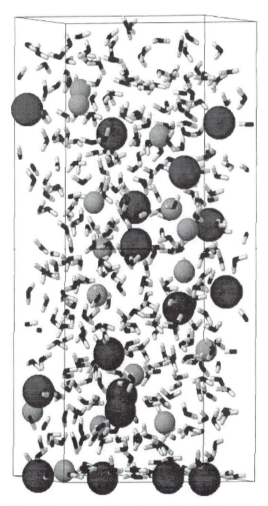

*Figure 2.3.* Sodium chloride solution near a metal surface (at bottom). Single time-step results of molecular dynamics simulation for a positive surface charge, From E. Spohr (1999). Molecular simulation of the electrochemical double layer. *Electrochimica Acta* **44**, 1697-1705, reproduced with permission from Elsevier.

Consider a metal surface (electrode or catalyst) with a regular lattice atomic structure. As an electrode in an electrolyser, the external voltage applied to the system causes a build-up of charges on the metal surface. In the classical description of an electrochemical device, charges of opposite sign, derived from ions in the electrolyte, are drawn towards the electrodes to form a double layer of opposite charges. At the positively charged electrode, a layer of negative ions builds up in the electrolyte, along the electrode surface. This is then the environment in which the water splitting takes place. To model the structure in this double-layer region, molecular mechanics has been used.

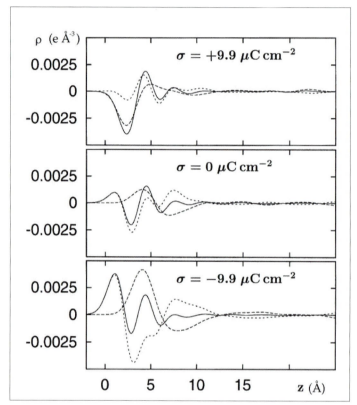

*Figure 2.4.* Charge density as function of distance from electrode, for the electrolyte modelled in Fig. 2.3. The three cases represent positive, zero and negative electrode charge. Solid lines are total charge densities, longer dashes are ionic charge densities, and shorter dashes are aqueous charge densities. The curves have been smoothed. From E. Spohr (1999). Molecular simulation of the electrochemical double layer. *Electrochimica Acta* **44**, 1697-1705, reproduced with permission from Elsevier.

"Molecular mechanics" is a classical mechanics approximation to a description of the motion of atomic or molecular centres, neglecting electrons entirely except for keeping track of charges, but retaining the possibility of a non-symmetric shape of each molecule, that may be characterised by its orientation in space. The molecules are assumed to interact via a heuristic force, given by parametrised potential functions such as the one depicted in Fig. 2.1. Figure 2.3 shows an example (frozen in time) of the size of problems that can be treated in this way: there are 400 water molecules and 32 ions (Na$^+$ or Cl$^-$) in a rectangular space with one charged side representing the electrode. The figure shows that the calculation provides patterns that look realistic. Averaging over the total simulation time of a few nanoseconds, the charge density variations as a function of distance from the electrode can be established, as shown in Fig. 2.4. The three cases shown are results of calculations for a positively, zero and negatively charged electrode. It is seen that as expected the Cl$^-$ ions accumulate near a positively charged electrode and the Na$^+$ ions accumulate near a negative charged electrode, but, in addition, the water molecules have become polarised so that the total charge as a function of distance from the electrode is oscillating as a result of interference between the ionic and water-related charges. The model is of course limited by its use of a phenomenological short-range interaction, which is the same for water and ions, by the molecular dynamics calculation and by not treating the specific quantum reactions that may take place at the electrode surface (Spohr, 1999).

In an electrolyser, the charged double layers (or more complex oscillatory charge densities) exist at both electrodes, and the ions and water molecules in the electrolyte have a thermal distribution of kinetic energies, depending on the operating temperature. The kinetic energy causes impact of the reactants on the electrodes, which again may change the rate of reactions. A quantum mechanical description thus entails first the determination of electron density distributions and potential energies along possible paths of reaction, and then a time-dependent solution of the Schrödinger equation or a semi-classical approximation to it, for reactant molecules in each of the states possible within their thermal distribution. These can be determined by finding the possible vibrational excited states of each molecule involved.

Figure 2.5 shows the potential energy surface of the simpler reaction (2.17) at the negative electrode, as a function of the distance $z$ of the hydrogen molecule from the surface and the distance $d$ between the two hydrogen atoms. Quantum calculations have been performed as a function of all the six hydrogen co-ordinates (three for each atom), and it has been found that a hydrogen-pair position parallel to the surface has the lowest energy. Near the surface, centres of mass for the hydrogen molecule may approach top surface metal atoms or the "holes" between them. In case of a platinum negative electrode, the first possibility seems to provide the easiest reaction path at larger distances (Pallassana *et al.*, 1999; Horch *et al.*, 1999), but once separated, the hydrogen atoms appear with their centre of mass at the "hole" site, i.e. the face centre in case of the cubic (fcc) Pt lattice pattern with Miller indices (111)[a]. In Fig. 2.5, the approach of the hydrogen pair is along a vertical line from the fcc position for a Ni(111) lattice.

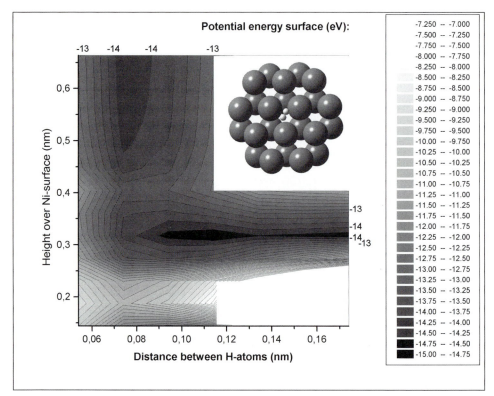

*Figure 2.5.* Potential energy surface in eV along two co-ordinates for hydrogen dissociation on a Ni catalyst surface. The abscissa is the distance $d$ between the two hydrogen atoms, and the ordinate is the distance $z$ of the hydrogen pair from the second layer of the Ni surface. The pair is assumed parallel to the surface. The quantum chemical calculation of the potential energies employs the density functional theory B3LYP (see section 3.2) and a set of 508 basis functions (SV) for 2×337 electrons in 26 atoms including two layers of Ni atoms and the two hydrogen atoms, positioned as shown in the top-view insert (Sørensen, 2004b).

Figure 2.5 shows that the hydrogen molecule may approach the surface at a fixed separation of $d = 0.074$ nm (equal to the experimentally determined separation in an isolated hydrogen molecule) down to a distance of about $z =$

---

[a] The three Miller indices are the reciprocal of the displacements along the three major axes defining the unit cell of a lattice structure (i.e. the distance to the position of the corresponding atom of the next unit cell in each direction). Cf. most any solid state physics textbook, e.g. Kittel (1971).

0.5 nm from the lower surface. There, it needs to pass through a saddle point in the potential energy (which is possible due to the vibrational energy associated with a finite temperature, kinetic energy or quantum tunnelling) in order to reach the potential energy valley associated with the dissociated state at $z = 0.317$ nm. A steeply rising potential energy prevents the hydrogen atoms from getting closer to the surface, although there are variations in the potential as a function of the position of the Ni atoms. The calculated saddle point (transition structure) barrier height is about 0.90 eV as seen from the hydrogen molecular state, and $d$ is only slightly increased. The final energy minimum once the hydrogen atoms have been separated is about 0.33 eV below the initial one. The separation between the hydrogen atoms at this minimum is around $d = 0.1$ nm, but the energy stays flat for increasing $d$.

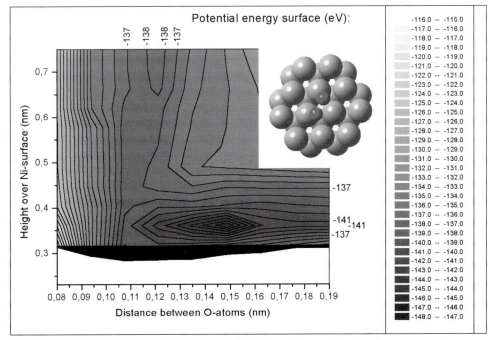

*Figure 2.6.* Potential energy surface in eV along two co-ordinates for oxygen dissociation on a Ni catalyst surface. The abscissa is the distance $d$ between the two oxygen atoms, and the ordinate is the distance $z$ of the oxygen pair from the 2nd layer of the Ni surface. The pair is assumed parallel to the surface. The quantum chemical calculation of the potential energies employ Hartree-Fock for the Ni atoms and adds correlations from density functional theory (B3LYP, see Section 3.2) for the oxygen atoms, a set of 508 basis functions (SV) for 2×344 electrons in 26 atoms including two layers of Ni atoms and the two oxygen atoms, positioned as shown in the top-view insert. Like in Fig. 2.5, the absolute value of the energy scale is taken arbitrarily, so that only potential energy differences matter (Sørensen, 2004b).

At the positive electrode, the pair of oxygen atoms seem to prefer separating over a face centre of a Pt-lattice, but the published quantum calculations are not able to reproduce the experimental energies well (Eichler and Hafner, 1997; Sljivancanin and Hammer, 2002). The precise nature of the surface is important, and the occurrence of steps in the crystal structure termination at the surface can greatly enhance the dissociation of, e.g., oxygen molecules on a platinum surface (Stipe *et al.*, 1997; Gambardella *et al.*, 2001). Similar effects have also been found for hydrogen adsorption (Kratzer *et al.*, 1998). Figure 2.6 gives the potential energy surface for two O-atoms over a Ni surface, along co-ordinates similar to those of Fig. 2.5. Dissociation at the surface is here clearly demonstrated. Further discussion of the quantum calculations for oxygen formation or dissociation taking place during water electrolysis or fuel cell operation will be given in sections 3.1 and 3.2.

Finite temperature effects may be calculated on top of the stationary quantum calculations of electronic structure, either by classical molecular dynamics calculations (with the argument that quantum effects are less important for the atomic nuclei than for the electrons) or by Monte Carlo calculations based on sampling path integrals (Weht *et al.*, 1998) or plane wave density functional methods (Reuter *et al.*, 2004). The path integral method is a general method of treating many-body systems in a quantum statistical way. It involves calculating a two-point density matrix, which under certain conditions may be expressed in terms of effective potentials from calculations such as those used in Figs. 2.5-2.6. The Monte Carlo technique is then used to determine the average thermal behaviour of the system.

The overvoltage introduced in connection with (2.20) can now be interpreted in terms two identifiable deviations from the potential determined by overall reaction energies. One is caused by the non-static situation of an electrochemical device, in which a current flows through the electrolyte, causing ions to interact with the molecules they pass. The other, occurring at each electrode, is connected with the polarisation build up, as illustrated in Fig. 2.4 and involving both water and ions in the electrolyte. The dynamic nature of both these effects imply that the overpotential could be defined as the difference between the real value and that for a completely static situation without any currents across the device (Hamann *et al.*, 1998). With interactions between the primary reactants and the surrounding molecules, the simple potentials depicted in Figs. 2.1, 2.5 and 2.6 could strictly speaking no longer be drawn as a function of a single variable.

The effect of the solvent can be modelled at various levels of sophistication. The simplest is to consider the molecule studied as surrounded by an infinite, homogeneous dielectric medium (the solvent) and interacting only via its total dipole moment. To avoid infinities in the numerical calculations, this also involves invoking a void (cavity) around the molecule studied, so that the distance between a point in the solvent and a point belonging to the molecule is never zero. This dipole model was first proposed by Onsager (1938). Refinements include replacing Onsager's fixed spherical cavity by the unification of spheres around each atom of the molecule and introducing various parametrisations of the dipole interaction. At the other extreme, each molecule of the solvent might be included in the quantum chemical calculations, which of course would then exclude use of the most detailed models.

Considering current commercial alkaline water electrolysers with 25-30% KOH and Ni electrodes, typical overpotentials are $V_a+V_c = 0.32$ V and $Rj = 0.22$ V at $j = 0.2$ A m$^{-6}$, to be added to the reversible cell potential of 1.19 V (Andreassen, 1998).

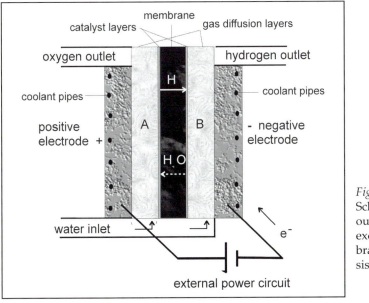

Figure 2.7. Schematic layout of proton exchange membrane electrolysis cell.

Improvement of electrolyser efficiency involves the design of catalysts able to promote high rates of water dissociation at the positive electrode and high rates of hydrogen recombination at the negative electrode. The liquid electrolyte may be replaced by solid polymer membrane electrolytes, offering the possibility of higher stability and longer life than the liquid electrolytes. Overpotentials as low as 0.016 V have been obtained for currents up to $10^4$ A m$^{-2}$ (Rasten *et al.*, 2003). It is also possible to provide the water feed at only one place (the negative electrode side in contrast to the example illustrated in Fig. 2.2), choosing a membrane that allows water penetration to the positive electrode side. The water penetration in one direction and the H$^+$ migration in the opposite direction can be balanced so that the oxygen is produced free of water (Hamilton Sundstrand, 2003). The hydrogen gas still contains water that has to be separated through a cleaning stage.

As mentioned, higher process temperatures may also increase efficiency. This is not possible with simple liquid electrolytes such as the KOH solution, but could be achieved with suitable solid oxide membranes similar to those used in solid oxide fuel cells and accepting water input in the form of steam (Dutta *et al.*, 1997).

An interesting idea investigated by Cheng *et al.* (2002) is to subject the entire electrolysis device to a constant, fast rotation. The centrifugal acceleration lowers the overvoltage and improves efficiency more than it is lowered by having to spend energy on maintaining the rotation.

Since hydrogen is often to be distributed at elevated pressures (e.g. in containers for vehicle use), a compression step may be avoided by carrying out already the electrolysis at high pressure. This requires suitable protection against explosion risks (Janssen *et al.*, 2004).

All types of fuel cells may be operated in reverse mode to split water. Proton exchange membrane fuel cells are becoming available both for electrolysis and for two-way operation, at efficiencies for electrolysis from 50 to 95% for one- or two-way systems (Shimizu *et al.*, 2004; Agranat and Tchouvelev, 2004; Proton Energy Systems, 2003; cf. section 3.5.5). Optimisation for hydrogen production involves the use of larger membrane areas, a low number of cells in a stack, and efficient hydrogen removal ducts (Yamaguchi *et al.*, 2000). Figure 2.7 gives an overview of the layout of a proton exchange membrane electrolyser. Recently, new catalysts have been developed, allowing high efficiency for operation in either direction (Ioroi *et al.*, 2002). Fuel cells, including reverse operation, will be discussed in more detail in Chapter 3.

## 2.1.4 Gasification and woody biomass conversion

An emerging technology for producing hydrogen from natural gas or heavy fuel oil, albeit with substantial electricity inputs, is high-temperature plasma-arc gasification, based on which a pilot plant operates on natural gas at 1600°C at Kvaerner Engineering in Norway (Zittel and Würster, 1996). The resulting products are in energy terms: 48% hydrogen, 40% carbon and 10% water vapour. Since all the three main products are useful energy carriers, the conversion efficiency may be said to be 98% minus the modest amounts of energy needed for the process. However, conversion of natural gas to carbon is not normally desirable, and the steam can be used only locally, so quoting a 48% efficiency is more meaningful. Concentrating solar collector energy input to decomposition of natural gas is investigated as an alternative, but expensive, route (Hirsch and Steinfeld, 2004; Dahl *et al.*, 2004).

Gasification is seen as a key pathway towards hydrogen when starting from coal or lignin-containing biomass (wood, wood scrap or other solid-structure plant material). The gasification takes place by heating with steam:

$$C + H_2O \rightarrow CO + H_2 - \Delta H^0, \tag{2.23}$$

where $\Delta H^0$ is 138.7 kJ/mol for water already in the gas phase. In air, competition may derive from the oxygen gasification (combustion) process

$$C + O_2 \rightarrow CO_2 - \Delta H^0, \tag{2.24}$$

with $\Delta H^0 = -393.5$ kJ/mol for $CO_2$ in the gas phase. Additional processes include the Boudouard process (2.6) and the shift reaction (2.2). For biomass, the carbon is initially contained in a range of sugar-like compounds, such as the cellulose materials listed in Table 2.2. Without catalysts, the gasification takes place at temperatures above 900°C, but use of suitable catalysts can bring the process temperature down to around 700°C. If additional hydrogen is to be produced by the shift reaction, this has to take place in a separate reactor operating at a temperature of about 425°C (Hirsch *et al.*, 1982).

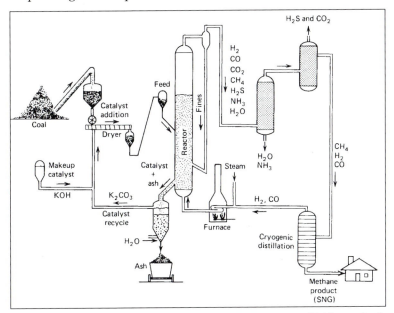

*Figure 2.8.* Schematic diagram of catalytic gasification process (SNG: synthetic natural gas). From Hirsch *et al.* (1982). *Science* **215**, 121-127, reprinted with permission. Copyright 1982 American Association for the Advancement of Science.

Coal may be gasified *in situ*, i.e. before extraction. If the coal has already been mined, the traditional methods include the Lurgi fixed-bed gasifier (providing gas under pressure from non-caking coal at a conversion efficiency as low as 55%) and the Koppers–Totzek gasifier (oxygen input, gas at ambient pressure, also of low efficiency). By using suitable catalysts, it is possible to shift the temperature regimes and combine all the processes within a single reactor, an example of which is shown in Fig. 2.8.

Peat and wood can be gasified in much the same way as coal. Wood gasification has a long history. The processes may be viewed as "combustion-like" conversion, but with less oxygen available than needed for burning. The ratio of oxygen available and the amount of oxygen that would allow complete burning is called the "equivalence ratio". For equivalence ratios below 0.1 the process is called "pyrolysis", and only a modest fraction of the biomass energy is found in the gaseous product, with the rest being in char and oily residues. If the equivalence ratio is between 0.2 and 0.4, the process is called a proper "gasification". This is the region of maximum energy transfer to the gas (Desrosiers, 1981). Table 2.2 lists a number of reactions involving polysaccharidic material, including pyrolysis and gasification. In addition to the chemical reaction formulae, the table gives enthalpy changes for idealised reactions (i.e., neglecting the heat required to bring the reactants to the appropriate reaction temperature). The specific heat of the material is 3 kJ $g^{-1}$ wood at the peak of energy in the gas, increasing to 21 kJ $g^{-1}$ wood for combustion at equivalence ratio equal to unity. Much of this sensible heat can be recovered from the gas, so that process heat inputs for gasification can be kept low. The energy content of the gas produced peaks at an equivalence ratio of about 0.25 and falls fairly rapidly both below and above this ratio (Reed, 1981; Sørensen, 2004a).

| Chemical reaction | Energy consumed (kJ $g^{-1}$) [a] | Products / process |
|---|---|---|
| $C_6H_{10}O_5 \rightarrow 6C + 5H_2 + 2.5\,O_2$ | 5.94[b] | elements, dissociation |
| $C_6H_{10}O_5 \rightarrow 6C + 5H_2O(g)$ | −2.86 | charcoal, charring |
| $C_6H_{10}O_5 \rightarrow 0.8\,C_6H_8O + 1.8\,H_2O(g) + 1.2\,CO_2$ | −2.07[c] | oily residues, pyrolysis |
| $C_6H_{10}O_5 \rightarrow 2C_2H_4 + 2CO_2 + H_2O(g)$ | 0.16 | ethylene, fast pyrolysis |
| $C_6H_{10}O_5 + \tfrac{1}{2}O_2 \rightarrow 6CO + 5H_2$ | 1.85 | synthesis gas, gasification |
| $C_6H_{10}O_5 + 6H_2 \rightarrow 6"CH_2" + 5\,H_2O(g)$ | −4.86[d] | hydrocarbons, −generation |
| $C_6H_{10}O_5 + 6O_2 \rightarrow 6CO_2 + 5\,H_2O(g)$ | −17.48 | heat, combustion |

*Table* 2.2. Energy change for idealised cellulose thermal conversion reactions. (Source: T. Reed (1981)., *Biomass Gasification*, reproduced with permission. Copyright 1981, Noyes Data Corporation.)
[a] Specific reaction heat.
[b] The negative of the conventional heat of formation calculated for cellulose from the heat of combustion of starch.
[c] Calculated from the data for the idealised pyrolysis oil $C_6H_8O$ ($\Delta H_c = -745.9$ kcal $mol^{-1}$, $\Delta H_f = 149.6$ kcal $g^{-1}$, where $H_c$ = heat of combustion and $H_f$ = heat of fusion).
[d] Calculated for an idealised hydrocarbon with $\Delta H_c$ as above. $H_2$ is consumed.

Figure 2.9 gives the equilibrium composition calculated as a function of the equivalence ratio. By equilibrium composition is understood the composition of reaction products occurring after the reaction rates and reaction temperature have stabilised adiabatically. The actual processes are not necessarily adiabatic; in particular the low-temperature pyrolysis reactions are not. For the cellulosic wood considered in Table 2.2, the average ratios of carbon, hydrogen and oxygen are 1:1.4:0.6. Figure 2.10 shows three examples of wood gasifiers: the updraft, the downdraft and the fluidised bed types.

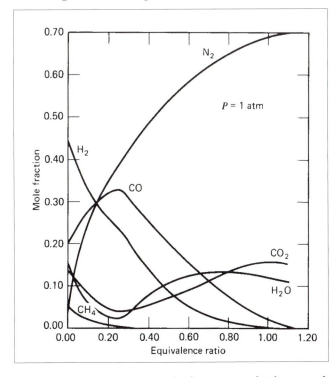

*Figure 2.9.* Calculated gas composition resulting from equilibrium processes between air and biomass, as a function of equivalence ratio. (From T. Reed (1981). *Biomass Gasification,* reprinted with permission. Copyright 1981, Noyes Data Corporation.)

The drawback of the updraft type is a high rate of oil, tar and corrosive chemical formation in the pyrolysis zone. This problem is solved by the downdraft version, where such oils and other matter pass through a hot charcoal bed in the lower zone of the reactor and become cracked to simpler gases or char. The fluidised bed reactor may prove superior for large-scale operations because passage time is smaller. The drawback of this is that ash and tars are carried along with the gas and have to be removed later in cyclones and scrubbers. Several variations on these gasifier types have been suggested (Drift, 2002; Gøbel *et al.*, 2002). Ni-based catalysts are used for both steam and dry reforming (Courson *et al.*, 2000). Practical ways of continuous hydrogen production by fluidised bed gasification are under investigation, e.g. within the Japanese energy programme (Matsumura and Minowa, 2004). For complex types of biomass, such as mixed urban waste or agricultural residues, gasification often leads to a gas of low hydrogen content (25-50%), requiring additional reaction or purification steps (Mérida *et al.*, 2004; Cortright *et al.*, 2002).

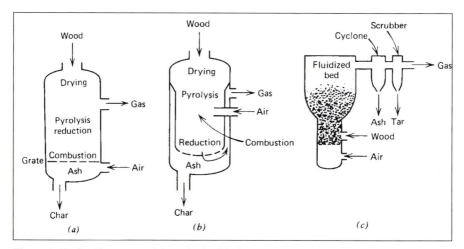

*Figure 2.10.* Gasifier types: (a) updraft; (b) downdraft; and (c) fluidised bed (from B. Sørensen, *Renewable Energy*, 2004, used with permission from Elsevier.)

The gas produced by gasification of biomass is itself a "medium-quality gas" with a burning value in the range of 10-18 MJ m$^{-3}$. It may be used directly in Otto or diesel engines or to drive heat pump compressors. However, in order to serve as a source of useful hydrogen, it has to be upgraded to pipeline-quality gas (about 30 MJ m$^{-3}$) and a hydrogen purity depending on the application envisaged. If methanol is the desired fuel, as it might be for some fuel cell operations, the following reaction is carried out at elevated pressure:

$$2H_2 + CO \rightarrow CH_3OH. \tag{2.25}$$

The mixture on the left-hand side of (2.25) is called "synthesis gas". More details on the methanol production from biomass may be found in Sørensen (2004a).

There are environmental impacts to consider in connection with biomass production, collection (e.g. by forestry) and transport to gasification site, from the gasification and related processes and finally from the use made of the gas. The gasification residues, ash, char, liquid wastewater, and tar have to be disposed of. Char may be recycled to the gasifier, while ash and tars could conceivably be utilised in the road or building construction industry.

Figure 2.11 gives an overview of the conversions of various types of biomass, including the gasification and pyrolysis processes described in this section. Biological production by fermentation and other bacterial processes will be described in section 2.1.5, while direct splitting by light or by heat will be covered in sections 2.1.6 and 2.1.7. The fluid "end" products such as ethanol or methanol may be further converted into hydrogen, although it is presumably more appropriate to use them directly, e.g. as fuels in the transportation sector. Of course, methane can also be used in high-temperature fuel cells such as solid oxide or molten carbonate fuel cells, without further conversion into hydrogen.

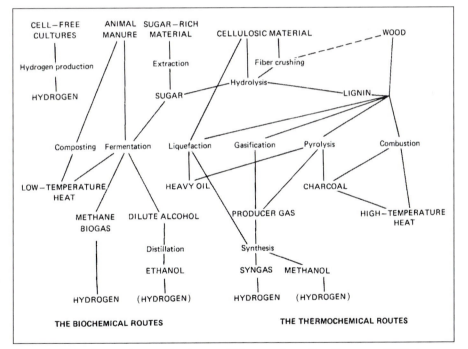

*Figure 2.11.* Pathways for non-food uses of biomass (based on Sørensen, 2004a.)

## 2.1.5 Biological hydrogen production

Production of hydrogen from biomass may be achieved by biological fermentation or by other bacterial or algae decomposition of water or another suitable substrate. The conversion processes may proceed in the dark or with the assistance of light. Growing the biological substance in the first place requires energy input, normally from sunlight, and there are thus several conversion efficiencies involved: from primary energy source to biological material, from energy in biological material to energy in hydrogen produced, or the overall efficiency from solar radiation to hydrogen. Because production of molecular hydrogen is rarely the purpose of natural biological systems, they have to be modified in order to serve this purpose, e.g. by genetic engineering.

Some primary sources ("substrates") for biological hydrogen production include those listed in the top area of Fig. 2.11. To these should be added water in case of the direct photolysis process and as the primary substrate for photosystem II discussed below. In these cases the biological system directly produces hydrogen (which then has to be prevented from subsequent use by the biological organism itself). Some organisms instead use an indirect route to produce hydrogen, starting not from water, but from an organic compound such as sugar, that requires less energy to complete the conversion. In organic waste fermentation, a succession of decomposition processes precede the hydrogen forming process. Many of the reactions depend on enzymes in order to work. There are enzymes in biological systems tailored to catalyse nitrogen to ammonia conversion (nitrogenases), with hydrogen as a possible byproduct and hydrogen oxidation enzymes (uptake hydrogenases) removing any hydrogen produced if not prevented. The are also bi-directional enzymes performing either function and, of course, enzymes catalysing other processes.

## Photosynthesis

The overall process of splitting water,

$$2H_2O \rightarrow O_2 + 4H^+ + 4e^- - \Delta H^0, \qquad\qquad (2.26)$$

requires an energy input of $\Delta H^0 = 590$ kJ/mol or $9.86 \times 10^{-19}$ J $= 6.16$ eV for the molecular process (2.26) splitting two water molecules. The solar spectrum, shown in Fig. 2.12, and the plant absorption spectrum, shown in Fig. 2.13, is such that in most cases four light quanta are required for delivering enough energy. Because of this, biological systems have had to develop an intricate system for collecting light, for transferring it to the site of water splitting and for further collecting and storing the energy in a form useful to subsequent biological processes. All of these processes take place in a photosynthesis complex, consisting of two photosystems and a number of auxiliary components, organised in and around a membrane structure similar to what would be needed in non-biological water splitting systems for separating the oxygen and hydrogen produced. The risk of recombination is usually prevented by not producing free molecular hydrogen at all, but rather some other energy-containing substance. Alternatively, oxygen and hydrogen should be on different sides of the membrane.

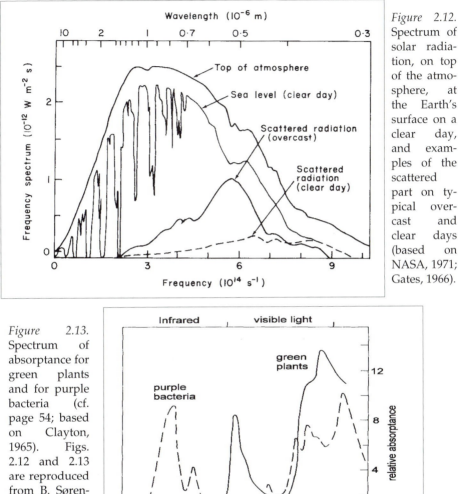

Figure 2.12. Spectrum of solar radiation, on top of the atmosphere, at the Earth's surface on a clear day, and examples of the scattered part on typical overcast and clear days (based on NASA, 1971; Gates, 1966).

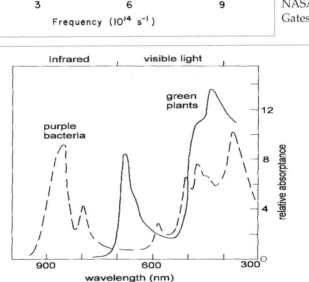

Figure 2.13. Spectrum of absorptance for green plants and for purple bacteria (cf. page 54; based on Clayton, 1965). Figs. 2.12 and 2.13 are reproduced from B. Sørensen, *Renewable Energy*, 2004, by permission from Elsevier.

Figure 2.14 shows the components of the photosystems residing along the thylakoid membranes. Insights into the structure of the membrane and component architecture has recently been obtained by atomic force microscopy (Bahatyrova *et al.*, 2004). The space inside the membranes is called the *lumen*,

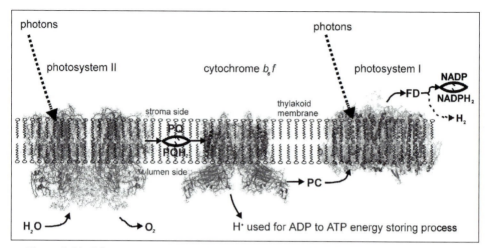

*Figure 2.14.* Schematic picture of the components of the photosynthetic system in green plants and cyanobacteria, see text for explanation (Sørensen, 2004c; with use of Protein Data Bank IDs 1IZL (Kamiya and Shen, 2003), 1UM3 (Kurisu *et al.*, 2003) and 1JBO (Jordan *et al.*, 2001)).

while on the outside there is a fluid, the *stroma*, rich in dissolved proteins. The whole assembly is in many cases enclosed within another membrane defining the chloroplasts, entities floating within the cytoplasma of living cells, while in some bacteria there is no outer barrier: these are called "cell-free".

Although some of the basic processes responsible for photosynthesis have been inferred long ago (see Trebst, 1974), the molecular structure of the main components have only recently been identified. The reason is that the huge molecular systems have to be crystallised in order to perform spectroscopy such as X-ray diffraction that allows identification of atomic positions. Crystallisation is achieved by some gelling agent, typically a lipid. Early attempts either damaged the structure of the system of interest, or the agent used to crystallise showed up in the X-ray pictures in a way difficult to separate from the interesting atoms. As a result, very slow progress has finally resulted in finding suitable agents and then improving the resolution from initially being over 5 nm down to first some 0.8 nm by 1998 and recently to about 0.2 nm. This allows most of the structure to be identified, although not the precise position, e.g. of hydrogen atoms. In the future, it is expected that the actual splitting of water may be followed as a result of time-slice experiments under pulsed illumination.

Figure 2.14 places the main photosystems along the thylakoid membrane, shown schematically but with the three systems exhibiting extrusions on the lumen or stroma side according to experimental findings. The splitting of water by solar radiation takes place in photosystem II, with the hydrogen emerging in the form of $pqH_2$, where $pq$ is plastoquinone. The cytochrome $b_6f$ (resembling the cytochrome $bc_1$ at work in the mitochondria of higher animals) transfers the energy in $pqH_2$ to plastocyanin ($pc$) and recycles $pq$ to photosystem II. The plastocyanin migrates on the lumen side of the membrane to photosystem I, where more trapping of sunlight is needed to transfer the energy to the stroma side of the membrane, where it is picked up by ferredoxin ($fd$). Ferredoxin is the basis for the transformation in the stroma, of NADP to $NADPH_2$, which is able to assimilate $CO_2$ by the Benson-Bassham-Calvin cycle (see Sørensen, 2004a) and thereby form sugar-containing material such as glucose and starch. However, a few organisms are capable of alternatively producing molecular hydrogen rather than $NADPH_2$. Possibly, the selection of the hydrogen-producing process can be stimulated by genetic engineering. In order to highlight the points where intervention may be possible, the individual processes will be described in a little more detail.

The structure of photosystem (PS) II, similar for different plants and bacteria, helps to reveal the mechanism of light trapping, water splitting and transfer between sites. Figure 2.15a and 2.15b, based on a thermophilic cyanobacterium (*Thermosynechoccus vulcanus*) show the overall structure of PS II, as seen from the side as in Fig. 2.14 and from the top[a]. The structure is that of a dimer with two nearly identical parts. Removing one part and pivoting in order to face directly the "inner" surface where the interface between the two parts was, the remaining monomer looks as shown in Fig. 2.16 with a number of amino acid chains indicated as solid helices drawn over the molecular structure. In Fig. 2.17, the view is tilted in order to better identify the reaction centre details shown in Fig. 2.18. At the 0.37 nm resolution of the X-ray data used, 72 chorophylls are identified in the photosystem II dimer. The zoomed monomer picture shows the 4 central chlorophyll rings (the experiment did not identify the side chains that allow the chlorophylls to attach to the membrane, cf. Fig. 2.19) and the unique manganese cluster of 4 atoms (there are another 4 in the other monomer). The role of Mn is like that of a catalyst for the water splitting, as will be discussed below. The many chlorophylls are part of an antenna system capable of absorbing solar radiation at the lower wavelength (higher energy) of the two peaks shown in Fig. 2.14 for green plants. The side chains shown in Fig. 2.19 and the long molecules of β-carotene probably also contribute to light absorption, with a subsequent transfer of energy from the initial chlorophyll molecules capturing radiation to the ones located in the reaction centre region near the Mn atoms, as indicated in Fig. 2.18.

---

[a] The sketch of amino acid chains and helices in molecular structure pictures is schematic (using the software package MSI (2000) for all pictures of this kind).

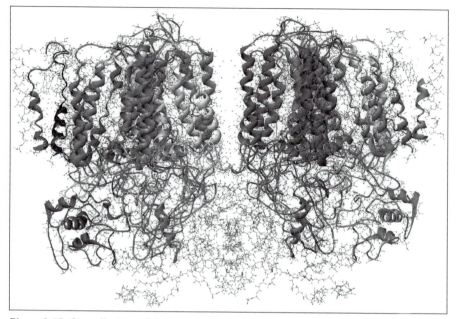

*Figure 2.15.* Overall view of the structure of photosystem II (*a,* above: seen from side, with stroma at top and lumen below; *b,* below: seen from top, i.e. looking down from stroma perpendicular to the view in *a*). Based on Protein Data Bank ID: 1IZL (Kamiya and Shen, 2003). Helices of amino acid chains are shown (cf. Fig. 2.16).

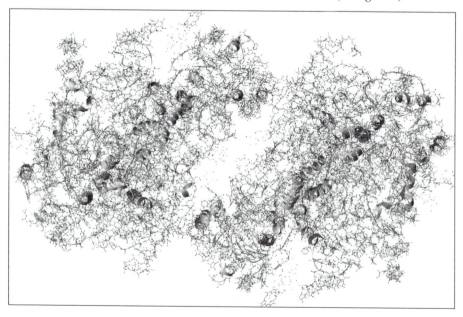

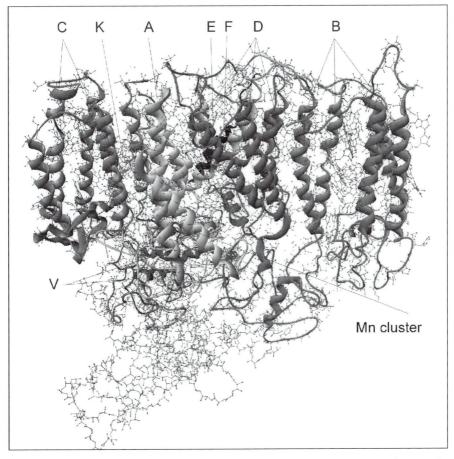

*Figure 2.16.* Protein arrangements in one monomer of photosystem II. A is the protein chain with the second quinone acceptor $pq_B$, B is a protein active in primary light harvesting, C and D are the subunits CP43 and D2, E and F are the cytochrome *b559* α and β helices, K is the protein K of the reaction center, and V is the cytochrome *b550*. The position of the 4 $Mn^{2+}$ atoms of the reaction centre is indicated. Based on Protein Data Bank ID: 1IZL (Kamiya and Shen, 2003).

The present understanding of the working of photosystem II is that solar radiation is absorbed by the antenna system including the large number of chlorophyll molecules such as those of region B in Fig. 2.16 and possibly some of the other molecules including as mentioned the β-carotenes. When absorbed, a light quantum of wavelength λ = 680 nm delivers an energy of

$$E = hc/\lambda = 2.9 \times 10^{-19} \text{ J} = 1.8 \text{ eV}, \tag{2.27}$$

where $c$ is the vacuum speed of light ($3 \times 10^8$ m s$^{-1}$) and $h$ is Planck's constant ($6.6 \times 10^{-34}$ Js). The excited chlorophyll state is a collective state involving electrons in all parts of the central ring structure (not the central Mg atom in particular, as demonstrated by experiments substituting other elements in this position). The excited state may decay and re-emit the energy as a photon (fluorescence), but the probability of transferring the energy to a neighbouring chlorophyll molecule is higher. In this way, the energy is handed from one chlorophyll molecule to another until it reaches one of the central chlorophyll-*a* molecules pointed out in Fig. 2.18b. The particular arrangement in this region favours a particular set of reactions, summarised in Fig. 2.20.

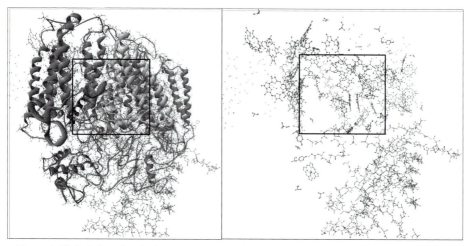

*Figure 2.17.* Left-hand side (as in Fig. 2.16) of photosystem II, but as seen from an angle of about 60° left rotation relative to that of Fig. 2.16 and from about 30° below the horizontal view of Fig. 2.16. To the left, amino acid chains are shown, while they are left out on the right-hand side. The squares roughly indicate the part enlarged in Fig. 2.18. Based on Protein Data Bank ID: 1IZL (Kamiya and Shen, 2003).

By transferring an electron to pheophytin (one of the two molecules shown in Fig. 2.18b), the donating chlorophyll-*a* is left in a positively charged state. The pheophytin passes on the electron to a plastoquinone, from which it is passed to a second plastoquinone believed to be better able to form $pcH_2$ and move around, after taking two protons from two of the processes shown in Fig. 2.20, and subsequently migrate from photosystem II to the cytochrome $b_6f$ complex for the next step shown in Fig. 2.14. The plastoquinone structure is shown in Fig. 2.21, taken from a recent study of the cytochrome $b_6f$ complex (Kurisu *et al.*, 2003); only the ring of one of the $pq$'s is seen in the photosystem II X-ray studies of Kamiya and Shen (2003) or Zouni *et al.* (2001). Considerable attention has been paid to identification of the processes, by which hydrogen ions are actually removed from water molecules, using 4 quanta of energy from the central chlorophyll absorbers.

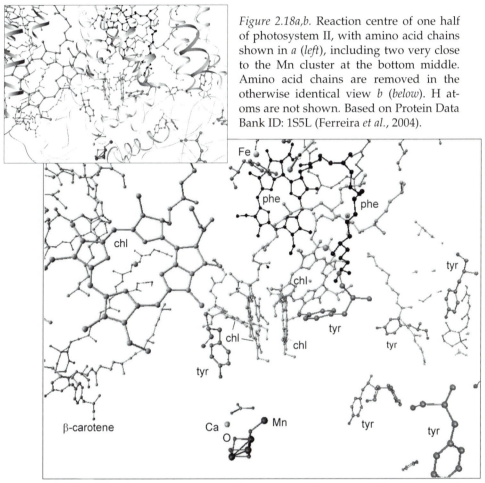

*Figure 2.18a,b.* Reaction centre of one half of photosystem II, with amino acid chains shown in *a* (*left*), including two very close to the Mn cluster at the bottom middle. Amino acid chains are removed in the otherwise identical view *b* (*below*). H atoms are not shown. Based on Protein Data Bank ID: 1S5L (Ferreira *et al.*, 2004).

Figure 2.18*b above* shows the same area as Fig. 2.18*a*, with amino acid backbones and side chains left out except tyrosines (labelled *tyr*). These have a special role in water splitting. The 4 chlorophyll-*a*'s (sometimes denoted $chl_{D1}$, $P_{680,D1}$, $P_{680,D2}$ and $chl_{D2}$) finally receiving the captured light energy are indicated (*chl*), as are the two pheophytin rings (*phe*) and a β-carotene being part of the antenna system. The two middle-left chlorophylls rings and one pheophytin ring belong together with the left tyrosine amino acid (sometimes denoted *tyr161* or $Y_Z$) to the subunit *D1*, the right-hand ones to *D2*. A non-heme $Fe^{3+}$ atom is enlarged at the top, and the Mn cluster below has 4 O and 4 Mn atoms identified, with one Mn away from the centre molecule. A $Ca^{2+}$ atom is suggested as part of the cluster, and a molecule with the structure $H_xCO_3$ is nearby. Not all side chains of the ring molecules have been identified.

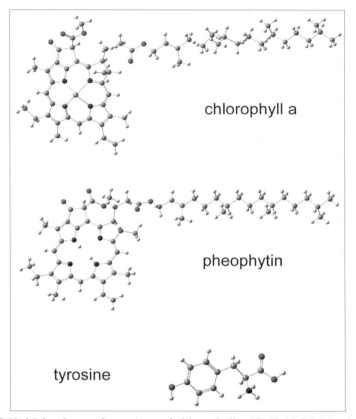

*Figure 2.19.* Molecular configurations of chlorophyll-*a* ($C_{55}H_{72}N_4O_5Mg$), pheophytin ($C_{55}H_{74}N_4O_5$) and tyrosine ($C_9H_{11}NO_3$). For clarity the detailed three-dimensional structure is not followed in detail. The surrounding environment would influence it. The tail components of the chlorophyll and pheophytin molecules serve to stabilise the placement of the molecules in the thylakoid membrane. Chlorophyll-*b*, a kind found among the chlorophylls of the antenna system in some photosystems, differs from *chl a* by having one $CH_3$ group replaced by CHO.

It has long been suspected that the Mn cluster serves to facilitate this process, but first, the energy has to get from the excited and positively charged chlorophyll molecule into the neighbourhood of the Mn atoms. The Mn atoms have double positive charges and the cluster is believed to be held in place by weak forces tying them to some of the amino acids of the nearby A-chain (Fig. 2.18a). The chemical jargon is that the Mn atoms are being "co-ordinated" by the polypeptides of the *D1* subunit. The amino acid most likely to donate an electron to the *chl*⁺ ring is the ring structure of a tyrosine molecule identified as *tyr161* (using the label numbering of the recent experiments quoted). The most recent X-ray experiment has a resolution of 0.35 nm, so there is considerable room for additional structure (and atoms) being identified as resolution approaches typical atomic distances around 0.1 nm.

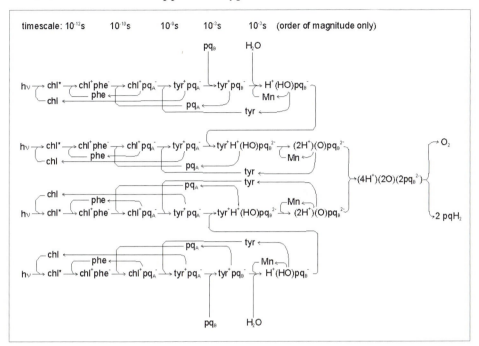

*Figure 2.20.* Photosystem II reaction centre conversions of the energy furnished by 4 light quanta (hv) originally captured in the antenna system, under the catalysing influence of the cluster of 4 $Mn^{2+}$ atoms. Order of magnitude reaction times are listed at the top (Barber, 2002). The notation used is dictated by chemical balancing and does not in itself imply nearness of any pair of reagents, although some reaction mechanisms would require such physical nearness (Sørensen, 2004c).

In Fig. 2.20, all four chlorophyll molecules, having received solar energy, transfer an electron via pheophytin to plastoquinone-$A$ and take one from a tyrosine molecule. In the first set of two reactions (top and bottom in Fig. 2.20), a hydrogen atom in a neighbouring water molecule is made to transfer an electron to the tyrosine(-161), presumably under influence of the Mn atoms. In the second set of reactions, the electron transfer from plastoquinone-$A$ to an already negatively charged plastoquinone-$B$ makes this doubly charged, which may also help to rip an electron from the second hydrogen atom in the water molecule. There are now two $H^+$ ions and a $pq_B$ molecule with a charge of –2, which combine to form $pqH_2$. The subscript $B$ is now dropped, as it is regarded as an indication of the site of reaction and therefore irrelevant for a plastoquinone that can move along the membrane to the following process step (in the cytochrome $b_6 f$ complex). Finally, the two oxygen atoms left behind by the process pairs will form an oxygen molecule, possibly again under the wings of the Mn catalyst. This understanding of the water splitting process is consistent with the recent structure data, but the matter cannot yet be considered as completely settled. Early theoretical explanations use sets of oxides $MnO_x$ (with x = 2 or more) that change their charge or charge co-ordination state relative to the trapped water molecules, in consort with the one-by-one stripping and release of 4 protons ($H^+$) and 4 electrons ($e^-$) (Kok *et al.*, 1970; Hoganson and Babcock, 1997).

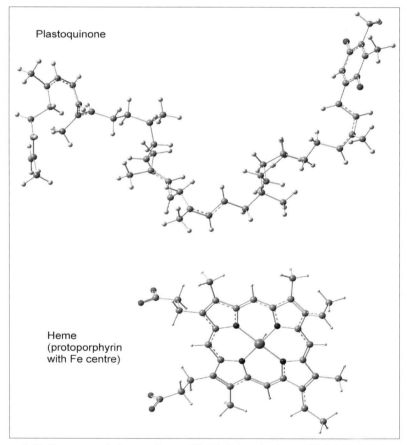

*Figure 2.21*. Structure of plastoquinone ($C_{53}H_{80}O_2$) and a heme ($C_{34}H_{32}N_4O_4Fe$), as they appear in cytochrome $b_6 f$.

The scheme of Fig. 2.20 does also allow an interpretation where the plastoquinones at site *A* and *B* are on equal footing: one is reached by the first reaction line in Fig. 2.20. If the second reaction involves the same pheophytin as the first, it will follow the reaction scheme in line 2 of Fig. 2.20, otherwise as in line 4. The remaining two steps are now fixed, provided that the plastoquinones have little probability for harbouring more than two negative charges. With the positions of most atoms determined, it should be possible to perform more realistic theoretical calculations of the reaction processes in the near future and thereby verify the status of the proposed reaction paths.

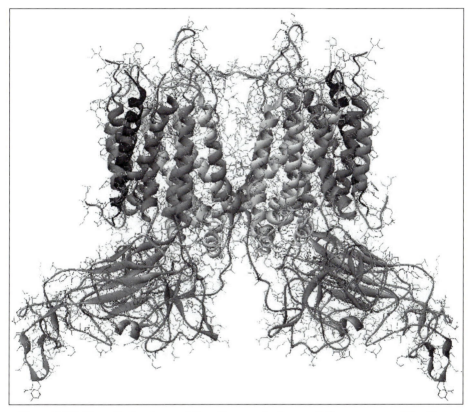

*Figure 2.22.* Overall structure of the dimer cytochrome $b_6f$, as measured by X-ray scattering for the thermophilic cyanobacterium *Mastigocladus laminosus*. Its position relative to the thylakoid membrane is indicated in Fig. 2.14. Resolution: 0.3 nm. Based on Protein Data Bank ID: 1UM3 (Kurisu *et al.*, 2003).

Once the $pq\text{H}_2$ reaches cytochrome $b_6f$ and gets close to a protoporphyrin heme molecule, the two protons are released and plastoquinone is recycled to the photosystem II. Figure 2.22 shows the overall structure of the cyanobacterium *Mastigocladus laminosus* cytochrome $b_6f$ (Kurisu *et al.*, 2003), which together with the structure derived from the alga *Clamydomonas reinhardtii* (Stroebel *et al.*, 2003) is the most recent element in completion of the photosystem structure. Similar functions are believed to be performed in animal mitochondriae by cytochrome $bc_1$ (Iwata *et al.*, 1998) and in purple bacteria by cytochrome $c_2$ (Camara-Artigas *et al.*, 2001). Figure 2.23 gives details of the $b_6f$ components, including three iron-containing hemes in the cytochrome $b_6$ and one in cytochrome $f$, plus the inorganic $Fe_2S_2$ molecule serving as an intermediary between the two cytochrome components.

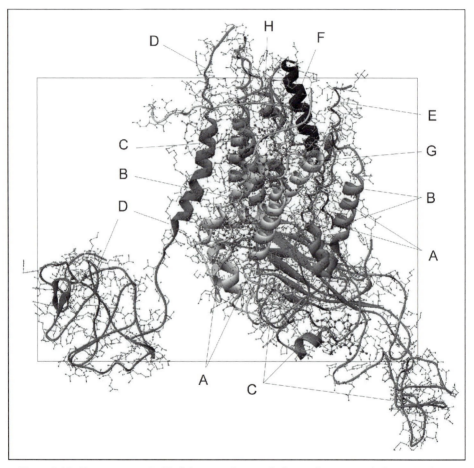

*Figure 2.23.* Shown is one half of the cytochrome $b_6f$ complex, rotated about 40° horizontally and moved about 20° upward relative to the view in Fig. 2.22. The structure at the lower right (C) is the cytochrome $f$ and the structure at the lower left (D) is called the Rieske protein. The upper structure (A) is the cytochrome $b_6$. The other chains identified are a subunit (B) and 4 proteins (E-H). The area within the rectangle is enlarged in Fig. 2.24. Based on Protein Data Bank ID: 1UM3 (Kurisu *et al.*, 2003).

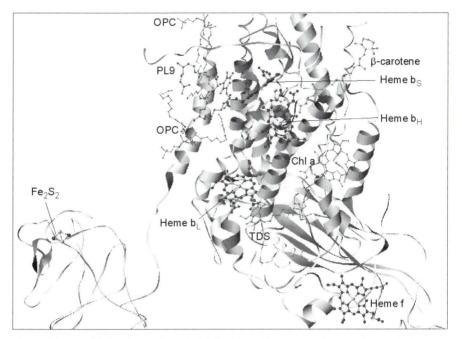

*Figure 2.24a* and *b* (without chains). Molecules relevant for the reaction path in cytochrome $b_6f$. See text for explanation. Protein Data Bank ID: 1UM3 (Kurisu *et al.*, 2003).

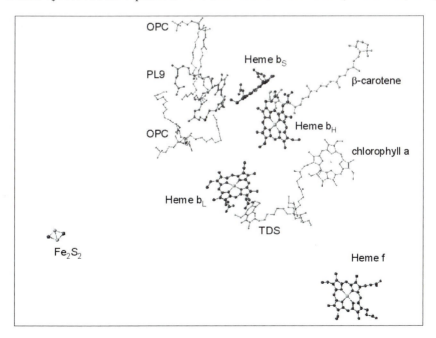

*Figure 2.25.* Structure of plastocyanin from the blue-green alga *Anabena variabilis*, with amino acid chains and central Cu atoms indicated. Based on Protein Data Bank ID: 1NIN (Badsberg *et al.*, 1996).

Although as judged from Fig. 2.23 and the blow-up in Fig. 2.24, the $Fe_2S_2$ molecule seems too far away to be important, in fact the Rieske protein on which it sits can be turned around the hinge connecting it to the membrane part of the structure and thereby bring the $Fe_2S_2$ molecule close to the hemes $b_L$ (L for lumen side) and $f$. The two hemes $b_H$ and $b_S$ closer to the stroma side are adjacent to a plastoquinone molecule (*PL9* in Fig. 2.24) and are thought to accomplish the recycling of $pq$ after the hydrogen ions (protons) have been separated.

The path via heme $f$ is taken by electrons on their way to the large plastocyanin (*pc*) molecule occupying a cavity on the lumen side before it moves with the extra electrons to photosystem I. The rather large *pc* molecule shown in Fig. 2.25 is from the blue-green alga (another word for cyanobacterium) *Anabaena variabilis* (Badsberg *et al.*, 1996). Smaller units with only one central Cu atom have been identified, for example, in the plant *Arabidopsis*, where the role of *pc* as an electron donor to photosystem I has been established by measurements indicating a large negative surface electrostatic charge (Molina-Heredia *et al.*, 2003).

Figure 2.24 also shows a few other molecules identified: TDS ($C_{25}H_{38}O_5$) and OPC ($C_{45}H_{87}NO_8P$), as well as more distant molecules of chlorophyll $a$ and β-carotene. Amino acids from the protein chains (indicated in Fig. 2.24a) help to keep the heme molecules in place and also may facilitate the electron transfer. The transfer of electrons to the lumen side *pc* with corresponding positively charged hydrogen ions implies the creation of a proton gradient across the membrane. This is where most of the energy transferred from photosystem II goes, and this energy may be stored for later use by the organism by the process

$$ADP^{3-} + H^+ + HPO_4^{2-} \rightarrow ATP^{4-} + H_2O, \qquad (2.28)$$

where ADP and ATP are adenosine di- and triphosphates (structure shown in Fig. 2.26). ATP is formed on the stroma side of the membrane, implying a transport of protons across the membrane.

As the next step in photosynthesis, the photosystem I shown in Fig. 2.27 receives the negatively charged plastocyanin $(pc)^{-1}$ from cytochrome $f$, but has to spend energy in transferring electrons across to the stroma side where they are picked up by ferredoxin for use either in molecular hydrogen formation or for the $CO_2$ assimilating processes of the organism. In order to provide the energy required, photosystem I contains an antenna system of chlorophyll molecules capable of capturing further solar radiation, albeit at a slightly lower energy than that characterising the central capture in photosystem II, corresponding now to wavelengths of about 700 nm.

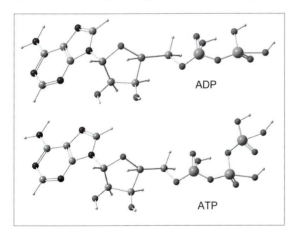

Figure 2.26 (left). Molecular structure of $(H_3)ADP$ = $(C_{10}H_{15}N_5O_{10}P_2)$ and $(H_4)ATP$ = $(C_{10}H_{16}N_5O_{13}P_3)$.

Electrons are brought into PS I from the lumen side, either by plastocyanine $pc$ or, in some cyanobacteria, by the cytochrome $c_6$. It has been suggested that plants developed the route with $pc$ and its Cu core at a time in development history, when Fe was less available in the areas of plant growth (Rosa *et al.*, 2002). Referring to Fig. 2.27, a cavity is seen on the lumen (lower) side, associated with two horizontal helical pieces of subunit $F$. This cavity, quite close to the chlorophylls *chl-A1* and *chl-B1* at the beginning of the electron transfer chain (Fig. 2.29), is supposed to be the docking site of *cyt-b$_6$* or *pc* (Fromme *et al.*, 2003). The somewhat surprising presence of a $Ca^{2+}$ atom on the lumen side (Fig. 2.28a) may have something to do with the coordination of the three monomers that make up the total PS I (Fromme *et al.*, 2001). The three antenna systems, each with two branches (A and B in Fig. 2.27), consist of a total of 270 chlorophylls (mostly type *a*, cf. Fig. 2.19), to which comes the $3 \times 6$ central ones (Fig. 2.29). This very large number (also compared with the 72 *chl*'s in photosystem II) is dictated by the fact that PS I must collect solar energy for all the requirements of the organism, whether a plant or a bacterium.

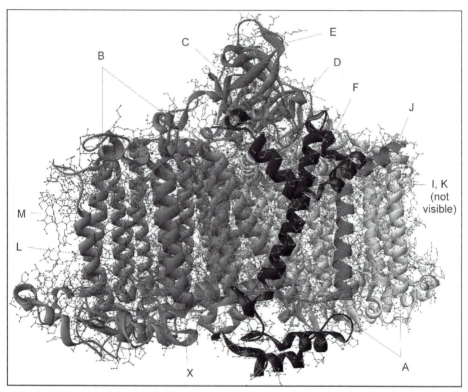

*Figure 2.27.* Structure of one photosystem I monomer (a total of three are at approximately 120° spacing) from the thermophilic cyanobacterium *Synechococcus elongatus*. A and B are two chlorophyll antennae, and C is the docking site for ferredoxin, harbouring two of the three iron-sulphur centres (cf. Fig. 2.28, seen from an angle 40° to the right). F to M are membrane spanning protein subsystems, while D, E and partly X extend upwards into the stroma side. Resolution: 0.25 nm. Based on Protein Data Bank ID: 1JBO (Jordan *et al.*, 2001).

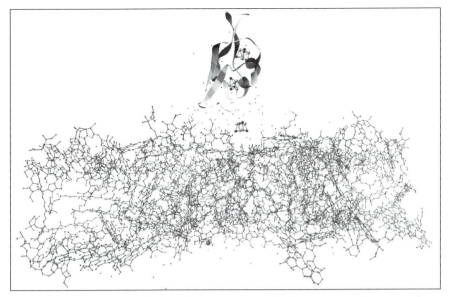

*Figure 2.28.* Photosystem I side *(a: above)* and top *(b: below)* views with protein chains omitted except for chain C on the stroma side near the three $Fe_4S_4$ clusters. A Ca atom is indicated on the lumen side. Protein Data Bank ID: 1JBO (Jordan *et al.*, 2001).

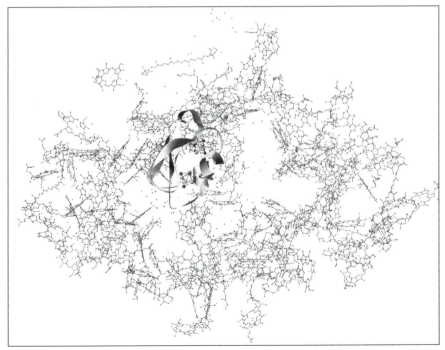

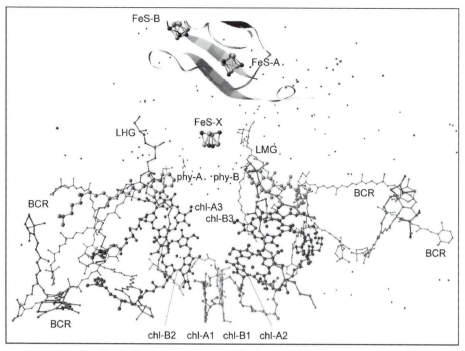

*Figure 2.29.* Detail of photosystem I, seen from the side, with the division between units A and B (see Fig. 2.27) in the middle. Chlorophyll molecules (*chl*) other than the six central ones involved in transfer of energy to the three $Fe_4S_4$ clusters (*FeS*) are omitted. Chlorophylls, phylloquinones (*phy*, also known as vitamin K) and iron-sulphur clusters are further indexed with the subunit label A, B or X (cf. Fig. 2.27), and the chlorophylls are shown with an index number 1 to 3. Other identified molecules include LHG ($C_{38}H_{75}O_{10}P$), LMG ($C_{45}H_{86}O_{10}$) and a number of β-carotenes (BCR). The scattered dots are oxygen atoms of water molecules (no hydrogens are shown). Based on Protein Data Bank ID: 1JBO (Jordan *et al.*, 2001).

The energy captured from solar energy is transferred from *chl* to *chl* in the antennas A and B, until it reaches the central pair [*chl-A1, chl-B1*] (Fig. 2.29). These two chlorophylls (usually denoted P700) are slightly different in shape and it is not known if both participate in the same manner. An electron is quickly (about 1 ps) passed along one of the vertical paths to *chl-A3* or *chl-B3*. It is not known if it passes through *chl-B2* or *chl-A2*, or if these are parts of the primary collection site P700. In a following rather fast step (some 30 ps), the electron is passed to either *phy-A* or *phy-B*, from where somewhat slower processes (20-500 ns) carry the electron successively to the three $Fe_4S_4$ molecules *FeS-X*, *FeS-A* and *FeS-B*. The surrounding amino acid chains play an important role in these transfers (Chitnis, 2001; Brettel and Leibl, 2001). Although charge recombination can occur, the primary electron donor is normally left in a positively charged state P700$^+$ throughout the successive electron transfers to *FeS-B*. The P700$^+$ (presumable *chl-A1* or *chl-B1)* is brought back to its neutral state by the electron brought into PS I by *pc* (or *cyt-c$_6$*) and is then ready for handling the next electron transfer from lumen to stroma.

Unlike PS II, where 4 light quanta were required in a co-ordinated fashion, due to the energetic requirements for water splitting, the PS I processes may proceed with just a single light quantum of 700 nm or $2.8 \times 10^{-19}$ J (1.7 eV). However, this observation does not invite any simple explanation for the dual structure of the central energy transfer chains A and B starting from P700. It could perhaps be that a redundancy was built in during evolution and its use later abandoned.

While the protein chain C surrounding the iron-sulphur clusters (Fig. 2.28) is similar in structure to ferredoxin, it needs a real ferredoxin protein docked on the stroma side to transfer the electron from *FeS-B* away from PS I . The subsystems C, D and E on the stroma side of PS I (Fig. 2.27) allow docking not only of the fairly small ferredoxin system (an example of which is shown in Fig. 2.30 – there are variations in structure between species such as plants and cyanobacteria), but also of the larger flavodoxin protein (Fromme *et al.*, 2003), an example of which is shown in Fig. 2.31.

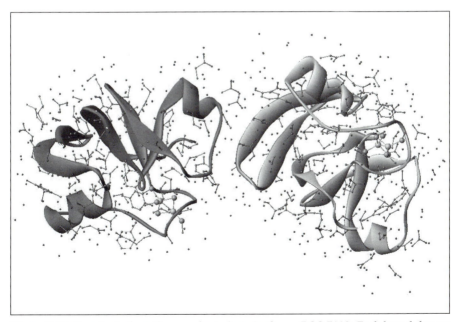

*Figure 2.30. Ferredoxin from the cyanobacterium Anabaena PCC 7119. Each branch has an Fe$_2$S$_2$ cluster. Proteins are shown schematically as ribbons, and side chains are shown as atoms. Surrounding dots are oxygen atoms from water molecules (hydrogens not shown). Based on Protein Data Bank ID: 1CZP (Morales et al., 1999).*

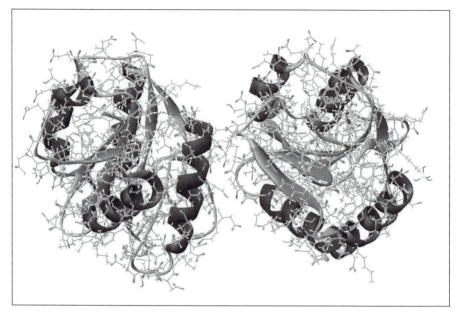

*Figure 2.31.* Flavodoxin from *Anabaena PCC 7119.* Based on Protein Data Bank ID: 1OBO (Lostao *et al.*, 2003).

Brought to the stroma side by the succession of photosystems, one now finds a number of electrons carried by the proteins ferredoxin and flavodoxin. Furthermore, protons (positively charged hydrogen ions) have been separated and moved across the thylakoid membrane (thereby creating an $H^+$ gradient from lumen to stroma). This makes room for two types of recombination processes on the stroma side. One is the "normal" reaction in plants and most bacteria:

$$NADP + 2e^- + 2H^+ \rightarrow NADPH_2, \qquad (2.29)$$

where $NADPH_2$ combines with $ATP^{4-}$ from (2.28) and water to assimilate $CO_2$ from the atmosphere for the synthesis of sugars and other molecules needed for the organism in question (see Sørensen, 2004a). The second possibility is the subject of interest here, namely to produce molecular hydrogen by some process equivalent to

$$2e^- + 2H^+ \rightarrow H_2, \qquad (2.30)$$

where just as in the inorganic case (2.17) and Fig. 2.5 an "agent" (in the inorganic case a catalyst) is needed for promoting the reaction. In the organic case, such an agent may be hydrogenase.

## Bio-hydrogen production pathways

Except for some rare bacteria, few organisms in nature produce hydrogen through (2.30). Clearly, the path in (2.29) is more profitable for the organism itself. To achieve bio-hydrogen production, it may therefore be necessary to trick suitable organisms into producing hydrogen by removing the subsequent steps of the "natural" pathway, e.g. by genetic engineering. The problem is that this would also remove the life-support system of the organism and prevent it from growing or reproducing. It is therefore most likely that a compromise will have to be struck, where only a part of the solar energy collected by the organism will be used for hydrogen production, while the rest is left for the purposes of the organism itself. This has implications for the overall efficiency of bio-hydrogen production. While the photosynthetic processes are efficient as judged by the requirements of the organism, they are low when the organism is seen as a solar collector, intercepting on average (over day and night) an energy flow of some 100-200 W m$^{-2}$. The energy efficiency of photosynthetic production of biomass relative to incident solar radiation is about 0.2% as a global average (Sørensen, 2004a), rising to at most 2% (on coral reefs) except for special energy-subsidised agri- or aquaculture. This implies maximum theoretical efficiencies for bio-hydrogen production of about 1% and the ensuing economic consequences discussed in Chapter 6.

Cyanobacteria (also called blue-green algae) convert nitrogen to ammonia, with the use of a catalyst called *nitrogenase*. The structure of a nitrogenase is shown in Fig. 2.32. Important components are an iron protein and a molybdenum-iron protein, the latter exhibiting an interesting mid-point atom (Fig. 2.33), assumed to be nitrogen and only recently discovered (Einsle *et al.*, 2002). Nitrogen fixing bacteria live in symbiosis with a range of green plants, such as legume crops, and account for about half of the nitrogen needed by human society agriculture. The other half is provided by chemical fertilisers based upon the Haber-Bosch ammonia synthesis route. The interesting feature of the nitrogenase Mo-Fe centre is its similarity to the tools used in the industrial ammonia synthesis process. Two amino acids keep the Mo-Fe cluster in place: a histidine attached to the Mo end and a cysteine attached to the single Fe atom at the opposite end. The middle atom stabilises the two Fe-triangles, but the detailed working of the catalyst is still debated (Smith, 2002). The overall reaction for ammonia synthesis, with ATP as the energy source and an inorganic phosphate ion as a byproduct, can be written

$$N_2 + 8\,H^+ + 8\,e^- + 16ATP^{4-} + 16H_2O \rightarrow 2NH_3 + H_2 + 16\,ADP^{3-} + 16\,H_2PO_4^-.$$
(2.31)

The presence of oxygen, from the exterior or the interior of the organism, has a deleterious effect on nitrogenase, and cyanobateria separates nitrogen fixation and oxygen evolution, either spatially or temporally (Berman-Frank *et al.*, 2001).

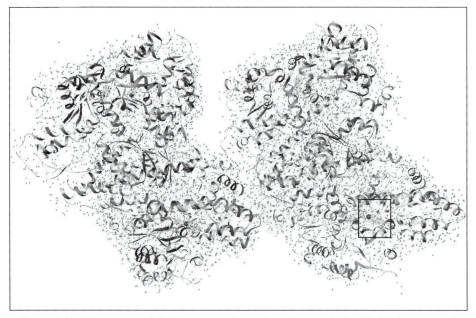

*Figure 2.32.* Nitrogenase Mo-Fe protein from *Azotobacter vinelandii.* Hydrogen and amino acid atoms are not shown. Based on Protein Data Bank ID: 1M1N (Einsle *et al.*, 2002). There are four sets of $Fe_8S_7$ and $MoFe_7S_9N$ clusters, with one enlarged in Fig. 2.33.

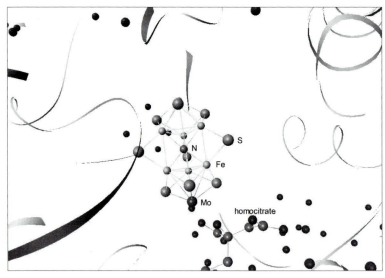

*Figure 2.33.* The highly specific Mo-Fe cluster of nitrogenase with centre ligand suggested to be N (blow-up of rectangle in Fig. 2.32). Based on the Protein Data Bank ID: 1M1N (Einsle *et al.*, 2002). Resolution: 0.116 nm.

As hydrogen evolution is a side effect of nitrogen conversion, the energetic efficiency is modest. Under normal circumstances, cyanobacteria use the hydrogen byproduct of nitrogen fixation to fuel energy-requiring processes, by a conversion process catalysed by the enzyme *uptake hydrogenase* (Tamagnini *et al.*, 2002). The active site of uptake hydrogenase contains a nickel-iron complex, the structure of which was first identified in *Desulfovibrio gigas* (Volbeda *et al.*, 1995) and later in *D. vulgaris* Miyazaki F (Ogata *et al.*, 2002). The extraction of hydrogen for human uses thus requires genetically modifying the cyanobacteria to suppress the action of uptake hydrogenase (Happe *et al.*, 1999).

Under anaerobic (oxygen-free) conditions, green algae can either use hydrogen as an electron donor in the $CO_2$ assimilation process or can, in the absence of oxygen, produce molecular hydrogen by combining protons with electrons from ferredoxin. These processes involve an enzyme (catalyst) called *reversible hydrogenase*, probably similar to the *bi-directional hydrogenase* seen in certain cyanobacteria (Tamagnini *et al.*, 2002; Pinto *et al.*, 2002). Use of genetic engineering to transfer some of the properties of specific hydrogenases such as the "Fe-only" type from the bacterium *Clostridium pasteurianum* (Peters *et al.*, 1998; shown in Fig. 2.34), which can only produce hydrogen in dark, oxygen-free environments, to photosynthetic cyanobacteria such as *Synechococcus elongatus* has been carried out successfully (Asada *et al.*, 2000), with $H_2$ production capability as a result.

The persistent role of Fe-S molecules in electron transfer is quite remarkable. In photosystem I (Figs. 2.28-2.29), in ferredoxin (Fig. 2.30) and now in the nitrogenase and hydrogenase enzymes, similar but not identical clusters of typically 2-5 iron-sulphur molecules are performing the energy transfer processes of a variety of organisms.

For natural cyanobacterial species, one of the highest rates of hydrogen production measured is just under 1%[a] (1.6% relative to the photochemical active part of the solar spectrum, cf. Figs. 2.12 and 2.13) for the *Anabaena*

---

[a] Only efficiencies relative to the radiation energy received are mentioned here. In biochemical literature, one may find yields of hydrogen production quoted as mol per kg of chlorophyll-*a* per hour, or per kg of culture, and the light input quoted as *einsteins* (E) per square metre and second. These data are useful only if the geometry of the reactor (area exposed to light and depth) is known and the light source used emits monochromatic light. The non-SI unit *einstein* is defined as Avodagro's number times the energy of a single light quantum, $h\nu$, and is thus not strictly a unit but linearly proportional to the frequency of the light and therefore useless for a distribution of light frequencies as found in solar radiation or in laboratory white light sources such as tungsten lamps. The nearest SI unit is *lumen*, which assumes the specific frequency distribution of black-body radiation at the temperature 2040 K. This distribution peaks at a wavelength of 555 nm (where 1E = 0.214 MJ) and resembles white light of certain lamps, but is not useful for the solar spectrum, as shown in Fig. 2.12.

*variabilis* over short periods of time (about a half hour), low irradiation (50 W/m$^2$) and in a pure argon atmosphere (Masukawa *et al.*, 2001, 2002). Average efficiencies over 24 h of outdoor solar radiation exposure are only about 0.05% (Tsygankov *et al.*, 2002a). The biomass addition to the bacterial culture itself is typically 0.5 to 1% of incident radiation accumulated over a number of growth days. When solar radiation is intense, most bacteria and algae harvest more photons through their photosystems than the metabolic and hydrogen-producing processes can handle, and the conversion efficiencies goes down. The ensuing fluorescence and heating further reduces hydrogen production, and to improve efficiencies, genetic manipulation truncating the chlorophyll antenna size has been used, e.g., for the unicellular algae *Chlamydomonas reinhardtii* (Polle *et al.*, 2002). Obviously, this does not in itself remedy the poor utilisation of available sunlight, but experiments have been made with two-layer collectors, where the first layer is genetically modified and the second is not (Kondo *et al.*, 2002; using bacteria of the genus *Rhodobacter sphaeriodes* and obtaining about 2% efficiency at 500 W/m$^2$ irradiation for 24 h). The same effect can be achieved by sulphur deprivation, but still obtaining only low levels of hydrogen production (Tsygankov *et al.*, 2002b).

*Figure 2.34.* Iron-only hydrogenase from *Clostridium pasteurianum*, showing a string of five Fe-clusters: one $Fe_2S_2^+$, three $Fe_4S_4^{2+}$ and a peculiar $C_5H_4O_7S_2Fe_2$ molecule. Based on Protein Data Bank ID: 1FEH (Peters *et al.*, 1998). Resolution: 0.18 nm. Hydrogen and amino acid atoms are not shown.

In understanding the efficiencies of hydrogen production by algae and bacteria, it should be kept in mind that there are fundamental limitations to the fraction of solar radiation that can be retrieved in the form of hydrogen. The solar spectrum on clear and overcast days is not the same (see Fig. 2.12), implying that a single collection system with a spectrally varying sensitivity cannot work well at all times. In addition to the losses also found in photovoltaic devices (due to the effects of bandgap and internal resistance, cf. Sørensen, 2004a), the photosystem in plants and bacteria have much more narrow frequency bands of light acceptance, as seen from Fig. 2.13, implying that substantial parts of the solar spectrum cannot be used. Furthermore, the radiation harvested has to be used for the growth and life support of the organism (respiration) before some of it can be diverted to producing hydrogen not used by the organism. The behaviour of hydrogen production efficiencies are typically uneven, with an initial delay while the organism switches from light absorption and respiration to hydrogen production, and later diminishing due to saturation or lack of feedstock. It is therefore important to distinguish between the peak efficiencies that may be experienced for short intervals and the average efficiencies over long periods of time. For comparison with other energy systems, an averaging time of a year should be used, due to the seasonal effects of solar radiation. As pointed out by Bolton (1996), the efficiency may be factorised to reflect incoming solar radiation and the fraction of solar radiation captured and useful to the hydrogen-producing processes, thus involving the quantum yield after losses of light quanta below the threshold energy and excess energy exciting various quantum states and eventually emerging as heat, as well as the hydrogen transport losses within the organic system, similar to an internal resistance of a semiconductor solar cell device.

For cyanobacteria, the distribution of radiation energy use between biomass and hydrogen production was in one experiment found to be about 10:1 (Tsygankov et al., 2002a). The smallness of the hydrogen production efficiency for direct bio-production is thus an inescapable condition, both for direct hydrogen production by the organism and for alternate production of biomass during the day and hydrogen in the dark. The implications of this are important for cost, as there has to be either a reactor device with cover, if hydrogen is to be retrieved from the same system as the one receiving solar radiation, or two separate systems. For alternate biomass/hydrogen production one may have an open large-area system for growth under sunlight and then transfer the biomass to a less expensive reactor with smaller surface area for the hydrogen-producing process. This points to the third possibility, where the biomass is just a feedstock (possibly in the nature of a residue, of discarded refuse or from some other inexpensive biomass source) which is collected and fed into some sort of reactor converting it into hydrogen. This route may employ several conversion techniques, such as the gasification schemes mentioned in section 2.1.4, or fermentation, a biological process that does not involve light input for its bacterial processes (see fermentation section below).

## Hydrogen production by purple bacteria

Purple bacteria do not have the two photosystems depicted in Fig. 2.14, but a single one incapable of water splitting but capable of delivering the energy required for $CO_2$ assimilation, with acetic acid or hydrogen disulfide as the electron donor, in the case of non-sulphur purple bacteria. Purple sulphur bacteria use sulphur or sulphur compounds as a substrate. The working of the single photosystem is similar to that of photosystem I (Minkevich *et al.*, 2004). Purple bacteria also contain nitrogenase and hydrogenases and may in the absence of nitrogen develop molecular hydrogen, without the problems encountered in organisms with both photosystems, related to the damage to nitrogenase caused by oxygen. For direct light-dependent hydrogen production, efficiencies are low (the absorption spectrum is shown in Fig, 2.13). The non-sulphur purple bacteria are considered for use by fermentation of organic waste, with hydrogen as one possible output. Some purple non-sulphur bacteria such as *Rubrivivax gelatinosus* can produce hydrogen by the water-gas shift reaction (2.2), while at the same time using organic substrates by means of an oxygen-tolerant hydrogenase (Levin et al., 2004).

## Fermentation and other processes in the dark

The production of an energy-rich gas from an organic substrate under oxygen-free and dark conditions is called *fermentation*. It is becoming a conventional energy conversion tool in biogas reactors, where the gas emerging is chiefly methane and carbon dioxide (Sørensen, 2004a). However, direct production of hydrogen is possible by selection of suitable bacteria for the degradation process. In biogas production, the bacterial cultures are often not controlled, and there may be hundreds of different bacteria working in different temperature regimes (from the mesophilic 25-40°C to the thermophilic 40-65°C region or even higher temperatures). For obtaining a high hydrogen content of the gas produced, it is necessary to use specific bacterial cultures, and even so, the $CO_2$ contamination is unavoidable and some amounts of $CH_4$, CO and $H_2S$ are usually present and require the hydrogen to be extracted by a subsequent purification process.

The feedstock may be carbohydrates such as glucose, starch or cellulose or more complex waste streams including industrial or habitat liquid and solid wastes of plant or animal origin, including food waste and animal dung. The cost of feed may be lower for the less specific types, but hydrogen yields are higher for transformation of glucose to acetic or butanoic acid (butyrate):

$$C_6H_{12}O_6 + 2H_2O \rightarrow 2CH_3COOH + 2CO_2 + 4H_2 + 184.2 \text{ kJ/mol}, \qquad (2.32)$$

$$C_6H_{12}O_6 \rightarrow CH_3CH_2CH_2COOH + 2CO_2 + 2H_2 + 257.1 \text{ kJ/mol}. \qquad (2.33)$$

In praxis, yields of 2 to over 4 mol $H_2$ per mol of glucose have been reported, corresponding to realising the stoichiometric ratios for the processes listed above (Ueno *et al.*, 1996; Hawkes *et al.*, 2002; Lin and Lay, 2004; Han and Shin, 2004). Hydrogen build-up inhibits production, so continuous removal of produced hydrogen is essential (Lay, 2000). Among the bacterial cultures used in current laboratory experiments are several strains of *Clostridium* (*pasteurianum, butyricum, beijerinkii*, etc.). The naturally occurring bacteria cultures found in cow dung and generally in sewage sludge also seem dominated by *Clostridia*. These bacteria are suitable for pure sugar fermentation and in proper mixtures also for the more complex types of feedstock mentioned. Like in biogas plants, the ability to handle varying compositions of feedstock as a function of time requires the presence of several types of bacteria, each of which should be able to grow to an important culture when the right feed becomes present.

The fermentation processes depend on temperature, usually producing more hydrogen in the thermophilic region (about 55°C) than in the mesoplilic region (for starch; Zhang *et al.*, 2003). They also depend on pH, with optimum production in acidic environments (pH = 5.5 for cow dung; Fan *et al.*, 2004). The production in some cases varies considerably with time, while in other cases a rather stable production is obtained with continuous feedstock input. Stable production seems to require modest rates of new input (e.g. diluted sludge, used for olive mill waste using *Rhodobacter sphaeroides*; Eroglu *et al.*, 2004). High performance is found for fermentation of granular sludge high in sucrose, producing a gas of 63% hydrogen and a volume yield of 280 litres of hydrogen per kg of sucrose, over 90 days (which is about 100% of the ratio in (2.33)), at a rate of 0.54 litres of $H_2$ per hour and per litre of sludge (using culture with 69% *Clostridium* and 14% *Sporolactobacillus racemicus*; Fang *et al.*, 2002). Also, the unicellular cyanobacterium *Gloeocapsa alsicola* has given high hydrogen yields in laboratory glucose fermentation experiments (Troshina *et al.*, 2002).

Figure 2.35 shows a generally accepted model for the main process steps in the enzymatic conversion of glucose, with energy and hydrogen transfers mediated by the ATP-ADT process as in plants and the $NADH_2$-NAD process (NAD = nicotinamide-adenine dinucleotide) replacing the $NADPH_2$-NADP process (2.29) of green plants and algae. The thick arrows indicate carbon flows, while fainter arrows are used to indicate electron and proton flows. Hydrogenase as well as several other enzymes is seen to play a role.

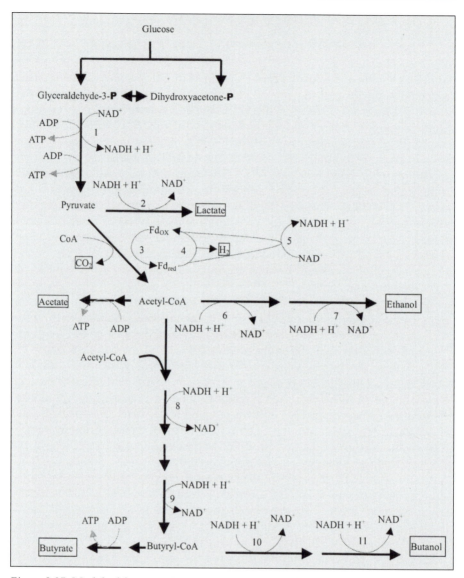

*Figure 2.35.* Model of fermentation processes in *Clostridium pasteurianum*. Enzymatic components are labelled: 1, glyceraldehydes-3-phosphate dehydrogenase; 2, lactate dehydrogenase; 3, pyruvate-ferredoxin oxidoreductase; 4, hydrogenase; 5, NADH-ferredoxin oxidoreductase; 6, acetaldehyde dehydrogenase; 7, ethanol dehydrogenase; 8, β-hydroxybutyryl-CoA dehydrogenase; 9, butyryl-CoA dehydrogenase; 10, butyraldehyde dehydrogenase; 11, butanol dehydrogenase. **P**, phosphate. From Lin and Lay, 2004 (used by permission from Int. Assoc. Hydrogen Energy), which is a modified version based on Dabrock *et al.* (1992), American Society for Microbiology.

## Industrial-scale production of bio-hydrogen

The two hydrogen production routes involving either photosynthetic algae/ cyanobacteria or fermentation bacteria converting earlier produced organic residues pose different challenges to a profitable industrial scale-up of the current laboratory-scale experimental work.

Several of the organisms invoked in experiments with hydrogen production are modifications of naturally occurring and often very abundant species. Some 40 years ago, the fairly large, filamentous photosynthetic cyanobacterium *Trichodesmium* was demonstrated to fix nitrogen by means of nitrogenase, and it was believed to be the main plankton species responsible for ammonia production in the open oceans. Subsequently, a number of unicellular cyanobacteria of a size under $10^{-5}$ m have been shown to be responsible for 20-60% of chlorophyll biomass and carbon fixation in the oceans (Zehr *et al.*, 2001; Palenik *et al.*, 2003; Sullivan *et al.*, 2003). Dominant species are *Synechococcus* and *Prochlorococcus*, and abundance may exceed $10^9$ per litre of sea water. In addition, large quantities of bacterial rhodopsin have been found, e.g., in *Escherichia coli* (Béja et al., 2000). This rhodopsin is capable of converting solar radiation and forming an electric proton gradient across the cell membrane and was earlier believed to exist only in archaea. Finally, the abundance of viruses in ocean water has been estimated as high as $10^{10}$ per litre, being important pathogens both for fish and for marine phytoplankton, assuming responsibility for the toxic algae blooms that have created severe problems both for holiday swimmers and for commercial aquaculture and fisheries (Culley *et al.*, 2003; Azam and Worden, 2004). Similarly, freshwater cyanobacteria are known to host microcystins responsible for outbursts of extreme toxicity blooms in lakes and other freshwater reservoirs all over the world (see, e.g., Shen *et al.*, 2003). The cyanobacteria most common in freshwater systems include *Microsystis, Anabaena, Nostoc* and *Oscillatoria*. These facts must be taken into account in designing new uses of the hydrogen-producing variants of the organisms in question.

As mentioned, natural hydrogen-producing bacteria are unlikely to produce industrially interesting amounts of hydrogen, and direct photoproduction of hydrogen is thus dependent on genetically modified strains such as the ones described in the previous sections. Because a solar collector is required with a large area due to the modest efficiency of conversion, a possibility would be to place this collector on the ocean surface, presumable offering lower area prices compared to on-shore land prices even in marginal areas. The need for hydrogen recuperation to a pipeline demands a closed reactor system containing the cyanobacteria, a glass or otherwise transparent cover and inlet/outlet devices for hydrogen transport and for replenishing or replacing the cultures as needed. The need to avoid accumulation of additional cyanobacteria in the surrounding water, as well as the requirement of not allowing genetically altered species to mix with the wild type in oceans, calls for a closed system for bacterial supply, photoreactor and hydrogen handling. The layout could in schematic terms be like the one shown in Fig. 2.36. The off-shore reactor could, instead of the glass-covered flat-plate collector indicated (and used in the experimental ocean facility developed by RITE in Kyoto; Miyake *et al.*, 1999), have a reactor system consisting of tubular pipes, allowing a higher portion of solar radiation to hit the cyanobacterial substrate in case of non-perpendicular impact angle and easier hydrogen transport to the pipe ends (Akkerman *et al.*, 2002). This system is in use for thermal solar collectors, where it gives a modest increase in energy production (some 10%) at a cost, which, however, is more than 10% higher than that of flat-plate systems.

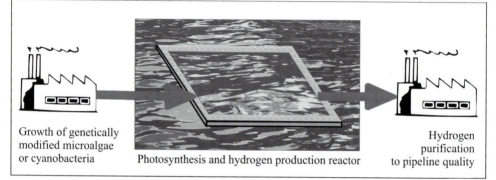

Growth of genetically modified microalgae or cyanobacteria

Photosynthesis and hydrogen production reactor

Hydrogen purification to pipeline quality

*Figure 2.36.* Schematic layout of photobiological hydrogen production plant with auxiliary plants for production of modified bacterial strains and for hydrogen gas purification (Sørensen, 2004c).

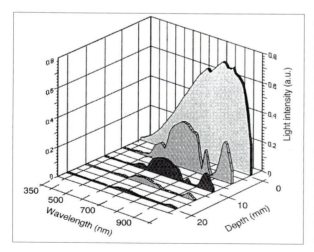

*Figure 2.37.* Attenuation of solar radiation as a function of penetration into a biohydrogen reactor containing partly modified cultures of *Rhodobacter spheroides*. Reprinted from J. Miyake, M. Miyake, Y. Asada (1999). Biotechnological hydrogen production: research for efficient light energy conversion. Journal of Biotechnology **70**, 89-101, with permission from Elsevier.

The algae or cyanobacterial cultures must occupy a shallow box because the solar radiation diminishes rapidly with depth of penetration (through water plus culture), as shown in Fig. 2.37. The penetration of radiation is highly wavelength-dependent, with low wavelengths of light being able to penetrate deeper (see Sørensen, 2004a; section 2.2.2). For an actual tank containing a photosynthetic purple bacterium *Rhodobacter spheroides RV* culture, Miyake *et al.* (1999) find an even more rapidly declining intensity, having dropped to 10% at 1 cm depth and to 1% at 2 cm depth. At the specific wavelengths of absorption by the two types of chlorophyll antennas present in the single photosystem, practically all light is absorbed within the first 0.5 cm. In accordance with the remarks on saturation in the previous sections, the efficiency of hydrogen production is found to be low in the first 0.5 cm layer and high in the deeper layers, but with negligible absolute production due to the near absence of light.

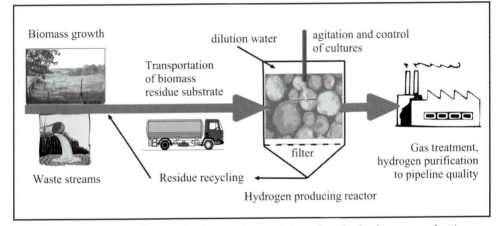

*Figure 2.38.* Schematic layout of substrate fermentation plant for hydrogen production, with infrastructure for residue collection/recycling and hydrogen gas purification. The blanket hydrogen reactor is similar to those used in biogas plants (Sørensen, 2004c).

Figure 2.38 shows a possible layout for industrial bio-hydrogen production based upon bacterial fermentation of already available biomass, such as harvested residues or wastes from other biomass uses (food, wood, etc.). The blanket-type reactor has been used in several experimental hydrogen fermentation studies (Chang and Lin, 2004; Han and Shin, 2004). The feedstock in this case has to be collected and transported to the site of the hydrogen production plant, which will generally have the appearance and components of biogas plants, either low-technology traditional ones or the large, industrial biogas plants currently used in most of the world in connection with waste water treatment plants and in a few cases in connection with biogas production from mixed household and (non-toxic) industry wastes (cf. Sørensen, 2004). After hydrogen extraction, the remaining residues may be returned to agriculture and, due to its high nutrient value, may replace or reduce use of fertiliser. Most fermentation plants use the thermophilic temperature regime, but going to slightly higher temperatures (70-80°C) may be possible without too severe an economic penalty from having to heat sections of the plant, because extensive recovery and recycling of heat is possible (Groenestijn *et al.*, 2002).

Recently, certain microbial systems have demonstrated the ability to generate electricity directly (with little or no hydrogen production) in fuel cell-like devices. These will be described in section 3.8. Production of hydrogen by photoelectrochemical devices may employ microbial sensitisers and enzymes or engineered substances derived from bacterial constituents. These will be mentioned in section 2.1.6 after the inorganic devices.

## 2.1.6 Photodissociation

In analogy to photovoltaic and photoelectrochemical devices aimed at electricity production (see Sørensen, 2004), efforts have been made to modify the devices in question to deliver hydrogen directly rather than electricity. Of course, there is always the alternative to convert the electricity to hydrogen in a second step, e.g. by conventional electrolysis. This is conveniently used as a backstop technology for economic comparison with new proposals.

The basic problem to overcome is that hydrogen production does not start until a sufficiently large cell voltage is reached, and further along the road, there is the familiar problem of separating oxygen- and hydrogen-evolving streams, for reasons including safety.

Figure 2.40a shows a hydrogen-producing cell, and Fig. 2.39a shows the corresponding electricity-producing cell. Figures 2.39b and 2.40b show current-voltage (IV) diagrams for the two systems, to be explained below.

The devices shown in Fig. 2.39a and 2.40a employ a solar cell with at least one semiconductor *p-n* junction. A *p*-layer contains lower atoms (i.e. atoms with a lower number of protons, $Z$) than the bulk of the material, manufactured by doping with foreign atoms, and an *n*-layer contains higher atoms (see, e.g., Sørensen, 2004a, section 4.A). Upon incident solar radiation, an electron is excited from the valence band of the semiconductor into its conduction band, and the potential behaviour across the junction will cause the excited electron to move towards the *n*-side terminal and the hole left behind in the valence band to move towards the *p*-side terminal. A current will flow through the external circuit connecting the terminals (in Fig. 2.39a), depending on the external load (resistance in the external circuit) as shown in Fig. 2.39b. $V_{oc}$ is the open circuit voltage corresponding to infinite external resistance (circuit cut off), and $I_{sc}$ is the maximum photo-induced current obtained with short-circuited external circuit (zero resistance). The maximum power that can be delivered to the external load is the maximum product $IV$. It deviates from the product of $I_{sc}$ and $V_{oc}$ by the fill factor $FF = IV/I_{sc}V_{oc}$.

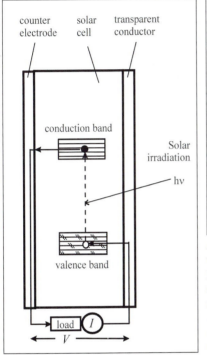

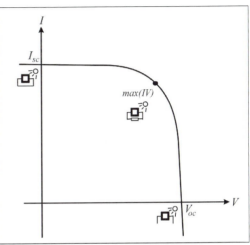

Figure 2.39a,b. Schematic picture of a solar cell with electron excitation and external circuit (a: left) and current $I$ vs. voltage $V$ diagram under illumination (b: above). $I_{sc}$ is short-circuit current, and $V_{oc}$ is open-circuit voltage. The solar cell may consist of one or more $p$-$n$ semiconductor junctions or of $p$-$i$-$n$ amorphous junctions.

In Fig. 2.40a, one of the terminal electrodes is separated from the solar cell with the other terminal electrode, to make room for an electrolyte containing water and some ion-conducting medium (a salt or redox couple). This design is known from photoelectrochemical solar electricity cells, where the photoelectrode is usually made of nanocrystalline $TiO_2$ covered by a metal- or pure-organic colour dye acting as a photo-sensitiser. In case of the modified device aimed at producing hydrogen (and oxygen), the outer electron circuit is shortcircuited in order to obtain a maximum current through the electrolyte. The voltage across the electrolyte is not a simple quantity, but depends on the chemical reactions taking place within the electrolyte, notably the water splitting and hydrogen (oxygen)-evolving processes. The $IV$-curve is in this case a hysteresis curve, i.e. having a different shape depending on whether the voltage is increasing or decreasing. The simple curve shown in Fig. 2.40b is just a limited part of the increasing voltage curve for large $V$, out of a total voltammogram, a technique further discussed below. It shows a photocurrent for device A that indicates activity in producing hydrogen, while for device B, the current only appears after an external bias voltage, i.e. an additional electromotive force, has been added to the external circuit. The indication is that a considerable "internal resistance" has to be overcome before hydrogen production can take place. This is equivalent to stating that the voltage difference between the electrodes must exceed the amount needed for the hydrogen splitting reaction, given in (2.21), based on the reactions (2.16) and (2.17) and also indicated in Fig. 2.40a, plus any internal electrode losses and resistance terms, as formally given by (2.18).

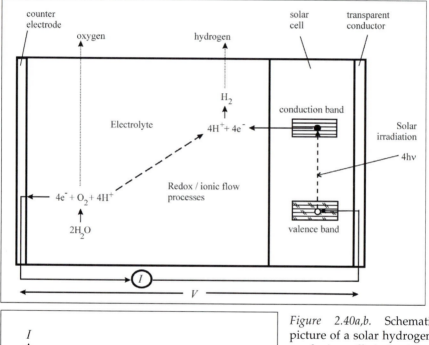

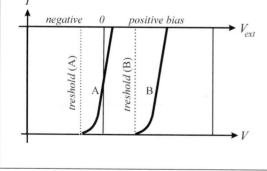

*Figure 2.40a,b.* Schematic picture of a solar hydrogen-producing cell aiming to excite a sufficient number of electrons for creating the voltage difference needed for water splitting (a: above) and current-voltage curves for such devices with sufficient (A) or insufficient (B) voltage for unassisted water splitting (b: left).

As an example, Khaselev and Turner (1998) used a high-efficiency tandem solar cell consisting of first a *p-n* junction (for electricity production as in Fig. 2.39) or a Schottky-type junction (for hydrogen production as in Fig. 2.40) made of GaInP$_2$ with a bandgap of 1.83 eV (absorbing the central part of the visible solar spectrum) and then a tunnel diode interconnect and a second semiconductor layer consisting of an *n-p* GaAs junction with a bandgap of 1.42 eV (suitable for the near-infrared portion of the solar spectrum, passing unhindered through the first layer). Solar irradiation may excite electrons in either layer, and if one electron is excited in GaInP$_2$ while another is already in the conduction band of the GaAs layer, then one of these electrons (most likely the one from GaAs, as it has the lowest excitation energy) may combine with the hole left in the valence band of the other layer, so that the net effect is to raise one electron to a state with energy more than 3.2 eV over the initial state. Two such excited electrons provide energy enough for the water splitting process (2.18). Like in the photosynthetic water splitting systems described in section 2.1.5 (see also Chapter 3, Fig. 3.3), it takes four light quanta of typical solar radiation (only the most energetic tail of the distribution could do with less) to supply the 6.16 eV needed (equal to 590 kJ/mol or 2 times $4.93 \times 10^{-19}$ J for the two water molecules) to split water into hydrogen and oxygen.

The excited electrons will emerge on the electrolyte side of the semiconductor junctions (see Fig. 2.40a), due to the potential gradient created, with holes in the valence band being filled by electrons from the counterelectrode, moving through the external circuit. The full potential gradient now makes its presence felt across the electrolyte, completing the positive and negative electrode reactions (2.16) and (2.17), provided that the electrolyte can provide transport for the protons $H^+$, either directly (similar to the flow processes shown in Chapter 3, Fig. 3.7) or through charge transfers involving redox couples present in the electrolyte. Hydrogen will then evolve near this negative electrode, while oxygen emerges at the positive counterelectrode, which in the experimental device is made of metallic platinum. The efficiency of the Khaselev-Turner device, illuminated by a tungsten lamp with about 10 times the energy flow typical of solar light, is quoted as 12.4%, with an efficiency they define as

$$\eta_C = 1.23 \, I/P_{in}, \tag{2.34}$$

in terms of the photocurrent $I$ (A/m$^2$) and the incident power of radiation $P_{in}$ (W/m$^2$). This corresponds to considering the electrolysis to be 100% efficient, as 1.23 V is the lower heating value of hydrogen at 25°C. A better method would obviously be to measure the volume of hydrogen produced per unit of time and calculate the output power from that. This device provides enough voltage to produce hydrogen without any energy subsidy, i.e., according to the scheme (A) in Fig. 2.40b.

In the general case, where an external bias voltage $V_{bias}$ may provide additional energy for the production of hydrogen, the solar-to-hydrogen efficiency may be defined by (Bolton, 1996)

$$\eta = (R \, \Delta G - I \, V_{bias})/P_{in}. \tag{2.35}$$

Here, $R$ is the rate of hydrogen production, $\Delta G$ is the corresponding free energy, $I$ is the photocurrent, and $P_{in}$ is the incident solar radiation. Both numerator and denominator have to be inserted in the same units, e.g. $Wm^{-2}$. In case it is necessary to apply a bias voltage, the hydrogen production can be said to by made by a solar-assisted hybrid generator.

At present, only laboratory-scale demonstrations have been made. While the system discussed above uses a high-efficiency tandem solar cell to obtain energy, there are many other possibilities. The system shown in Fig. 2.41 uses an amorphous solar cell with three $p$-$i$-$n$ layers as the negative electrode (Yamada *et al.*, 2003). The $p$-$i$-$n$ structure of an amorphous solar cell functions similarly to the $p$-$n$ junction of a crystalline solar cell (Sørensen, 2004a), with an intrinsic $i$-layer having interstituent hydrogen atoms (a-Si:H) in order to enhance light absorption. Instead of expensive platinum or Ru-based electrodes, simpler materials are used: a Co-Mo compound at the hydrogen-evolving electrode and a Ni-Fe-O compound at the oxygen-evolving electrode. The electrolyte is a strongly alkaline (pH = 13) $Na_2SO_4$ plus KOH solution. A solar-to-hydrogen conversion efficiency of 2.5% is claimed.

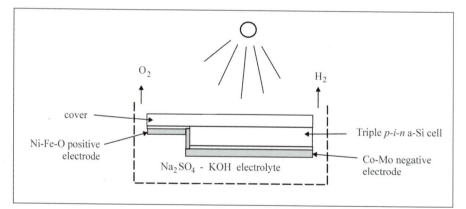

*Figure 2.41.* Single-chip photoelectrochemical electrolyser, with electrode assembly simply immersed into the electrolyte, according to the laboratory-scale design of Yamada *et al.* (2003).

An obvious device modification is to separate the two electrodes, so that the gases evolving may be collected easier through separate channels. Typically, a separator in the form of a perfluorised membrane would be used in analogy to proton exchange membrane fuel cells (see section 3.7).

The electrode reactions may be weakened by back-reactions such as conduction band electrons from the positive electrode recombining with protons formed in the electrolyte or electrons from the electrolyte filling vacancies in the electrode valence band. These reactions depend on the pH of the electrolyte, and it may therefore seem desirable to be able to control the alkalinity of the electrolyte near the positive electrode independently from that at the negative counter-electrode. Milczarek *et al.* (2003) use a two-compartment photoelectrochemical electrolyser separated by a perfluorised membrane to perform this separate optimisation. The device, which is illustrated in Fig. 2.42, uses 2.5 molar $Na_2S$ at the positive electrode and 1 molar $H_2SO_4$ at the negative electrode. The negative electrode is made of a Pt slab covered with a film of a perfluorinated Nafion (trademark of Dupont de Nemours) and a further layer of electrodeposited Pt, while the positive electrode consists of Ti covered by a CdS film. The two electrolytes are separated by the membrane (tradename Aldrich Nafion-417), allowing proton transport. The efficiency, based on the photocurrent as in (2.34), was claimed to be 7% on a sunny day, rising to 12% on an overcast day with a radiation level of 200 $Wm^{-2}$. However, the use of two electrolytes implies a difference in chemical potential, which acts like an external bias voltage, so the real efficiency is lower and will degrade with time.

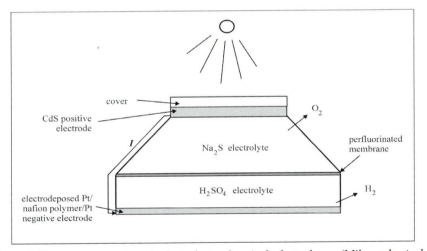

*Figure 2.42.* Two-compartment photoelectrochemical electrolyser (Milczarek *et al.,* 2003), suggesting optimum performance for an $H^+$ transporting membrane area 2-3 times larger than that of the light-exposed electrode. Gas outtakes are only indicated.

Many photoelectrochemical hydrogen-producing devices described in the scientific literature since the first one by Fujishima and Honda (1972) are unable to deliver enough potential energy difference for water splitting and therefore apply an additional external bias voltage in order to enable the system to draw the missing energy (e.g., Kocha *et al.*, 1991; Mishra *et al.*, 2003; Radecka, 2004). There are also suggestions to use concentrators to increase the amount of solar radiation reaching the hydrogen-producing device (Aroutiounian *et al.*, 2004), but again this is unlikely to be compatible with the low cost needed because of the modest efficiency achievable.

Whereas the device with incorporation of what is close to a full photovoltaic cell necessarily leads to a high hydrogen production cost, devices such as that illustrated in Fig. 2.41 are likely to be limited by low efficiency.

The behaviour of the electrolyte can be understood in terms of the description of redox processes, which are processes by which the **ox**idation level of a substance can increase and decrease (**red**uction, hence "red-ox"), whereby ionic charges can be transported through the electrolyte. The relation between the potential difference across the electrolyte and the Gibb's free energy of the redox reactions

$$X \Leftrightarrow X^+ + e^-, \quad e^- + Y \Leftrightarrow Y^-; \quad \text{or} \quad X + Y \Leftrightarrow X^+ + Y^- \tag{2.36}$$

is given by *Nernst's equation* (see, e.g., Bockris *et al.*, 2000; Hamann *et al.*, 1998), relating the reversible potential (2.21) to the concentrations $c_i$ of the substances entering into (2.36) at a given temperature $T$ (in degrees Kelvin)

$$-z\mathscr{F}V_r = \Delta G = -z\mathscr{F}V^0 + \mathscr{R} T \, log \, (c_{X+} \, c_{Y-} \, / \, c_X \, c_Y), \tag{2.37}$$

where $z$ is the number of electrons (one in (2.36)), $\mathscr{F}$ is Faraday's constant given in connection with (2.21), $\mathscr{R}$ is the gas constant (8.315 J K$^{-1}$ mol$^{-1}$), and the constant $V^0$ is called the standard potential. It is taken at some reference temperature, normally 298 K, and may be looked up in tables such as those found in CRC (1973).

A way to gain insight into the complicated response of an electrolyte to the application of a voltage across a pair of electrodes enclosing the electrolyte is to perform a two-way voltage sweep measurement, the result of which can be displayed in a *voltammogram* (Bard and Faulkner, 1998). The voltage is first increased from zero to a maximum value and then reduced back to zero. The outcome may look as shown in Fig. 2.43. It shows different regions of behaviour determined for small applied voltage by electrode processes, then by the transformation of the reactants such as those described by the hydrogen forming reactions indicated in Fig. 2.40a, and finally, a region where the process is limited by diffusion of reactants into the electrode region. This assumed a fairly quick voltage sweep. If the voltage is increased so slowly that there is always time for diffusion to the active region, this behaviour disappears. On the way back, one observes a lower current, due to the possibility that the oxidised substance may experience other reactions (Wolfbauer, 1999).

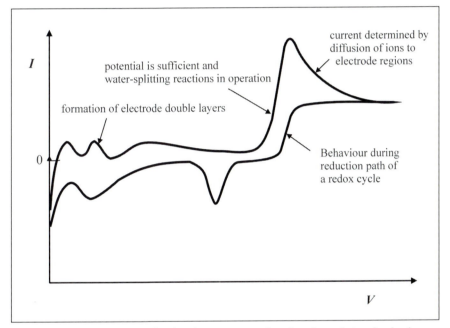

*Figure 2.43.* Schematic sketch of voltammogram for the electrolyte of a hydrogen-producing photoelectrochemical cell.

Some of the biological systems considered for hydrogen production in section 2.1.5 bear resemblance to the photoelectrochemical systems. The possibility of making hydrogenase produce hydrogen molecules rather than hydrogen bound in organic substances has spurred efforts to insert components such as the hydrogenase into inorganic systems, in order to improve control over the hydrogen-evolving processes. Also the photosynthesis could be "industrialised" by inserting light collecting biological antennas such as chlorophylls (Fig. 2.19) or porphyrins (Fig. 2.21) into electrochemical devices in the place of the (expensive) solar cells used in the devices described above.

Hydrogenase purified from the bacterium *Pyrococcus furiosus* has been combined with the enzymes of the pentose phosphate cycle to produce hydrogen from glucose-6-phosphate and $NADP^+$ (Woodward *et al.*, 2000). Similar experiments have been made using hydrogenase from *Thiocapsa ro-*

*seopersicina* (Wenk *et al.*, 2002; Qian *et al.*, 2003) immobilised in viologen polymer matrices. These experiments used an applied bias voltage. More close analogies to the photoelectrochemical cells have been studied by Saiki and Amao (2003) and Tomonou and Amao (2004). Here a light absorbing system based on tetraphenyl-porphyrin tetrasulphate or Mg-chlorophyll-*a* (from *spirulina*) was coupled partly to an NADH or NADPH system able to extract energy from glucose (or with a further enzyme from polysaccharides) and partly a methylviologen system with a platinum catalyst capable of producing hydrogen from the protons transferred to it. Absence of oxygen is essential to all of these systems.

## 2.1.7 Direct thermal or catalytic splitting of water

Another route contemplated for hydrogen production from water is thermal decomposition of water. As the direct thermal decomposition of the water molecule requires temperatures exceeding 3000 K, which is not possible with presently available materials, attempts have been made to achieve decomposition below 800°C by an indirect route using cyclic chemical processes and catalysts. These thermochemical or water splitting cycles were originally designed to reduce the required temperature to the low values attained in nuclear reactors, but could of course be used with other technologies generating heat, say at around 400°C. An example of the processes considered in early studies is the three-stage reaction (Marchetti, 1973)

$$6FeCl_2 + 8H_2O \rightarrow 2Fe_3O_4 + 12HCl + 2H_2 \ (850°C),$$
$$2Fe_3O_4 + 3Cl_2 + 12HCl \rightarrow 6FeCl_3 + 6H_2O + O_2 \ (200°C), \tag{2.38}$$
$$6FeCl_3 \rightarrow 6FeCl_2 + 3Cl_2 \ (420°C).$$

The first of these reactions still requires a high temperature, implying a need for external energy to be supplied in addition to problems related to the corrosive substances involved. A similar scheme using $CaBr_2$ involves three reactions at 730, 550 and 220°C (Doctor *et al.*, 2002). This research is still a long way from having created a practical technology.

At temperatures of 700-900°C, a number of mixed metal oxides of perovskite structure ($ABO_3$) have been studied as proton conductors, e.g., for use in solid oxide fuel cells. Water vapour introduced at one electrode of such a proton conductor sustaining a sufficiently strong electric current (with the use of externally provided electric energy) is able to split water and produce some amounts of hydrogen (Matsumoto *et al.*, 2002; Schober, 2001). Introducing a membrane to separate hydrogen as it is produced, thermal water splitting can be augmented, but the production rates are still very minute at the temperatures in question (Balachandran *et al.*, 2004).

Other reaction schemes are considered in section 5.4.2.

# 2.2 Issues related to scale of production

## 2.2.1 Centralised hydrogen production

Different technologies are suited for application on a scale determined both by the type of application and by the characteristics of the technology itself. If the cost of a given technology exhibits an economy of scale, it is preferred to use that technology in a centralised fashion, with large individual units. If the technology is cheapest in smaller units (for fixed overall production), the situation is of course opposite, but this is rare, and a more common situation is that the cost is insensitive to scale of production. This allows the kind of applications aimed for to determine the scale employed. However, there may also be specific scale requirements set by the type of usage. For example, technology for passenger cars must have a size and weight suitable for typical motor vehicles. Here one finds little flexibility, due to existing infrastructure such as roads, size of garages and parking spaces, etc. Generally, if a new technology requires changes in infrastructure, both the cost and the inconvenience associated with changing to it must be taken into consideration.

Traditional hydrogen production technologies, such as steam reforming or electrolysis, do exhibit some economy of scale, e.g., for heat exchangers or associated with environmental control options. For production of hydrogen by photosynthesis, the economy of scale resembles that of agriculture, while for dark fermentation, known economy of scale advantages for biogas plants should be transferable. Photodissociation by devices resembling solar cells or ambient temperature fuel cells are likely not to show any marked economy of scale, while high-temperature direct splitting of water may derive advantages from a large scale of operation.

## 2.2.2 Distributed hydrogen production

The identification of hydrogen production by reverse operation of low-temperature fuel cells as a technology not suffering by small scale of operation explains the interest in developing this technology for dispersed (decentralised) employment. Visions of hydrogen production in individual buildings (see section 4.5) may be based on reversible proton exchange membrane fuel cells taking over the function of existing natural gas burners, while at the same time offering to produce hydrogen for vehicles parked in the building and to regenerate electricity from stored hydrogen in situations of insufficient electricity supply from the outside (e.g. in connection with the desire of electric utility suppliers to perform load-levelling or manage intermittent renewable energy resources). If solar collector-based hydrogen generators become developed to a state of acceptable efficiency and cost, they can also be used in a decentralised fashion. Present experience from electricity-producing photovoltaic installations is that the economy of scale obtained for centralised plants (e.g. reducing installation cost) is often offset by the possibility of saving conventional roof and facade elements when such solar panels are installed.

The question of suitability for decentralisation becomes more critical in the case of vehicle-integrated hydrogen production systems, which must be economic at small scale.

## 2.2.3 Vehicle on-board fuel reforming

The basis for on-board production of hydrogen in principle could be fuels such as biofuels or fossil fuels, notably gasoline, as well as methanol, ethanol and similar intermediate stages between fuels produced from natural organic or from more artificial industrial primary materials. Of these, only methanol can be obtained by a reformation process similar to that of natural gas, (2.1), at moderate temperatures of 200-300°C. Reformation of other hydrocarbons usually requires temperatures above 800°C. Methanol is also interesting because of its similarity to gasoline in terms of fuelling infrastructure. The methanol energy content of 21 MJ $kg^{-1}$ or 17 GJ $m^{-3}$ is lower than that of gasoline, but because fuel cell cars are more efficient than gasoline cars, the fuel tank size will be similar.

Compressed and liquefied hydrogen are limited for use in automobiles by low energy density and safety precautions for containers and by the requirement of creating a new infrastructure for fuelling, so there is a strong interest in converting conventional fuels to hydrogen on-board. It is primarily to avoid having to make large changes to the current gasoline and diesel fuel filling stations that schemes based on methanol as the fuel distributed to the vehicle fuel tank have been explored. The energy density of methanol given above corresponds to 4.4 kWh $litre^{-1}$, which is roughly half that of gasoline. Hydrogen is then formed on-board by a methanol reformer before being fed to the fuel cell to produce the electric power for an electric motor. The set-up is illustrated in Fig. 2.44. Prototype vehicles with this set-up have been tested in recent years (cf. Takahashi, 1998; Brown, 1998).

Methanol, $CH_3OH$, may eventually be used directly in a fuel cell without the extra step of reforming to $H_2$. Such fuel cells are similar to proton exchange membrane fuel cells and are called direct methanol fuel cells (cf. section 3.7). As methanol can be produced from biomass, hydrogen may in this way be eliminated from the energy system. On the other hand, handling of surplus production of power systems with intermittent production may still conveniently involve hydrogen as an intermediate energy carrier, as it may have direct uses and thus may improve the system efficiency by avoiding the losses in methanol production. The electric power to hydrogen conversion efficiency is currently about 65% in high-pressure alkaline electrolysis plants (Wagner *et al.*, 1998; with prospects for higher efficiencies in the future, cf. section 2.1.3), and the efficiency of the further hydrogen (plus CO or $CO_2$ and a catalyst) to methanol conversion is up to around 70%. Also, the efficiency of producing methanol from biomass is about 45%, whereas higher efficiencies are obtained if the feedstock is methane (natural gas) (Jensen and Sørensen, 1984; Nielsen and Sørensen, 1998).

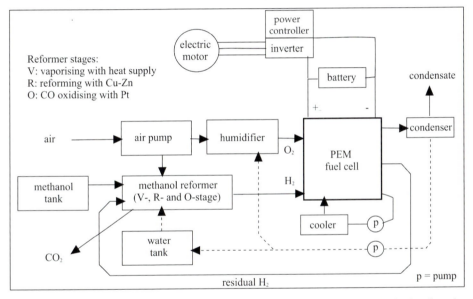

*Figure 2.44.* Layout of methanol-to-hydrogen vehicle power system with fuel cell and electric motor. The power controller allows shift from direct drive to battery charging (from B. Sørensen, *Renewable Energy*, 2004, used with permission from Elsevier).

## Production of methanol

As methanol may serve as a substitute for hydrogen in fuel cells and an intermediate fuel that can be used to produce hydrogen, its own production is of relevance and will be discussed in brief. Methanol may be produced from fossil sources such as natural gas or from biological material.

Conventional steam reforming of natural gas may produce methanol through a reaction of the form

$$2CH_4 + O_2 \rightarrow 2CH_3OH, \qquad\qquad (2.39)$$

as one among a number of possible reactions, of which several were mentioned in section 2.1.1 (Wise, 1981). The initial production of a synthesis gas by (2.1) is a likely pathway for the methanol synthesis (2.39). The process typically yields 0.78 mol of methanol per mol of methane at a thermal efficiency of 64% including considerations of heat inputs (Borgwardt, 1998). A direct synthesis may be obtained by partial oxidation (cf. section 2.1.2), at temperatures in the range of 425-465°C (Zhang *et al.*, 2002) and with similar efficiencies as for steam reforming.

There are various ways of producing methanol from biomass sources, as indicated in Fig. 2.11. Starting from wood or isolated lignin, the most direct routes are by liquefaction or by gasification. The pyrolysis alternative gives only a fraction of the energy in the form of a producer gas (Güllü and Demirbas, 2001). By high-pressure hydrogenation, biomass may be transformed into a mixture of liquid hydrocarbons suitable for further refining or synthesis of methanol (Chartier and Meriaux, 1980).

Current biomass-to-methanol production schemes use a synthesis gas (a mixture of $H_2$ and CO) derived from wood gasification by a process similar to that of coal gasification. The low-quality "producer gas" resulting directly from the wood gasification (extensively used in cars throughout Europe during World War II) is a mixture of carbon monoxide, hydrogen gas, carbon dioxide and nitrogen gas, varying with reaction temperature. If air is used for gasification, the energy conversion efficiency is about 50%. If pure oxygen is used instead, some 60% efficiency is possible, and the gas produced would have less nitrogen content (Robinson, 1980).

Gasification could conceivably be performed with heat from (concentrating) solar collectors, for example, in a fluidised bed gasifier maintained at 500°C. A probably more economic alternative is to use some of the biomass to produce the required heat, although this would entail environmental emissions that would have to be controlled.

Once the producer gas has been cleaned, removing $CO_2$ and $N_2$ (e.g. by cryogenic separation) as well as impurities (the nitrogen), methanol is generated at elevated pressure by the reaction

$$2H_2 + CO \rightarrow CH_3OH. \qquad\qquad (2.40)$$

Changing the $H_2/CO$ stoichiometric ratio to suit (2.40) may be achieved by the "shift reaction" (2.2) discussed in section 2.1.1. Steam is to be added or removed in the presence of a catalyst (e.g. iron oxide or chromium oxide).

For woody biomass consisting of 46% C, 7% H and 46% O plus minor constituents, the process may yield 16 mol of methanol per kg of biomass, with a thermal efficiency of 51% when accounting for the necessary input of electric power for wood drying, etc. (Borgwardt, 1998).

Non-woody biomass may be converted into methane contained in biogas (cf. Sørensen, 2004a), followed by the methane-to-methanol conversion process mentioned above, e.g. using steam reformation.

The conversion efficiency of the synthesis gas-to-methanol step is about 85%, and it is usually assumed that improved catalytic gasification techniques are required to raise the overall conversion efficiency above the 51% mentioned above (Faaij and Hamelinck, 2002). The total efficiency will become lower if all life-cycle contributions involving energy inputs, e.g. for collecting and transporting biomass, are included (EC, 1994).

The octane number of methanol is similar to that of ethanol, but the heat of combustion is less, as mentioned above. Methanol can be mixed with gasoline in standard engines or used in specially designed Otto or diesel engines. An example would be a spark ignition engine run on vaporised methanol, with the vaporisation energy being recovered from the coolant flow (Perrin, 1981). Methanol uses generally are similar to those of ethanol, but several differences exist in the assessment of environmental impacts, from production to use (e.g. toxicity of fumes at filling stations).

Historical costs of methanol have varied between US$ 6 and 16 per GJ (Lange, 1997), and estimated future costs with production stimulated by demand from the transportation and power utility sectors could be in the range of US$ 5.5 to 8 per GJ (Lange, 1997; Faaij and Hamelinck, 2002). In regard to environmental considerations, the gasification would be made in closed environments, where all emissions are collected, as well as ash and slurry. Cleaning processes in the methanol formation steps will recover most catalysts in reusable form, but other impurities would have to be disposed of along with the gasification products. Precise schemes for waste disposal have not been formulated, but probably only a subset of nutrients could be recycled to agri- or silviculture (like in the case of ethanol fermentation; SMAB, 1978).

Production of ammonia by a process similar to the one yielding methanol is an alternative use of the synthesis gas. Production of methanol from *eucalyptus* rather than from woody biomass has been studied in Brazil (Damen *et al.*, 2002). Extraction of methanol from tobacco cells (*Nicotiana tabacum*) has been achieved by genetic engineering, incorporating a fungus (*Aspergillus niger*) to decompose cell walls and infect the plant with a mosaic virus promotor (Hasunuma *et al.*, 2003). More fundamental studies aiming to better understand the way in which conventional methanol production relies on degradation of lignin are ongoing (Minami *et al.*, 2002).

## Methanol-to-hydrogen conversion

The steam reforming of methanol to hydrogen involves the reaction analogous to (2.1),

$$CH_3OH + H_2O \rightarrow 3H_2 + CO_2 - \Delta H^0, \tag{2.41}$$

where $\Delta H^0 = 131$ kJ mol$^{-1}$ (liquid reactants) or 49 kJ mol$^{-1}$ (gaseous reactants) has to be added as heat. The process works at a temperature of 200-350°C. The water-gas shift reaction or its reverse may be operating as well,

$$CO_2 + H_2 \leftrightarrow CO + H_2O + \Delta H^0, \tag{2.42}$$

with $\Delta H^0 = -41$ kJ mol$^{-1}$. This could lead to CO contamination of the hydrogen stream, which is unacceptable for fuel cell types such as the proton exchange membrane or alkaline fuel cells (above ppm level) and only little acceptable (up to 2%) for phosphorous acid fuel cells. Fortunately, the CO production is low at the modest temperatures needed for steam reforming, and adjusting the amount of surplus steam ($H_2O$) may be used to force the reaction (2.42) to go towards the left at a rate achieving the desired reduction of CO (Horny et al., 2004). At the high end of the temperature regime, this control of CO becomes more difficult. However, use of suitable membrane reactors with separate catalysts for steam reforming and water-gas shift allows this problem to be overcome (Lin and Rei, 2000; Itoh et al., 2002; Wieland et al., 2002). Figure 2.45 shows a typical layout of a modest-scale tubular reactor for on-board automotive applications. A typical thermal efficiency of hydrogen formation by this method is 74%, with near 100% conversion of the methanol feed.

It is possible to make the process autothermal, i.e. to avoid having to heat the reactants, by adding the possibility of the exothermic partial oxidation process (cf. section 2.1.2),

$$CH_3OH + \frac{1}{2}O_2 \rightarrow CO_2 + 2H_2 - \Delta H^0. \tag{2.43}$$

Here $\Delta H^0 = -155$ kJ mol$^{-1}$. By suitable combination of (2.41) and (2.43) the overall enthalpy difference may become approximately zero. There are still problems in controlling the temperature across the reactor, because the oxidation reaction (2.43) is considerably faster than the steam reforming (2.41). Proposed solutions include the use of a catalyst filament wire design leading to near-laminar flow through the reactor (Horny et al., 2004).

Catalysts traditionally used for the steam reforming process include a Cu-Zn catalyst containing mole fractions of 0.38 CuO, 0.41 ZnO and 0.21 Al$_2$O$_3$ (Itoh et al., 2002; Matter et al., 2004) and a metallic Cu-Zn catalyst in case of the wire concept (Horny *et al.*, 2004). For the oxidation reaction, a Pd catalyst is suitable, e.g. forming the membrane illustrated in Fig. 2.45 (using $Pd_{91}Ru_7In_2$; Itoh *et al.*, 2002). Also Pt has been used, as indicated in Fig. 2.44.

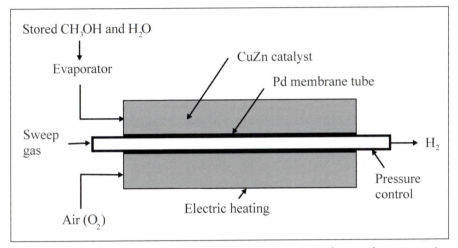

*Figure 2.45.* Schematic cross-section of membrane steam reformer for automotive uses (Lin and Rei, 2000; Itoh *et al.*, 2002).

For higher hydrocarbons, such as gasoline, steam reforming has to be performed at high temperature. Using conventional Ni catalysts, the temperature must exceed 900°C, but addition of Co, Mo and Re or use of zeolites allows a reduction of the temperature by some 10% (Wang *et al.*, 2004; Pacheco *et al.*, 2003).

Small-scale applications of fuel cells, aimed at increasing operation time before recharge over that of current battery technology, have stimulated the development of miniature reformers for both methanol and other hydrocarbons (Palo *et al.*, 2002; Presting *et al.*, 2004; Holladay *et al.*, 2004). Although the thermal efficiency of these devices aiming at use in conjunction with fuel cells rated in the range of 10 mW to 100 W is lower than for larger units, system efficiencies compare favourably with those of existing devices in the same power range. Applications will be discussed in section 4.6.

A theoretical understanding of the diffusion of hydrocarbons through the porous catalyst layer (see Fig. 2.45) may be obtained by simulations using semi-classical molecular dynamics (as in Fig. 2.3). Such calculations have been performed for the penetration of various hydrocarbons through $Al_2O_3$ catalysts with and without Pt insertions (Szczygiel and Szyja, 2004). As indicated in Fig. 2.46, it is found that fuel transport depends on both cavity structure and the adsorption on internal catalyst walls.

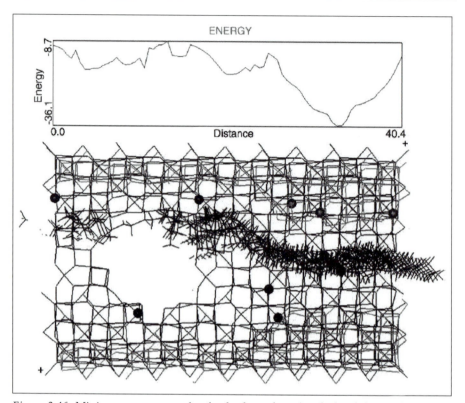

*Figure 2.46.* Minimum energy path of a hydrocarbon (methylcyclohexane) moving through an $Al_2O_3$ catalyst cavity channel with additional Pt atoms. In the upper part, energy in kcal/mol is shown as a function of distance along path in Å. The lower part gives a top view of the hydrocarbon trail through the catalyst. A molecular dynamics model was used. Reproduced from J. Szczygiel and B. Szyja (2004). Diffusion of hydrocarbons in the reforming catalyst: molecular modelling. *Journal of Molecular Graphics and Modelling* **22**, 231-239. With permission from Elsevier.

For high-temperature fuel cells (such as solid oxide fuel cells), reforming of methanol and certain other hydrocarbons to hydrogen occurs naturally within the fuel cell itself, in the presence of a catalyst such as Ru, but with an efficiency varying somewhat according to feed (Hibino *et al.*, 2003).

# 2.3 Hydrogen conversion overview

This section gives an overview of current and contemplated uses of hydrogen: for combustion, further conversion into liquid or gaseous fuels and fuel cell production of electricity, possibly with associated heat production. Details of a number of the technologies employed, specifically those involving fuel cells, are the subject of subsequent chapters. The non-energy uses and energy uses not involving fuel cells are only described in the present section.

## 2.3.1 Uses as an energy carrier

The use of hydrogen as an energy carrier may be contemplated without necessarily assuming it also to be an energy source. Energy forms difficult to transport or store for time-displaced uses could be converted into hydrogen and the hydrogen transported in containers or pipelines, possibly combined with hydrogen stores, and later converted to another energy form suited for the final energy uses. The low energy density by volume (see Table 2.1) sets certain limits to such uses of hydrogen, but there are many applications where density is not a big issue.

Advantages of hydrogen as an energy carrier include low environmental impacts and versatility in usage. A possible disadvantage might be the high requirements for tightness of containers and pipelines needed to avoid leakage.

Particular areas where an intermediate energy carrier is desirable include intermittent primary energy sources such as solar and wind energy, but also other energy sources linked to not easily storable electricity, e.g., nuclear energy. If use of fossil energy is extended, the quest for avoiding $CO_2$ emissions could lead to extended conversion of, e.g., coal to hydrogen, in order to be able to control greenhouse gas emissions at the primary stage (e.g. by forming carbonates and sending them to a repository such as old natural gas wells). A closer inspection of scenarios for fossil-hydrogen, nuclear-hydrogen and renewables-hydrogen energy supply systems is made in sections 5.3-5.5.

## 2.3.2 Uses as an energy storage medium

Seen as a storage medium, gaseous hydrogen is convenient in underground stores such as aquifers or flushed-out salt dome extrusions, requiring only a better lining compared to present use of the same geological formations for natural gas storage. The low volume density makes hydrogen storage in manufactured containers somewhat expensive, but compressed hydrogen storage is still considered a convenient solution for many applications in industry and for at least the first generations of fuel cell hydrogen vehicles and household-size generators. These and other storage options will be described further in section 2.4, including liquefied and molecularly trapped hydrogen stores.

It is possible for hydrogen to play a role as storage medium in connection with several types of energy systems, independent on whether or not hydrogen is also used as a general energy carrier. Renewable energy systems require energy storage in order to become self-contained solutions, and hydrogen satisfies a range of the storage requirements of such systems, particularly if an affordable fuel cell becomes available for recovery of energy in the versatile form of electricity. As mentioned above, just about any future energy system would derive benefits from access to hydrogen energy storage.

## 2.3.3 Combustion uses

Hydrogen can be used as a fuel in conventional spark-ignition engines such as the Otto and Diesel engines used in motorcars. The engine efficiency is as high as for gasoline or diesel fuels, and the hydrogen flame expands rapidly from the kernel of ignition (cf. Table 2.3). However, due to the lower energy density at the pressures suitable for piston cylinders, the volume of displacement must be 2–3 times as large as for gasoline engines, causing problems of space within engine compartments of passenger vehicles. One motorcar manufacturer active in developing hydrogen fuelled passenger cars has used huge 8 or 12 cylinder engines with displacement volumes of over 4 litres in order to approach an acceptable performance (BMW, 2004). Efficient conventional gasoline or diesel cars have total displacement volumes of around 1.2 litres, distributed on 3-4 cylinders (VW, 2003).

| Property | Hydrogen | Methanol | Methane | Propane | Gasoline | Unit |
|---|---|---|---|---|---|---|
| Minimum energy for ignition | 0.02 | – | 0.29 | 0.25 | 0.24 | $10^{-3}$ J |
| Flame temperature | 2045 | | 1875 | | 2200 | °C |
| Auto-ignition temperature in air | 585 | 385 | 540 | 510 | 230–500 | °C |
| Maximum flame velocity | 3.46 | – | 0.43 | 0.47 | | m s$^{-1}$ |
| Range of flammability in air | 4–75 | 7–36 | 5–15 | 2.5–9.3 | 1.0–7.6 | vol.% |
| Range of explosivity in air | 13–65 | | 6.3–13.5 | | 1.1–3.3 | vol.% |
| Diffusion coefficient in air | 0.61 | 0.16 | 0.20 | 0.10 | 0.05 | $10^{-4}$ m$^2$ s$^{-1}$ |

*Table 2.3.* Safety-related properties of hydrogen and other fuels (with use of Dell and Bridger, 1975; Zittel and Wurster, 1996).

Figure 2.47 shows the results of a computer simulation of the hydrogen combustion process in air (in a combustion chamber that could represent an engine cylinder or perhaps rather a gas turbine), substantiating the remark on the rapid consumption of the hydrogen injected. $H_2$ enters from the left side. The oxygen distribution shows "unused" oxygen along the chamber, into which air is drawn through numerous holes in the outer surface. At the bottom of Fig. 2.47 the distribution of nitrogen oxides formed is shown. It is similar to the distribution of high temperatures shown above, as the formation of nitrogen oxides increases dramatically for temperatures above 1700 K. High levels of $NO_x$ formation are a problem for combustion uses of hydrogen, which in this case no longer is pollution-free.

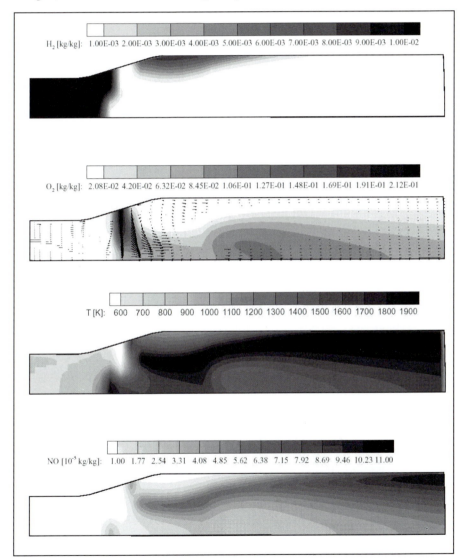

*Figure 2.47.* Simulation of a hydrogen combustion chamber. Top to bottom: $H_2$, $O_2$, temperature and NO distributions (From Weydahl *et al.* (2003). Proc. 14th World Hydrogen Conf. Used with permission from the Canadian Hydrogen Association).

The behaviour of hydrogen in thermodynamical engines has been established both by simulations such as the one shown in Fig. 2.47 and also by measurements, e.g., to establish the influence of compression ratios (Verhelst and Sierens, 2003; Karim and Wierzba, 2004). The wide range of flammability and easy ignition has two sides. One is the smooth operation at low loads. However, at high loads there are a number of problems with pre-ignition, backfiring or knocking that have to be addressed (Shoiji *et al.*, 2001). There are significant safety aspects, due to the wide flammability range of hydrogen (see Table 2.2). These have to be controlled, e.g., by double-valve systems on each cylinder (NFC, 2000). Work has also been done on engines fuelled by mixtures of hydrogen with gasoline or with natural gas (Fontana *et al.*, 2004; Akansu *et al.*, 2004).

Due to the large amounts of hydrogen by volume needed for use in internal combustion engines, a decent operational range for a passenger car of conventional dimensions requires much higher $H_2$ storage densities than possible on gas form. Therefore, liquid hydrogen is used, implying cooling to a temperature of 20 K (with an energy use that further reduces the efficiency) and using very special filling stations. These technologies have been developed and used for the prototype hydrogen combustion cars (Fischer *et al.*, 2003). Special tanks for containing liquid hydrogen have been constructed, using many layers of metal cylinders with high-grade insulation between them (Michel *et al.*, 1998). Even so, there is an escape of heat and a hydrogen-leaking problem to handle.

Hydrogen combustion is somewhat easier to incorporate into buses, because more generous sizing of the engine compartment is a smaller fraction of the overall size and – due to smaller operating speeds (in urban use) – buses can accommodate stored hydrogen on the roof, as it is done in the prototype hydrogen buses manufactured by MAN (Knorr *et al.*, 1998). Also, ships could accommodate hydrogen fuel instead of diesel in conventional engines.

The suggestion of using hydrogen for direct combustion has received a good deal of attention in the case of airplanes. Stores of liquid hydrogen may be incorporated into wings or fuselage, and near-conventional gas turbine engines can be used. Early tests of liquid hydrogen aircraft include the 1957 Boeing B57 bomber and in 1988 the Tupolev TU-154, and recently, a similar project named Cryoplane based on the Airbus A310 aircraft has been started (Pohl and Malychev, 1997; Klug and Faass, 2001). The more bulky fuel stores have lead to suggestions that these planes should operate at lower heights, which would have positive impacts on $NO_x$ emissions (Svensson *et al.*, 2004). This also makes the impact of water emissions less important, because the absorption of solar radiation by water (leading to greenhouse warming) is more pronounced above 10 km. Design for an airplane capable of operating from zero to Mach 5 (5 times the speed of sound, which is $1.22 \times 10^6$ m s$^{-1}$ at sea level) has been forwarded, with the vehicle using liquid hydrogen in the conventional gas turbine turbofan mode up to Mach 3 and a ramjet rocket mode above that speed (Qing and Chengzhong, 2001).

Liquid H$_2$ has been used in space vehicles for a number of years. Here the total mass to lift-off is an essential limiting parameter. Thus the high energy content by weight makes hydrogen preferable to any other fuel-based system, and the additional complication of cryogenic storage is deemed worthwhile. The engine may be a gas turbine or a rocket engine, depending on the mode of operating the space vehicle (cruising within the Earth's atmosphere or escaping the gravitation of the Earth). Both hydrogen and oxygen have to be carried in liquid form, as the space environment outside the Earth atmosphere does not contain air or oxygen. A basic feature of a rocket engine is the nozzle through which propellant gases are exhausted in order to provide the forward thrust. New developments in high-performance multiple nozzle design are still ongoing (Yu *et al.*, 2001), as are experiments with multiple fuels (typically liquid hydrogen and solid hydrocarbons) to suit the requirements at different stages of a near-Earth space journey with reusable shuttle-type space vehicles (Chibing *et al.*, 2001). The optimum relative location of liquid hydrogen and liquid oxygen injection sites has been studied by experiment and simulation (Kendrich *et al.*, 1999).

## 2.3.4 Stationary fuel cell uses

The uses of hydrogen in fuel cells for stationary markets have received considerable attention. Two areas of usage have been investigated. One is the electric power production sector, where the most efficient fuel cell types (solid oxide or molten carbonate) may have a future when they are sufficiently developed with regard to both technology and cost. The other area, which seems closer to the market introduction, is small, single-building units based on the proton exchange membrane fuel cell technology, aimed at replacing natural gas-fired heat burners and thereby making the house owner a provider of both heat and electricity. The details of these developments will be dealt with in section 4.5 and in the scenarios in Chapter 5.

## 2.3.5 Fuel cell uses for transportation

The most significant hope for introduction of hydrogen used in fuel cells is in the transportation sector, where the introduction of alternative energy sources such as renewables has been most elusive. While hydrogen produced from renewable energy sources is seen as the only stable solution for the long term, the use of hydrogen based upon natural gas may constitute a convenient transition solution. The most significant cost reduction needed for these scenarios to come true is for the fuel cell itself (assumed to be a proton exchange membrane type), but also the infrastructure change associated with distribution of hydrogen may constitute a significant cost. These issues will be dealt with in detail in sections 4.1-4.3 and in Chapter 5 on implementation.

## 2.3.6 Direct uses

It has been suggested that one of the first uses of hydrogen was for brick-making (Bao, 2001). Fired bricks replaced sun-dried ones for uses in luxury buildings more than 5000 years ago in Mesopotamia, and lead-glazed ones appeared from about 3000 years ago (Hodges, 1970). Only much later did kiln bricks become common in the Mediterranean region, due to the abundance of other building materials such as stone and marble (although the kiln technology was itself in widespread use for pottery). In China, pottery making seems imported from the West (of Asia). Little evidence for fired bricks exists before the construction of the Great Wall some 1000 years ago. The bricks used here are grey (rather than the red shades characterising many modern clay bricks), which is interpreted as them being heated with water and carbon monoxide in high-temperature kilns, where they are subjected to the water-gas shift reaction followed by reduction of clay,

$$CO + H_2O \rightarrow CO_2 + H_2$$
$$\downarrow$$
$$Fe_2O_3 + H_2 \rightarrow 2FeO + H_2O. \tag{2.44}$$

A 17th century Chinese source describes this process as "fire and water stimulate each other to make the best quality bricks" (Yinxing, 1637). The element hydrogen was unknown until the 19th century. Processes like (2.44) involving $H_2$ are in use today for the production of a range of materials.

The current main use of hydrogen is still in industry, where it is a standard commodity usually distributed in pressure containers in the form of gas flasks. In industrial areas such as Ruhr in Germany or London in England, hydrogen distribution through pipeline networks has been common.

Some 60% of the hydrogen use in industry is for ammonia production. The process is

$$N_2 + 3H_2 \rightarrow 2NH_3 - \Delta H^0 \tag{2.45}$$

with $\Delta H^0 = -107$ kJ mol$^{-1}$ at 425°C and 21 MPa. An iron catalyst is required. Since the process is exothermal, the temperature is as low as consideration of reaction velocity allows. At this temperature, the conversion efficiency is 15-20%, depending on the catalyst used (Superfoss, 1981; Zhu *et al.*, 2001).

Hydrogen is further used in hydrocracking and -refining, methanol synthesis (section 2.2.3) and aldehyde production. Future coal liquefaction also relies on hydrogen input.

# 2.4 Hydrogen storage options

The form of hydrogen storage most suitable depends on the application. Applications in the transportation sector require storage at a volume that can be accommodated within the vehicle and at a weight that does not limit the performance of the vehicle (for space applications one that still allows lift-off and disengagement from the Earth's atmosphere). Also, for building-integrated applications, storage volume must usually be restricted, while dedicated storage at power plants or remote locations may allow for more latitude. Examples of storage density by mass or by volume are given in Table 2.4. Below, hydrogen storage will be discussed in terms of compressed gas, liquefaction, cryo-adsorbed gas storage in activated carbon, metal hydride storage, carbon nanotube storage and reversible chemical reactions.

| Storage form | Energy density | | Density |
|---|---|---|---|
| | $kJ\ kg^{-1}$ | $MJ\ m^{-3}$ | $kg\ m^{-3}$ |
| Hydrogen, gas (ambient 0.1 MPa) | 120 000 | 10 | 0.090 |
| Hydrogen, gas at 20 MPa | 120 000 | 1 900 | 15.9 |
| Hydrogen, gas at 30 MPa | 120 000 | 2 700 | 22.5 |
| Hydrogen, liquid | 120 000 | 8 700 | 71.9 |
| Hydrogen in metal hydrides | 2 000–9 000 | 5 000–15 000 | |
| Hydrogen in metal hydride, typical | 2 100 | 11 450 | 5 480 |
| Methane (natural gas) at 0.1 MPa | 56 000 | 37.4 | 0.668 |
| Methanol | 21 000 | 17 000 | 0.79 |
| Ethanol | 28 000 | 22 000 | 0.79 |

*Table 2.4.* Energy density by weight and volume for various hydrogen storage forms, and mass density, including for comparison natural gas and biofuels (with use of Sørensen, 2004a; Wurster, 1997b).

## 2.4.1 Compressed gas storage

Storage of hydrogen in compressed gas form is the most common storage form today. Standard cylindrical flasks use pressures of 10-20 MPa, and fuel cell motorcar stores are currently in the range of 25-35 MPa. Tests are ongoing with pressure increased to 70 MPa in order to be able to store enough energy to obtain acceptable ranges for passenger cars, in particular the less efficient ones used in North America. While flasks for stationary uses are usually made of steel or aluminium-lined steel, weight considerations make composite fibre tanks more suitable for vehicle application. A typical design involves a carbon-fibre shell with inside polymer liner and outside reinforcement (in the US gunfire-proof). The first approved 70-MPa system holds 3 kg of hydrogen with an overall system weight of 100 kg (Herrmann and Meusinger, 2003).

Compression may take place at a filling station, receiving hydrogen from a pipeline. The energy requirement depends on the compression method. The work required for isothermal compression at temperature $T$ from pressure $P_1$ to $P_2$ is of the form

$$W = A\,T \log (P_2/P_1), \tag{2.46}$$

derived from the modified gas law

$$PV = AT \tag{2.47}$$

where $A$ is the gas constant $R = 4\,124$ J K$^{-1}$ kg$^{-1}$ times an empirical, pressure-dependent correction valid particularly for hydrogen and decreasing from 1 at low pressures to around 0.8 at 70 MPa (Zittel and Wurster, 1996; Herrmann and Meusinger, 2003). Adiabatic compression requires more energy input, and the preferred solution is a multistage intercooled compressor, which typically reduces the energy input to about half of the one for a single stage (Magazu *et al.*, 2003). For compression to about 30 MPa, the compression energy for the multistage concept is about 10 MJ/kg or some 10% of the energy in the hydrogen stored (Table 2.4). The further energy required for transferring the compressed gas to the tank in a vehicle is minute in comparison, and the transfer time for the flask sizes mentioned is under 3 min. Although this is some 60 times slower than the energy flow from a gasoline filling station to the car tank (Sørensen, 1984), it is still deemed acceptable.

Safety of compressed hydrogen tanks in proximity to vehicle passengers has been the subject of intense investigations, including crash tests and drops from considerable heights, as well as behaviour during fires, e.g., confined in long tunnels such as those found in many highway systems, leading traffic through mountains or under straits and other water bodies (Carcassi *et al.*, 2004; FZK, 1999). Incidences spanning from minor leaks to tank failure and explosion can usually not be handled by the safety distances used for stationary stores. Questions of safety standards will be discussed in section 5.2.

Power requirements of portable hydrogen equipment such as cameras, mobile phones or laptop computers currently covered by batteries, typically of lithium ion type, could with a small fuel cell and some 10-20 g of directly or indirectly stored hydrogen prolong operational time by a factor of 5-10. A discussion of such options using direct methanol fuel cells is made in section 4.6.

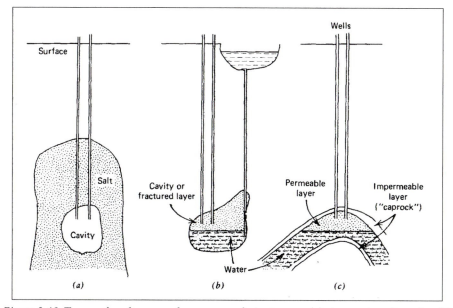

*Figure 2.48.* Types of underground compressed air storage: (a) storage in salt cavity, (b) rock storage with compensating surface reservoir and (c) aquifer storage (from B. Sørensen, *Renewable Energy,* 2004, used with permission from Elsevier).

For stationary hydrogen storage on a large scale, underground caverns or cavities are an appealing option, often offering storage solutions at a very low cost. Three possibilities of interest are salt dome intrusions, cavities in solid rock formations and aquifer bends.

Cavities in salt deposits (Fig. 2.48a) may be formed by flushing water through the salt. The process has already been used successfully in a number of cases in connection with storage of compressed natural gas. Salt domes are salt deposits extruding upwards toward the surface, therefore allowing cavities to be formed at modest depths. The former industrial hydrogen salt dome store at Teeside, UK, is currently being restored as a hydrogen technology demonstration site (Taylor *et al.*, 1986; Roddy, 2004).

Rock cavities (Fig. 2.48b) may be either natural or excavated, with walls properly sealed to ensure air-tightness. If excavated, they are considerably more expensive to make than salt caverns, but the latter only exists in a limited number of locations around the world.

Aquifers are layers of high permeability, permitting underground water flows along the layer. Gases may be stored in aquifers by displacing water (Fig. 2.48c), provided that the geometry includes upward bends allowing a gas pocket to be held in place by water on the sides. Above and below the aquifer must be layers of little or no permeability for the gas in question. This is often the case for clay layers. Aquifers that may be suitable for hydrogen storage are found in many parts of the world, except for areas dominated by rock all the way to the surface. These will have to use the more expensive rock storage shown in Fig. 2.48b.

The selection and preparation of sites for any of these gas stores is a fairly delicate process, because tightness can rarely be guaranteed on the basis of geological test drillings and modelling. The detailed properties of the cavity will not become fully disclosed until the installation is complete. The ability of the salt cavern to keep an elevated pressure may turn out not to live up to expectations. The stability of a natural rock cave, or of a fractured zone created by explosion or hydraulic methods, is also uncertain until actual full-scale pressure tests have been conducted. For the aquifers, the decisive measurements of permeability can only be made at a finite number of places, so surprises are possible due to rapid permeability change over small distances of displacement (Sørensen, 2004a).

The stability of a given cavern is further influenced by two factors: the temperature variations and the pressure variations. It is possible to keep the cavern wall temperature nearly constant, either by cooling the (possibly compressed) hydrogen before letting it down into the cavern or by performing the compression so slowly that the temperature only rises to the level prevailing on the cavern walls. The latter possibility (isothermal compression) is impractical for most applications because excess power must be converted at the rate at which it comes. Most systems therefore include one or more cooling steps. With respect to the problem of pressure variations, when the store holds different amounts of energy, the solution may be to store the hydrogen at constant pressure but variable volume (a possibility offered naturally by some aquifer stores). For the underground rock cavern, a similar behaviour may be achieved by connecting the underground reservoir to an open surface reservoir (indicated in Fig. 2.48b), so that a variable water column may take care of the variable amounts of hydrogen stored at the constant equilibrium pressure prevailing at the depth of the cavern. This kind of compressed energy storage system may alternatively be viewed as a pumped hydro storage system, with extraction taking place through gas-driven turbines rather than through water-driven turbines.

The aquifer storage system shown in Fig. 2.48c would have an approximately constant working pressure, corresponding to the average hydraulic pressure at the depth of the hydrogen-filled part of the aquifer. The stored energy $E$ may, in general, be expressed as

$$E = - \int_{V_0}^{V} P \, dV, \qquad (2.48)$$

considering the compressed gas cavern store at pressure $P$ as a cylinder with a piston at positions describing the volumes $V_0$ and $V$.

In the aquifer case, $E$ simply equals the pressure $P$ times the volume of hydrogen gas displacing water in the aquifer. This volume equals the physical volume $V$ times the effective porosity $p$, that is, the fractional void volume accessible to intruding hydrogen gas (there may be additional voids that the incoming gas cannot reach), so the energy stored may be written

$$E = pVP. \qquad (2.49)$$

Typical values are $p = 0.2$ and $P$ around $6 \times 10^6$ N m$^{-2}$ at depths of some 600 m, with useful volumes of $10^9$ to $10^{10}$ m$^3$ for each site. Such sites are already in use for storage of natural gas (cf. section 5.1.1).

An important feature of an energy storage aquifer is the time required for charging and emptying. This time is determined by the permeability of the aquifer. The permeability is basically the proportionality factor between the flow velocity of a fluid or gas through the sediment and the pressure gradient causing the flow. The linear relationship assumed may be written

$$v = - K (\eta\rho)^{-1} \partial P/\partial s, \qquad (2.50)$$

where $v$ is the flow velocity, $\eta$ is the viscosity of the fluid or gas, $\rho$ is its density, $P$ is the pressure, and $s$ is the path length in the downward direction. $K$ is the permeability. In SI units it has the dimension of m$^2$. The unit of viscosity is m$^2$ s$^{-1}$. Another commonly used unit of permeability is the *darcy*. One darcy equals $1.013 \times 10^{12}$ m$^2$. If filling and emptying of the aquifer storage are to take place in a matter of hours rather than days, the permeability has to exceed $10^{11}$ m$^2$. Sediments such as sandstone are found with permeabilities ranging from $10^{10}$ to $3 \times 10^{12}$ m$^2$, often with considerable variation over short distances.

In practice, there may be additional losses. The cap-rock bordering the aquifer region may not have negligible permeability, implying a possible leakage loss. Friction in the pipes leading to and from the aquifer may cause a loss of pressure, as may losses in the compressor and turbine. Typically, losses of about 15% are expected in addition to those of the power machinery. Large salt dome and aquifer stores in operation for natural gas are described in the implementation chapter (section 5.1.1), as they may have future uses for hydrogen, along with new installations in similar geological formations.

## 2.4.2 Liquid hydrogen storage

Liquid hydrogen storage requires refrigeration to a temperature of 20 K, and the liquefaction process requires an industrial facility expending a minimum of 15.1 MJ/kg. The actual energy spending is nearly three times higher for the refrigeration techniques available at the present. The liquefaction process requires very clean hydrogen, as well as several cycles of compression, liquid nitrogen or helium cooling, and expansion. The subsequent transfer to a filling station and from there to a store within a vehicle (assuming automotive application) uses little energy by comparison, and like the situation for compressed hydrogen filling can be accomplished in a few minutes. Pressures used are only slightly above the atmospheric, 0.6 MPa being a typical value. The liquid hydrogen storage technology was initially developed for space vehicles.

A vessel for storing liquid hydrogen consists of several layers (of metal) separated by highly insulating material, often involving low "vacuum-like" pressures of about 0.01 Pa and a resulting heat conductivity of 0.05 W $K^{-1}$ (Chahine, 2003). Including the entire storage device, the storage capacity by volume becomes roughly half the figures given in Table 2.4, for store sizes of the order of 130 litres. This means that liquefaction energy amounts to at least 30% of the stored energy.

A significant problem is the boil-off of hydrogen from the store caused by the need to control tank pressures (associated with the heat release of ortho- to para-$H_2$ conversion) by venting valves. Insulation properties influences the boil-off, which usually starts after a dormancy period of at most a few days and then proceeds at a level of 3-5% per day (Magazu *et al.*, 2003), with the lower values obtained by installing heat exchangers between outgoing cold hydrogen and air intake to regulate pressure. The boil-off has security implications, e.g. for cars parked in a garage. Boil-off limits the usefulness of liquid vehicle storage except where a fairly continuous driving pattern is involved.

## 2.4.3 Hydride storage

Molecular hydrogen is dissociated into hydrogen atoms in the neighbourhood of metals and certain compounds as illustrated in Fig. 2.5. If the lattice structure of the metal or alloy is suitable, there are interstitial positions capable of accommodating the relatively small hydrogen atoms. The energy associated with these changes can be quite modest, as demonstrated by the thermodynamical plot of pressure versus hydrogen concentration shown in Fig. 2.49. Heat is released when hydrogen enters the lattice and heat must be supplied to drive the hydrogen out of the lattice again. The mechanisms involved in Fig. 2.49 can be understood in terms of chemical thermodynamics (see, e.g., Morse, 1964).

### Chemical thermodynamics

Consider a chemical reaction, which may be described in the form

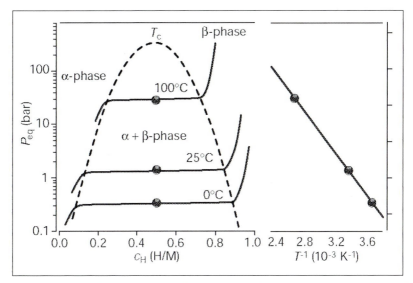

*Figure 2.49.* The left side of the figure shows the equilibrium pressure (1 bar ≈ 0.1 MPa) as a function of the hydrogen concentration (hydrogen to metal ratios, the metal being here LaNi$_5$) at three different temperatures. For temperatures below a critical value, the non-equilibrium α-phase (metal lattice intact) and β-phase (lattice expanded/ modified by the presence of hydrogen atoms) are separated by a pressure plateau region where the reactants remain in thermodynamical equilibrium during the absorption of more hydrogen. In the right-hand part of the figure, the logarithm of the plateau pressure is plotted against inverse temperature. From L. Schlapbach and A. Züttel (2001). Hydrogen-storage materials for mobile applications, *Nature* **414**, 353-358, reprinted by permission.

$$\sum_{i=1}^{N} v_i M_i = 0, \qquad (2.51)$$

where $M_i$ is the $i$th substance (reactant or product) and $v_i$ is its stoichiometric coefficient. As the reaction proceeds, the amounts of each substance changes, and eventually an equilibrium is formed. Introducing the concentration, $\chi_i$, defined as the mole fraction of the $i$th substance, the stoichiometry (2.51) requires that changes of the substances involved have the relationship

$$d\chi_i = v_i \, dx \text{ with the same } dx \text{ for all } i. \qquad (2.52)$$

For near-ideal multi-component gases, it is possible to use macroscopic average variables in a thermodynamical description, and write the change in the Gibb's free energy $G$ as

$$dG = \sum_{i=1}^{N} \mu_i v_i \, dx, \qquad (2.53)$$

where $\mu_i$ is the chemical potential of the $i$th substance. The equilibrium situation is characterised by a minimum of $G$. The quantity

$$K_P = \prod_{i=1}^{N} P_i^{v_i} \qquad (2.54)$$

is called the equilibrium constant. It is expressed in terms of the partial pressures $P_i$ of the $i$th substance raised to a positive power for the reaction products and to a negative power for the reactants. For ideal gases, the chemical potentials may be written as

$$\mu_i = \mu_i^0 + RT \log(P_i), \qquad (2.55)$$

where $R$ is the gas constant and the superfix "0" denotes a reference situation. Inserting (2.55) into (2.53), one gets for the equilibrium situation

$$\Delta G^0 = \sum_{i=1}^{N} v_i \mu_i^0 = -RT \sum_{i=1}^{N} \log(P_i) = -RT \log K_P . \qquad (2.56)$$

Introducing the enthalpy from the definition $H = G + TS$, the temperature dependence of the quantities in (2.56) can be expressed by van't Hoff's equation

$$\frac{d \log K_P}{dT} = \frac{\Delta H^0}{RT^2} . \qquad (2.57)$$

In integrated form this reads (using $\Delta H^0 = \Delta G^0 + T\Delta S^0$ to fix the integration constant)

$$\log K_P = -\frac{\Delta H^0}{RT} + \frac{\Delta S^0}{R} \qquad (2.58)$$

Finally, with use of $P_i = \chi_i P$ and the equilibrium conditions, one can rewrite (2.58) in the form (given in a slightly different way by Morse, 1964; Schlapbach and Züttel, 2001)

$$\log P - \log P^0 = -\frac{\Delta H}{RT} + \frac{\Delta S}{R} \tag{2.59}$$

valid for the equilibrium (plateau) values of $P$ and $P^0$ and at a given temperature $T$. This equation describes the linear curve in the right-hand side of Fig. 2.49, and it allows determination of the plateau pressures as a function of temperature illustrated in the left-hand side of Fig. 2.49. The dominant contribution to $\Delta S$ is the dissociation energy of hydrogen molecules, $-130$ J K$^{-1}$ mol$^{-1}$, and for the $T = 300$ K (25°C) curve to have its plateau pressure equal to a normal atmospheric sea-surface value of 0.1 MPa ($\approx 1$ bar), $\Delta H$ must be $-39$ kJ mol$^{-1}$ (of hydrogen; Züttel, 2004). Actual enthalpy values for different substances are discussed below.

**Metal hydrides**

Some metal alloys can store hydrogen at volume densities more than twice that of liquid hydrogen. However, if the mass storage density is also important, as is normally the case in automotive applications, then storage densities are still only 10% or less, compared to those of conventional fuels (see Table 2.4 and the summary in Fig. 2.54). This makes the concept doubtful for mobile applications, but interesting for decentralised hydrogen storage (as required, e.g., in one of the scenarios described in section 5.4), due to the positive aspects of near loss-free storage at near-ambient pressures (0.06–6 MPa) and high safety in operation with hydrogen transfer accomplished by adding or withdrawing modest amounts of heat, according to

$$Me + \tfrac{1}{2}\, x H_2 \leftrightarrow MeH_x, \tag{2.60}$$

where $Me$ stands for a metal or metal alloy, such as a binary $A_m B_n$ or higher order alloy. Among single-metal hydrides, $MgH_2$ and $PdH_{0.6}$ are the most well-studied ones, with hydrogen mass percentages of 7.6 and 0.6 and decomposition temperatures of 330 and 25°C (Grochala and Edwards, 2004). Binary and higher metal alloys generally store similar amounts of hydrogen by volume, but less hydrogen by mass (see Fig. 2.54).

The hydride of the composition $MgH_2$ seems the most interesting single-metal hydride, with its fairly high hydrogen fraction of 7.6% by mass. Although the desorption temperature is higher than ambient (about 330°C), it could be made available in motorcars, although this would lower the overall efficiency. The enthalpy of the exothermal transfer of hydrogen into the metal lattice is $-74.5$ kJ mol$^{-1}$ (Sandrock and Thomas, 2001), and Fig. 2.50 shows the tetragonal crystal structure of $MgH_2$. Quantum chemical calculations are used to follow the swelling of the lattice with incorporation of hydrogen in sheets or bulk materials (Shang et al., 2004; Liang, 2003a). The kinetic restraints at near-ambient pressures make the absorption and desorption processes very slow, and desorption times in the range of hours are unacceptable, at least for mobile applications (see also the modelling subsection at the end of this section).

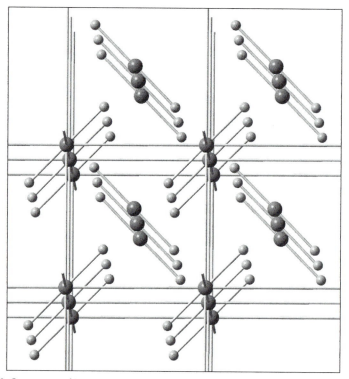

*Figure 2.50.* Structure of $MgH_2$ proposed on the basis of X-ray spectroscopy and quantum chemical calculations (see text). The larger atoms are Mg, and the smaller ones are hydrogen.

Other binary hydrides can be formed with light elements, without improvement in performance. The structure of NiH is shown in Fig. 2.51. For heavy metals, only palladium has received some attention, due to rapid absorption and desorption at ambient temperature and pressure. However, only 0.6% hydrogen by mass is stored, which together with high metal cost makes this option unattractive.

Hydrides with two metal constituents studied include a range of Mg alloys, including interesting ones with Ni and Fe, as well as alloys containing Al, Fe, Ni, Ti and La, as well as other metals. Of the total number of atoms, say $6 \times 10^{28}$ per $m^3$, the hydrogen fraction is usually below the value of 2 for $MgH_2$. The mass fraction achieved for these hydrides is often below 2%, with the $LaNi_5H_7$ illustrated in Figs. 2.52 and 2.53 as a typical example. Its volumetric storage density is about 115 kg $H_2$ per $m^3$, and the enthalpy evolution during absorption is $-30.8$ kJ mol$^{-1}$ (Sandrock and Thomas, 2001). While the maximum number of hydrogen atoms that can be accommodated within the LaNi$_5$ hexagonal structure is 7 (Fig. 2.53), sometimes only 6 are incorporated, possibly due to the swelling of those unit cells that already have accepted 7 H atoms. Attempts have been made to perform quantum chemical calculations to explore these effects and to determine the locations of absorbed hydrogen atoms, which is not at midpoints of the interstitial sites but must be strongly influenced by the Coulomb binding forces exerted by the lattice atoms, cf. Fig. 2.52 (Tatsumi *et al.*, 2001; Morinaga and Yukawa, 2002).

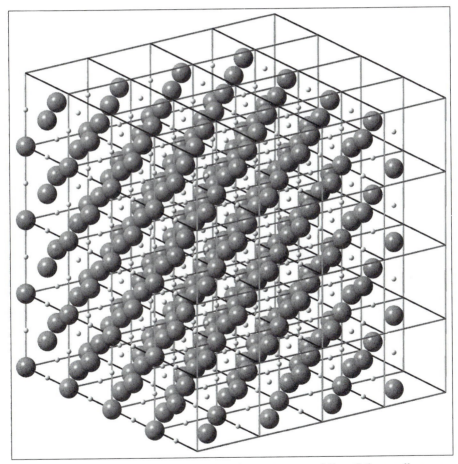

*Figure 2.51.* Structure of an NiH lattice. The large atoms are Ni, and the small ones are H. Note that the Ni lattice structure with Miller indices (111) is the same as that used in the catalyst reaction modelling shown in Figs. 2.5 and 2.6.

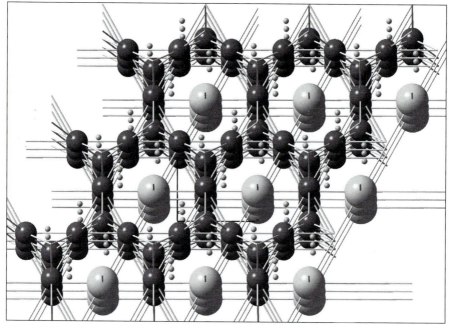

*Figure 2.52.* Structure of LaNi$_5$H$_7$ proposed on the basis of X-ray spectroscopy and quantum chemical calculations (see text). The atoms are illustrated with declining size from La over Ni to H (not all of the H's are visible, cf. Fig. 2.53).

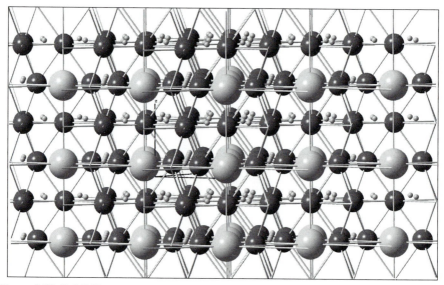

*Figure 2.53.* LaNi$_5$H$_7$ as in Fig. 2.52, but from an angle better showing the hydrogen atom positions. The atoms are illustrated with declining size from La over Ni to H.

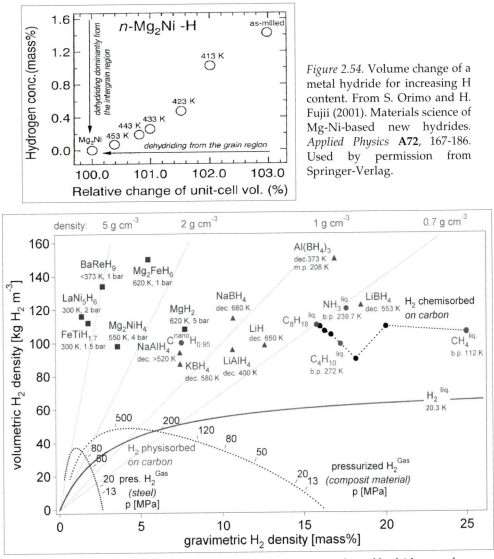

*Figure 2.54.* Volume change of a metal hydride for increasing H content. From S. Orimo and H. Fujii (2001). Materials science of Mg-Ni-based new hydrides. *Applied Physics* **A72**, 167-186. Used by permission from Springer-Verlag.

*Figure 2.55.* Overview of hydrogen storage properties. A number of hydrides are characterised by mass and volume $H_2$ densities (axes) as well as by simple density (scale on top). The circles in the right-hand part of the figure represent chemisorbed hydrogen on carbon, and for comparison, curves (in lower part of the figure) illustrate the properties of liquid and compressed hydrogen storage in steel and composite containers. From A. Züttel (2004). Hydrogen storage methods. *Naturwissenschaften* **91**, 157-172. Used by permission from Springer-Verlag. See also discussion in Züttel et al. (2004). and in Schlapbach and Züttel (2001).

The swelling as a function of hydrogen absorbed is shown in Fig. 2.54, for another well-studied hydride, $Mg_2NiH_4$, the storage capacity of which is seen in Fig. 2.55. The lattice absorption cycle also performs a cleaning of the gas, because impurities in the hydrogen gas are too large to enter the lattice.

The metal hydride with the highest mass percentage of hydrogen found so far is $Mg_2FeH_6$ (Fig. 2.55), but decomposing at a temperature near 400°C for ambient pressure (Fig. 2.56). Several higher alloys have been studied, e.g., $LaNi_{4.7}Al_{0.3}$ based on the alloy shown in Fig. 2.52 (Asakuma *et al.*, 2004). Increased stability is achieved by such substitution, but without definite improvements in storage properties (Züttel, 2004). The van't Hoff plots (as in the right-hand side of Fig. 2.49) for this and other alloys are shown in Fig. 2.56, again showing the difficulty in bringing the properties needed for automotive applications together.

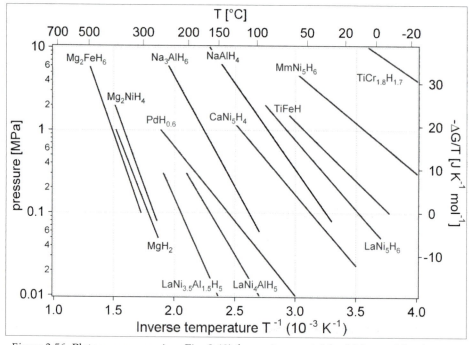

*Figure 2.56.* Plateau pressure (see Fig. 2.49) for various metal hydride combinations as a function of inverse temperature. From A. Züttel (2004). Hydrogen storage methods. *Naturwissenschaften* **91**, 157-172. Used by permission from Springer-Verlag.

The time needed for absorption depends not only on the alloy used, but also on its physical properties such as grain size. Figure 2.57 shows how dramatic such differences can be, using as an example the $LaNi_5$ alloy discussed above, with a catalyst helping the absorption process. The range of absorption and desorption times encountered in the alloys studied is illustrated in Fig. 2.58. The storage charge and discharge times are for most alloys in the range of 30-60 min, but there are ranges down in the area of 1 min, suitable for automotive applications. These require additives based on vanadium and carbon, as well as elevated temperatures and for absorption pressures of 10-15 times the ambient pressure.

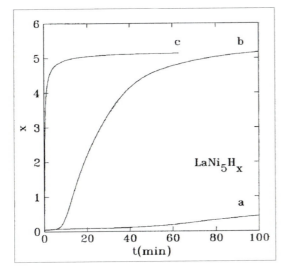

*Figure 2.57.* Hydrogen absorption in different forms of LaNi$_5$: a, polycrystalline; b, nanocrystalline; c, nanocrystalline with addition of catalyst. From A. Zaluska, L. Zaluski, J. Ström-Olsen (2001). Structure, catalysis and atomic reactions on the nano-scale: a systematic approach to metal hydrides for hydrogen storage. *Appl. Phys.* **A72**, 157-165. Used by permission from Springer-Verlag.

Models for following the hydrogen diffusion process may be based on Monte Carlo simulation or network simulation incorporating both a regular lattice structure and irregular grain regions (Herrmann *et al.*, 2001).

A maximum hydrogen storage capacity in chemical structures may be estimated from close packing of hydrogen atoms with a distance of about 0.2 nm, as closer distances are unlikely if lattice atoms are to be accommodated (cf. Problem 2.6.2 in section 2.6). Some of the metal hydrides in Fig. 2.55 are not far below the implied density, so emergence of novel miracle materials is not to be expected.

Although failure to meet all the criteria required for use of metal hydride stores in automobiles so far has barred their introduction, there are in fact manufacturers offering the technology for special stationary purposes, where, e.g., safety considerations disfavour use of compressed or liquid hydrogen tanks. The products offered have dealt with the problem of expansion in different ways. One company offers rolled up mats of hydride material (and heating coils) with generous room for expansion (the "Hy-Stor" trademark of Ergenics, Ringwood, NJ, owned by Hera in Quebec), while other products use stacks of boxes filled with granular hydride material (the simplest design using cylindrical boxes with a centre hole for heat application equipment).

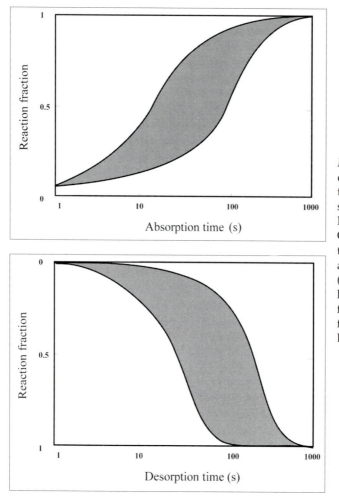

*Figure 2.58a,b.* Kinetics of hydrogen absorption (a: top) and desorption (b: below) in Mg alloys with V and C additives, at the temperature 350°C and pressure 1.4 MPa (absorption) and 10 kPa (desorption). The fastest reactions are for $MgH_2$. Based on Liang, 2003a.

## Complex hydrides

The term complex hydrides has been used for metal hydrides thought to be involved in complex bindings, but the distinction as a separate class is perhaps not entirely warranted, because the nature of the quantum chemical binding exhibits a sliding transition between the different classical binding types. The hydrides called *complex* typically involve light atoms, which in general terms would seem suitable for high hydrogen-to-metal ratios. Hydrogen is often not located in the most obvious interstitial sites, but rather near certain lattice atoms as a result of favoured binding to the atoms in these locations. This is also the case in the metal hydride shown in Fig. 2.52, exposing the gradual scale of assigning hydrides as metal or complex. An equally important parameter is the macroscopic structure of the material, from solid lattices through grains of different sizes to nanoporous structures with a high ratio of surface to bulk atoms.

Figure 2.55 shows mass percentages up to 18 for $LiBH_4$ at about 280°C, a substance first synthesised by Schlesinger and Brown (1940). A similar compound, $NaAlH_4$, shows reversibility with absorption and desorption at the more convenient temperatures of about 195°C, but with a considerably smaller mass percentage (Bogdanovic and Schwickardi, 1997; the 7.5% indicated in Fig. 2.55 includes H atoms not involved in the reversible storage reactions, as do the other complex hydride entries in the figure). The reactions involved for these substances called *alanates* are more complicated than simple water dissociation and diffusion into the lattice:

$$6NaAlH_4 \rightleftarrows 2Na_3AlH_6 + 4Al + 6H_2 \rightleftarrows 6NaH + 6Al + 9H_2 \qquad (2.61)$$

$$2Na_3AlH_6 \rightleftarrows 6NaH + 2Al + 3H_2 \qquad (2.62)$$

usually complemented by reactions for an accompanying Li-substituted alanate,

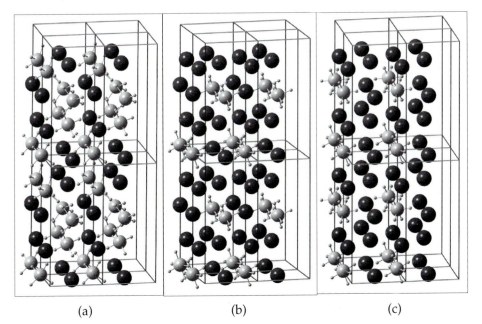

(a)             (b)             (c)

*Figure 2.59.* Molecular structure of $NaAlH_4$ (a) and $Na_3AlH_6$ in its α-form (b) and β-form (c). The small grey atoms are H, the large dark ones are Na, and the large lighter ones are Al.

$$2Na_2LiAlH_6 \rightleftarrows 4NaH + 2LiH + 2Al + 3H_2. \tag{2.63}$$

The four reactions above in isolation each contribute 3.7, 5.5, 3.0 and 3.5% to the hydrogen mass percentage, which for the combined reaction system is about 5.6%. Figures 2.59a-c show the molecular structure of the sodium alanates. In particular, $Na_3AlH_6$ has two forms of which the simple face centred cubic structure (Fig. 2.59c) only dominates at temperatures above 252°C (Arroyo and Ceder, 2004).

The complex sequences of reactions makes the filling and emptying of the storage system slow, and although speed can be increased by use of catalysts (typically Ti) in the reactions (2.61)-(2.63), hydrogen evolution still involves periods of 10-30 h (Bogdanovic *et al.*, 2000). This makes the complex hydride systems unsuitable for automotive uses, in addition to questions of stability (Nakamori and Orimo, 2004).

In the lithium borohydride case, the "desorption" process of regaining hydrogen is (Züttel *et al.*, 2003)

$$2LiBH_4 \rightarrow 2LiH + 2B + 3H_2, \tag{2.64}$$

a process proceeding at 300°C under influence of an $SiO_2$ catalyst, with a hydrogen mass yield of 13.8%. The choice of catalyst is not obvious. A Ti catalyst works for the reactions (2.61)-(2.63) and for $Li_3AlH_6$, but not for $LiAlH_4$. An explanation has been sought in terms of the energy changes following absorption of Ti atoms to a position well into the surface structure (Løvvik, 2004). Charging the system with hydrogen by the process reverse to (2.64) is possible, but complete reversibility has so far only been achieved at elevated pressure (10 MPa) and temperature (550°C) (Sudan *et al.*, 2004).

If the questions of efficient synthesis, catalyst use and stability are resolved, there are still issues of cost and safety to be addressed. The complex hydrides and related structures are relatively new research areas, where new ideas keep coming up. For example, hydrogen storage may occur in zeolites at a temperature of 77 K (Züttel *et al.*, 2004), and in metal-organic frameworks (Rosi *et al.*, 2003), with hydrogen mass adsorption on surface sites changing from 1 to 4.5%, as the temperature is decreased from ambient to 78 K. Another avenue of investigation is clathrate hydrates at 145 K (Mao *et al.*, 2002). Clathrates are water-ice structures, which form cages capable of accommodating extra hydrogen atoms (Sluiter *et al.*, 2003).

### Modelling metal hydrides

In order to understand (or predict) the energetic advantage for hydrogen of penetrating a metal lattice and forming a hydride, quantum calculations of the chemical structures involved may be performed.

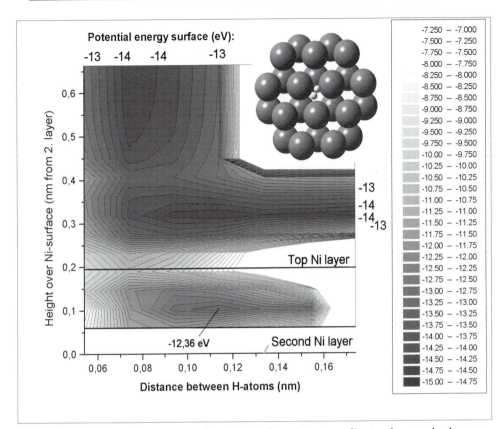

*Figure 2.60.* Potential energy surface in eV along two co-ordinates for two hydrogen atoms near a nickel surface: distance between hydrogen atoms $d$ and height $z$ of the centre of mass of the two H atoms over the (1,1,1) surface of the Ni-layer below the top layer. The quantum chemical calculation uses the density functional theory B3LYP and a set of basis functions called SV (see section 3.2). The energy scale has an arbitrarily chosen origin (Sørensen, 2004f).

Consider first the behaviour of hydrogen approaching a nickel surface. This was modelled in Fig. 2.5 in connection with the description of the ability of a metal catalyst such as Ni to split the hydrogen molecule into two hydrogen atoms. This process was shown to take place about 0.1 nm outside the Ni surface. It is easy to extend the calculation to see what happens if the hydrogen atoms are allowed to penetrate into the Ni lattice, simply by using the same method to calculate potential energies for H positions below the top Ni layer. The result is shown in Fig. 2.60.

In order to penetrate into the Ni lattice, the two hydrogen atoms must pass a barrier that is lowest when the two H atoms form a molecule ($d = 0.076$ nm).

The barrier is considerably higher than the dissociation barrier outside the top Ni surface. If the hydrogen atoms get inside the lattice, they can reach a fairly deep minimum in potential energy, located midway between the two Ni-layers and at a hydrogen atom distance of $d \approx 0.12$ nm. The position of the Ni atoms is further away. One Ni atom is responsible for the rise in potential energy at the right-hand side of Fig. 2.60 (white area in top Ni layer). The next Ni atom is at the left-hand side of Fig. 2.60, just to the left of the boundary but not in the same plane as the one spanned by the H atoms and the Ni atom in the white area. So the most favourable place for the H atoms inside the Ni structure is as expected midway between the layers and approximately midway between the Ni atom positions. However, the depth of the potential energy minimum is nearly 2 eV less than that of separating the H atoms outside the Ni surface. As a result, Ni is not a suitable material for absorption of hydrogen and forming hydrides. Ni has this property in common with the other metal catalysts in the same group, Pd and Pt.

The metals of group II in the periodic system comprise magnesium, the metal identified above in the metal hydride subsection as forming a metal hydride without needing a binary metal structure. It is thus interesting to see if a quantum chemistry calculation will show magnesium as better suited for absorbing hydrogen than nickel.

This calculation presented below is carried out using density functional theory with periodic boundary conditions and fast multipole techniques (Kudin and Scuseria, 1998, 2000), rather than explicitly treating the periodic lattice as a single large molecule in the way done for the Ni hydride above. Figure 2.61 (top) shows the optimised structure of $MgH_2$, with H atoms occupying positions within the lattice leading to a minimum in potential energy. In order to compare this energy to that of the same system but with hydrogen atoms outside the Mg lattice, a sequence of calculations were made, pulling the hydrogen atoms out of the lattice but preserving their relative positions among them. This allows the periodic boundary condition method to be used. Figure 2.61 (bottom) shows a situation with the hydrogen set pulled out 0.5 nm from the equilibrium position. Figure 2.62 shows the map of the potential energy surface, as function of $x$, the pull direction perpendicular to the Mg lattice surface, and $y$, the co-ordinate describing displacements parallel to the Mg surface. Figure 2.63 shows the potential energy curve along the $y = 0$ line of displacement (the potential minimum for fixed $x$ as a function of $y$, as seen in Fig. 2.62). For completeness, the potential energy of fully separated hydrogen atoms (calculated as the energy of the Mg lattice alone, plus the energy of the hydrogen atoms alone) is also indicated. All potential energies are given per unit cell of 2 Mg atoms and 4 H atoms, with arbitrary origin but the same for all displacement calculations.

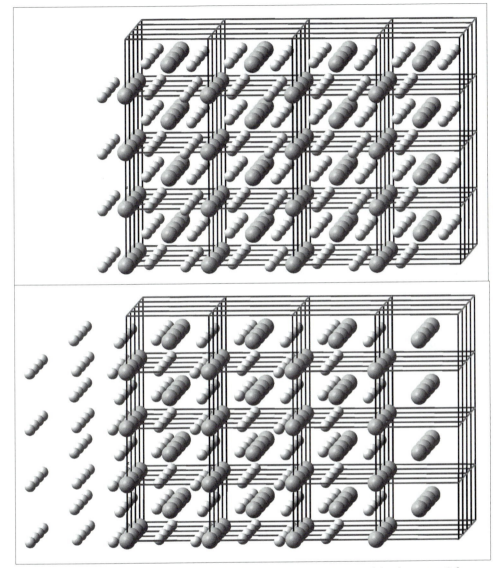

*Figure 2.61.* Two configurations of Mg-2H periodic structure used in the potential energy calculations of Figs. 2.62-2.63. The situation on top corresponds to the $MgH_2$ hydride equilibrium configuration (same as the one shown in Fig. 2.50), while the H atoms in the lower figure have been moved 0.5 nm out of the Mg lattice, to the left along the negative x-axis perpendicular to one of the lattice surface planes (Sørensen, 2004f).

It is seen, first of all, that the $MgH_2$ metal hydride indeed has a potential energy 3.5 eV (per unit cell) lower than that of the Mg and H atoms separated,

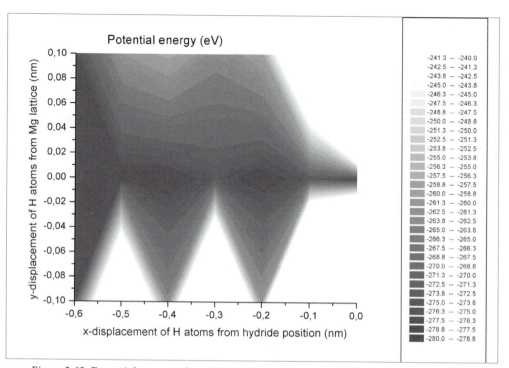

*Figure 2.62.* Potential energy plot of Mg-2H system as a function of displacement of the H atoms in two directions relative to the equilibrium position within the hydride. See text for further explanation (Sørensen, 2004f).

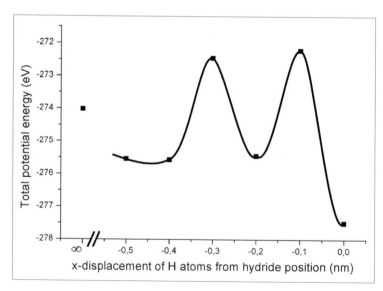

*Figure 2.63.* Potential energy plot for Mg-2H system as a function of $x$-displacement (as in Fig. 2.62 but with $y$ fixed at zero). See text for further explanation (Sørensen, 2004f).

thereby substantiating the ability of the Mg lattice to absorb hydrogen without energy added externally. On their way in, the hydrogen atoms make the potential energy oscillate in concord with the positions of Mg atomic centres, with an excursion range of 3 eV but at all times more than 2 eV higher than the final hydride energy (Fig. 2.63). The $y = 0$ path into the Mg lattice allows the most free passage, as seen by the effect of varying $y$ (Fig. 2.62), which is to rapidly increase the potential energy. The calculation gives the enthalpy of formation at standard temperature and pressure as $-71.93$ kJ mol$^{-1}$. The experimental value at 435°C is $-75.2$ kJ/mol (Bogdanovic *et al.*, 2000).

Details of the calculation are the use of the SV basis (Schaefer *et al.*, 1992) with automatic optimisation (Gaussian, 2003), and the PBEPBE parametrisation of the density functional exchange and correlation parts (Perdew *et al.*, 1996; cf. the discussion in section 3.2). The periodicity was allowed to prevail for about 1.6 nm in every direction, which was verified to be enough to ensure stability of calculated energies.

Figure 2.64 shows the electron wavefunctions in one unit cell of the system, illustrated by the density surface where the squared density has fallen to 0.05. It is seen that the "size" of the wavefunction parts surrounding Mg and H atoms is very similar, denouncing the common explanation of hydrogen absorption in metal lattices based on hydrogen being "such a small atom".

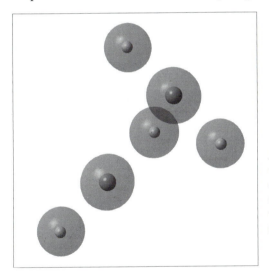

*Figure 2.64.* Isodensity surfaces (at the value of 0.05) for the Mg (darkest) and H (lighter shaded) atoms of a unit cell in the equilibrium structure of the metal hydride MgH$_2$ (Sørensen, 2004f).

## 2.4.4 Cryo-adsorbed gas storage in carbon materials

Hydrogen molecules may adsorb not just to metal surfaces, but also to surfaces of various solid materials including carbon, qualitatively following an adsorption and/or chemisorption curve of the type shown in Fig. 2.1. Adsorption of hydrogen takes place some 0.1 nm above the surface, in a single (mono) layer at a suitable low temperature (but above the hydrogen liquid phase transition temperature of 20 K). Too high a temperature leads to losses of adsorbed molecules by thermal disturbance, and for the only substance studied extensively so far, carbon, suitable temperatures are around the boiling point of nitrogen (77 K) and at pressures of about 10 MPa. Because only a monolayer is formed, the surface area must be as large as possible in order to get interesting storage parameters such as hydrogen mass or volume fraction. For a fully occupied monolayer, the hydrogen concentration is $1.2 \times 10^{-5}$ mol $m^{-2}$. The maximum amount of hydrogen adsorbed is thus directly proportional to the carbon surface area, and for thin sheets of graphene, activated carbon or graphite, the proportionality factor is about 2% by mass (H/(C+H)) per $m^2$ $kg^{-1}$ carbon surface area, with 1.5% established in practice (Nijkamp *et al.*, 2001; Züttel, 2004). Substances like activated carbon contain micropores that allow adsorption through surface areas 2-3 orders larger than the geometrical surface. Compared with storage in liquid hydrogen, the difference between the 20 and 77 K temperature should reduce filling costs and reduce problems with escaping hydrogen. The mass percentage seems too small for being attractive in mobile applications (although the company Zevco in the UK has proposed the concept in connection with its alkaline fuel cell taxi prototype). Theoretical modelling of the physisorption (i.e., not chemisorption, cf. Fig. 2.1) process has been done using Monte Carlo simulation (Williams and Eklund, 2000).

The use of nanotubes as the carbon material attracted some attention a few years back. However, it has become clear that nanotubes cannot store hydrogen gas in its voids, but only as adsorbed to the sides, implying a storage capability no greater than that of any other carbon surface (in fact smaller due to the curvature of the surface) (Zhou *et al.*, 2004).

Surface absorption of hydrogen on metal surfaces is a well-known phenomenon in the electrodes of batteries. It may also be at work for electrodes made of nano-structured carbon and accomplishes nearly the same mass percentage as cryogenic storage for suitable electrolytes, but at ambient temperature. Charging or discharging takes several hours or even days, so the concept is unsuitable for use in cars (Jurewicz *et al.*, 2004).

## 2.4.5 Other chemical storage options

Once again the distinction between different types of chemical hydrogen storage is a bit artificial, as reactions beyond dissociation were already involved in several of the forms described above and notably in the use of hydrides called complex. But generally, any reversible reaction scheme with different amounts of hydrogen on either side of the equation may be contemplated as a hydrogen storage device, and even a non-reversible device may serve the process if it can be manufactured with a hydrogen content that may be recovered under controlled circumstances, e.g., in a vehicle. Thus, the methanol reformer in a car storing methanol but converting it to hydrogen before using it in a fuel cell may be considered such a device, with hydrogen stored in the form of methanol.

A number of other chemical reactions have been contemplated for storage applications. Here only a single example shall be mentioned, involving the hydrocarbons decalin and naphthalene,

$$C_{10}H_{18} \rightleftarrows C_{10}H_8 + 5H_2, \tag{2.65}$$

which proceeds to the right at ambient pressure and a temperature of T ≈ 200°C, with a platinum-based catalyst and heating corresponding to an enthalpy supply of $\Delta H^0 = 8.7$ kWh kg$^{-1}$ of H$_2$ (Hodoshima *et al.*, 2001). The reaction proceeds to the left at ambient or slightly lower temperatures (5°C). The weight percentage of hydrogen in decalin is 7.3, which is favourable compared to most of the hydrides discussed above (although also here, only a part of the hydrogen is available as a gas on the right-hand side of (2.65)).

Comments on the practical options for use of various types of storage in hydrogen energy carrier systems are given below in section 2.4.6 and, for particular implementations, in Chapter 4 and section 5.1.

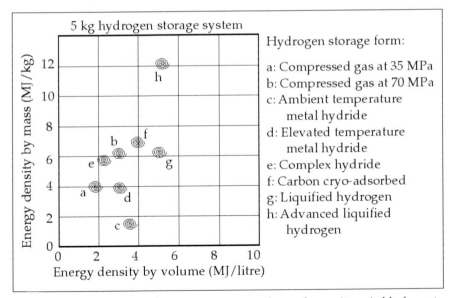

*Figure 2.65.* Energy density for storage systems of a total capacity suitable for automotive applications. Densities are estimated for a total package solution in each case.

## 2.4.6 Comparing storage options

The choice of a suitable hydrogen store depends on the application aimed for and should compare options at a system level.

Figure 2.65 summarises the energy density by volume and mass for systems considered for use in fuel cell passenger cars, based on the data from the preceding subsections, supplemented with estimates of system package overheads (containment, safety features, controls) according to Herrmann and Meusinger (2003). No presently available system is satisfactory on a mass basis, and only liquid hydrogen stores approach viability on a volume basis. At the same time, the latter systems have undesirable traits related to boil-off losses.

Similar assessments may be made for other applications, notably stationary uses, where the requirements in most cases are less stringent. An alternative tool in evaluation of system performance is a plot of energy density versus power density. Some of the systems under consideration (e.g., many of the metal hydride stores) have filling and delivery times leading to power densities too low for many applications.

# 2.5 Hydrogen transmission

## 2.5.1 Container transport

Hydrogen can be transported in containers of compressed or liquid hydrogen, as well as emerging storage forms in hydrides or other chemical substances, as described in section 2.4.

Intercontinental hydrogen transport, which may become required if conditions for hydrogen production are geographically distributed in a way different from that of demand, may make use of containers transported on ships, similar to those transporting liquefied natural gas today. Because the density of liquefied hydrogen is so much smaller than that of natural gas, the transportation costs are going to be higher. In addition, there are problems with the leakage from the containers (section 2.4.2), and safety in case of containment accidents from on-board causes, including those related to hydrogen loading and unloading or caused by ship collisions. Conceptual designs include spherical or cylindrical containers (Abe *et al.*, 1998). Intercontinental transport of hydrogen by ship has been estimated to cost around 25 US$/GJ or 3 US$/kg (Padró and Putsche, 1999).

Alternative materials for hydrogen storage during long-range transport are, as mentioned in section 2.4.5, methanol and higher hydrocarbons. The high-temperature reactions listed in Table 2.5 allow methane and other hydrocarbons to be transformed into producer gases with high hydrogen content, typically with more modest temperature requirements for higher hydrocarbons (such as the decalin mentioned in Eq. (2.65)).

| Closed-loop system | Enthalpy[a] $\Delta H^0$ (kJ mol$^{-1}$) | Temperature range (K) |
|---|---|---|
| $CH_4 + H_2O \leftrightarrow CO + 3H_2$ | 206 (250)[b] | 700–1200 |
| $CH_4 + CO_2 \leftrightarrow 2CO + 2H_2$ | 247 | 700–1200 |
| $CH_4 + 2H_2O \leftrightarrow CO_2 + 4H_2$ | 165 | 500–700 |
| $C_6H_{12} \leftrightarrow C_6H_6 + 3H_2$ | 207 | 500–750 |
| $C_7H_{14} \leftrightarrow C_7H_8 + 3H_2$ | 213 | 450–700 |
| $C_{10}H_{18} \leftrightarrow C_{10}H_8 + 5H_2$ | 314 | 450–700 |

Table 2.5. High-temperature, closed-loop chemical C-H-O reactions (Hanneman et al., 1974; Harth et al., 1981).
[a] Standard enthalpy for complete reaction.
[b] Including heat of evaporation of water.

The transport of hydrogen in the form of chemical compounds would seem to reduce losses as well as costs, relative to transport of liquid hydrogen or the in any case likely to be too bulky compressed gaseous hydrogen.

For shorter distances of transport, e.g., from central stores to filling stations, all the forms of transport can in principle be contemplated. The conversion or liquefaction/deliquefaction processes may carry too high energy losses and costs, so that despite the bulkiness of compressed hydrogen it may be a more acceptable solution.

## 2.5.2 Pipeline transport

The alternative for hydrogen transport over short or medium distances is pipelines, with the viable distances being determined by ease and cost of establishing pipelines, usually over land but conceivably also in some cases off-shore. Again, the technology is basically already developed for natural gas transport, and only modest alterations are believed to be required in order to contain the smaller hydrogen molecules with acceptable leakage rates.

Current polymer pipes used for transportation of natural gas at pressures under 0.4 MPa have sufficient strength to carry hydrogen, and diffusion losses arising mainly from pipe connectors have been estimated at three times higher for hydrogen than for natural gas (Sørensen et al., 2001). Transmission of natural gas at pressures up to 8 MPa uses steel pipes with welded connections. Most metals can develop brittleness after absorption of hydrogen, especially at locations contaminated with foreign substances (dirt or $H_2S$) (Zhang *et al.*, 2003). Adding a little oxygen (about $10^{-5}$ by volume) inhibits these effects. Inspection programmes similar to those carried out today should suffice to make the risk comparable to that for natural gas transmission. Auxiliary components such as pressure regulators and gauges are not expected to cause problems, although lubricants used should be checked for hydrogen tolerance. Like for natural gas pipelines, spark ignition is a potential hazard. If hydrogen is distributed to individual buildings, gas detectors must be installed, and outdoor emergency venting devices should be considered. Hydrogen has previously been extensively used in town gas systems. Recent test installations in the USA and Germany have been investigated (Mohitpour *et al.*, 2000). Hydrogen pipeline costs are estimated at about 625 000 US$/km (Ogden, 1999).

# 2.6 Problems and discussion topics

### 2.6.1
Try to estimate the amount of hydrogen storage required if all electricity demand in your region should be covered either from wind power alone or from photovoltaic power alone (produced at your latitude). Compare with the scenario results in section 5.4. How many days of full demand should the store be able to hold?

### 2.6.2
An estimate of the maximum amount of hydrogen that can be stored in a molecular lattice may be obtained by close packing of hydrogen atoms with a distance between H atoms determined from empirical values in, e.g., metal hydrides. Assuming such a distance to be not less than 0.2 nm, show that the total amount of hydrogen than can be stored is about 250 kg m$^{-3}$.

For comparison, the distance between the hydrogen atoms in a hydrogen molecule is 0.074 nm.

### 2.6.3
Give an estimate of how much land (or water surface) area should be set aside for allowing world-wide needs for hydrogen as a transportation fuel to be derived by microbial photosynthesis. How much arable land would be used if the same amount of hydrogen were to be produced by fermentation. Discuss the possibility for combining the land (or ocean/waterway) use with other uses.

**2.6.4**
Try to make a plot of energy density (by mass) versus estimated power density as suggested in section 2.4.6 for the energy storage systems discussed.

**CHAPTER**

**3**

# FUEL CELLS

## 3.1  Basic concepts

### 3.1.1 Electrochemistry and thermodynamics of fuel cells

The electrochemical conversion of energy is a conversion of chemical energy to electrical energy or vice versa. An electrochemical cell is a device either converting chemical energy deposited within the device, or chemical energy supplied through channels from the outside to the cell, into electricity or it is operating in reverse mode, converting electricity to either stored chemical energy or an outward-going mass flow of chemical energy. At all stages, associated heat may be generated or drawn from the surroundings. The general layout of such a device is shown in Fig. 3.1.

According to thermodynamics (cf. textbooks, e.g., Callen, 1960), the maximum amount of chemical energy of the system that in a given situation can be converted into a high-quality energy form such as electricity is given by the free energy $G$, also called Gibbs energy,

$$G = U - T_{ref} S + P_{ref} V, \tag{3.1}$$

where $U$ is the internal energy of the system (such as chemical energy in the cases considered here), $S$ is its entropy, and $V$ is its volume, while $(T_{ref}, P_{ref})$, the absolute temperature and pressure of the surroundings, defines what is meant by "in a given situation". The classical law of energy conservation (which is also the first law of thermodynamics) states that the increase in internal energy is given by the net energy added to the system from the outside:

$$\Delta U = \int dQ + \int dW + \int dM. \tag{3.2}$$

Here $M$ is the net energy of material flowing into the device, $W$ is the net amount of mechanical or electric work performed on the system by the surroundings, and $Q$ is the net amount of heat received from the environment.

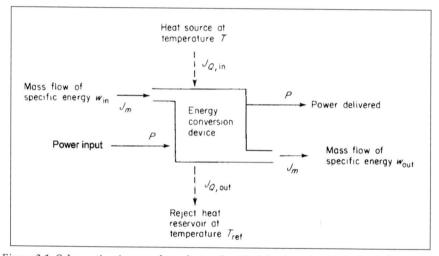

Figure 3.1. Schematic picture of an electrochemical device, showing mass (fuel), heat and power exchange to or from the device (from B. Sørensen, *Renewable Energy*, 2004, used by permission from Elsevier).

The amount of work performed by the device on its surroundings is $-W$. Integrated over a period of time, it may be expressed as

$$-\Delta W = -\Delta W_{elec} + \int P \, dV, \tag{3.3}$$

where the last term is zero if the system is held at a fixed volume. The electric work $-\Delta W_{elec}$, which the electrochemical system can deliver to a device (such as an electric motor) attached to an external circuit, may be expressed in terms of the electric potential difference between the positive and negative electrodes (see below) and the number of electrons travelling,

$$-\Delta W_{elec} = n_e \, N_A \, e \, \Delta\phi_{ext} = n_e \mathscr{F} \Delta\phi_{ext} \tag{3.4}$$

where the electron charge is $e = 1.6 \times 10^{-19}$ C (and $e\Delta\phi_{ext}$ thus the energy difference for a single electron), $n_e$ is the number of moles of electrons, $N_A = 6 \times 10^{23}$ Avogadro's constant (the number of particles – here electrons - per mole) and $\mathscr{F} = N_A\, e = 96\,400$ C mol$^{-1}$ is Faraday's constant. The energy (3.4) can also be expressed in terms of internal properties of the electrochemical device, namely the chemical potential difference $\Delta\mu$ across the device, eventually made up by more components $\Delta\mu_i$,

$$- \Delta W_{elec} = \sum_i n_{e,i}\, \Delta\mu_i. \tag{3.5}$$

The chemical potential for an electrolyte is an expression of the additional energy of its ions relative to that of the pure solvent. With the electrolyte components labelled as above by $i$, their contributions may be expressed as

$$\mu_i = \mu_i^0 + \mathscr{R}\, T\, log\, (f_i\, x_i), \tag{3.6}$$

where $\mathscr{R} = 8.3$ J K$^{-1}$ mol$^{-1}$ is the gas constant, $T$ is the temperature (K), and $x_i$ is the mole fraction of the particular component. $f_i$ is called the activity coefficient and may be regarded as an empirical constant. $\mu_i^0$ is the chemical potential of the $i$th component if it were the only one present, at the given pressure and temperature (see, e.g., Maron and Prutton, 1959).

In case the electrochemical device has a constant volume and the exchange of heat with the environment can be neglected, then the electricity produced must correspond to the loss (conversion) of free energy from the cell,

$$-\Delta G = -\Delta W_{elec}. \tag{3.7}$$

The free energy $G$, as given by (3.1), is the maximum work that can be drawn from the system under conditions where the exchange of (any kind of) work is the only interaction between the system and its surroundings. A system of this kind is said to be in thermodynamic equilibrium if its free energy is zero. Technically speaking, the expression of the free energy given in (3.1) implies dividing the total system into two subsystems, a small one (the electrochemical device) with extensive variables (i.e., variables proportional to the size of the system) $U$, $S$, $V$, etc. and a large one (the surroundings) with intensive variables $T_{ref}$, $P_{ref}$, etc. The reason for introducing the large system is to be able to consider its intensive variables (but not its extensive variables $U_{ref}$, $S_{ref}$, etc.) as constants, regardless of the processes by which the entire system approaches equilibrium.

This implies that the intensive variables of the small system, which may not even be defined during the process, approach those of the large system when the combined system approaches thermodynamical equilibrium ($G = 0$). The maximum work is found by considering a reversible process between the initial state and the equilibrium state. It equals the difference between the initial internal energy, $U_{init} = U + U_{ref}$, and the final internal energy, $U_{eq}$, and hence takes the form given in (3.1).

The approach of the entire system towards equilibrium, likely involving internal, irreversible processes, gives rise to a dissipation of energy given by

$$D = -dG/dt = T_{ref}\, dS(t)/dt, \tag{3.8}$$

assuming that entropy is the only time-dependent variable. It may not be possible to realise the approach to equilibrium in a finite span of time, and constraints such as confining the (small) system by walls to a finite volume may prevent reaching the true equilibrium state of zero free energy. If the extensive variables in such a constrained equilibrium state are denoted $U^0$, $S^0$, $V^0$, etc., then the available free energy becomes modified to an expression of the form

$$\Delta G = (U - U^0) - T_{ref}(S - S^0) + P_{ref}(V - V^0), \tag{3.9}$$

where chemical reaction energy is assumed included in the internal energy.

When the small system is constrained by walls, then the free energy reduces to the Helmholtz potential $U - TS$, and if the small system is constrained so that it is incapable of exchanging heat, the free energy reduces to the *enthalpy*, $H = U + PV$. The corresponding forms of (3.9) may be used to find the maximum work that can be obtained from a thermodynamic system with the given constraints. The general difference between enthalpy and free energy is made explicit through the relations (see, e.g., Sørensen, 2004a)

$$dU = T\, dS - P\, dV,$$

$$dH = T\, dS + V\, dP. \tag{3.10}$$

It should be emphasised that thermodynamics is a theory for the behaviour of systems near their equilibrium state, obtained from statistical treatment of the Newtonian laws of mechanics. In other words, thermodynamics tries to establish simple laws for the time development of averages of certain quantities (like temperature defined through averages of squared particle velocities). As weather forecasters know, it is not generally possible to find any simple behaviour for averaged quantities, or differently stated, no simple theory has yet been found for averages of thermodynamic quantities far away from equilibrium. Probably such basic laws simply do not exist. Since the systems of actual interest in electrochemistry are always away from equilibrium and are involved in complex, irreversible reactions or developments, then it should be clear that things like the maximum thermodynamical efficiency of a device can only be used as a theoretical reference point for comparing measured efficiencies, and in many cases there are no device designs that could bring the actual efficiency close to the thermodynamical maximum.

A description of the dynamical behaviour of the system must then go beyond the consideration of thermodynamics and try to establish relations between the rates of flows in the system, e.g.,

$$J_Q = dQ/dt \quad \text{(heat flow rate)},$$

$$J_m = dm/dt \quad \text{(mass flow rate)},$$ (3.11)

$$J_q = dq/dt = I \quad \text{(charge flow rate or electrical current)},$$

and the factors that may influence them, i.e. interactions (generalised forces) between the components of the system. From Fig. 3.1, the simple energy efficiency of an electrochemical device converting fuel and heat into electricity (thus in this case the electricity input is zero) is then at a given instant in time given by

$$\eta = \frac{J_{Q,in} - J_{Q,out} + J_m(w_{in} - w_{out})}{J_{Q,in} + J_m w_{in}}$$ (3.12)

where the specific energies $w_{in}$ and $w_{out}$ of the ingoing and outgoing substances (fuels) serve to convert the mass flow rates into energy flow rates. The efficiency of providing work (also called 2nd law efficiency or *exergy* efficiency) is defined as

$$\eta^{2.law} = \frac{W}{\max(W)},$$ (3.13)

where $W$ is the actual power output and $\max(W)$ the maximum work that might be produced, according to the theoretical understanding of the actual irreversible, not-close-to-equilibrium processes. The expression carefully avoids specifying which "theoretical understanding" one should use or if there exists a known, valid and applicable theory at all. This warning is meant to emphasise that although we believe there are valid theories on the basic level (classical and quantum mechanics, electromagnetic theory etc.), it is often not possible to pre-calculate the behaviour of an actual, macroscopic energy conversion system without having to introduce uncertainties associated with the approximate descriptions used to engineer systems too large, or having too many elementary constituents, to allow direct application of the fundamental theories available.

## Electrochemical device definitions

Electrochemical devices are named according to the following conventions. A device that converts chemical energy supplied as input fuels to the device into electric energy is called a *fuel cell*. If the free energy-containing substance is stored within the device rather than flowing into the device, the name *"primary battery"* is used. A device that accomplishes the inverse conversion (e.g. electrolysis of water into hydrogen and oxygen) may be called a *driven cell*. The energy input for a driven cell need not be electricity, but could be solar radiation, for example, in which case the process would be photochemical rather than electrochemical. If the same device can be used for conversion in both directions (or if the free energy-containing substance is regenerated outside the cell with addition of energy and recycled through the cell), it may be called a *regenerative* or *reversible fuel cell,* and finally, if the free energy-containing substance is stored inside the device, it may be called a *regenerative* or *secondary battery.*

The basic ingredients of all these electrochemical devices are two electrodes (positively and negatively charged) and an intermediate electrolyte layer capable of transferring (most often) positive ions in either direction, while a corresponding flow of electrons in an external circuit provides the desired power or uses power to produce fuels. I use the name *negative electrode* for the electrode where negative charges build up as a result of the presence of excess electrons and the name *positive electrode* for the electrode where positive charges build up as a result of a deficit of electrons. In contrast to the historical names *anode* and *cathode*, these definitions ensure that the electrodes retain their names, no matter whether electrons are transferred from electrode to electrolyte or the other way (i.e., the fuel or power production modes of a fuel cell) [o]. Use has been made of solid electrodes and fluid electrolytes, as well as fluid electrodes and solid electrolytes (e.g. conducting polymers). Before entering into the description of particular fuel cell types, a few general characteristics will be described with use of the theoretical concepts introduced above.

## Fuel cells

Figure 3.2 shows the general layout of a fuel cell, based on the free energy change $\Delta G = -7.9 \times 10^{-19}$ J for the reaction (left-to-right for fuel cell production of electricity, right-to-left for reverse fuel cell performing electrolysis):

---

[o] Some authors used to define the cathode as the electrode, to which electrons in the outer circuit flow and become transferred to the electrolyte. This gives rise to confusion, as the electrodes will change names each time the current is reversed (battery charge/discharge or electricity/fuel production in the case of fuel cells).

$$2H_2 + O_2 \leftrightarrow 2H_2O - \Delta G, \tag{3.14}$$

(cf. Chapter 2, (2.16)). For electricity production, a hydrogen gas is led to the negative electrode, to which it may lose electrons and thereby form $H^+$ ions capable of diffusing through the electrolyte (and membrane if present), while the electrons flow through the external circuit. The involved splitting of hydrogen molecules, $H_2$, into protons and electrons is typically speeded up by the presence of a catalyst (typically a metal, which could be the electrode itself or, e.g., a platinum film placed on the electrode surface). A reaction of the form

$$2H_2 \leftrightarrow 4H^+ + 4e^- \tag{3.15}$$

thus takes place at the negative electrode, where suitable catalyst choices may make it proceed at an enhanced rate (see, e.g., Bockris and Reddy, 1998; Bockris *et al.*, 2000; Hamann *et al.*, 1998).

Gaseous oxygen (or oxygen-containing air) is similarly led to the positive electrode, where a more complex reaction takes place, with the net result

$$O_2 + 4H^+ + 4e^- \leftrightarrow 2H_2O. \tag{3.16}$$

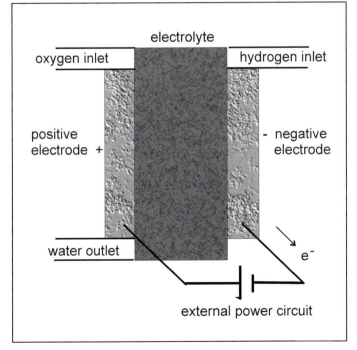

*Figure 3.2.* Schematic picture of a hydrogen–oxygen fuel cell. The fuel inlets are drawn so that they can illustrate inputs on either side of the electrodes.

This reaction is likely built up by simpler reactions, which could involve oxygen first picking up electrons or first associating with a hydrogen ion. Like in biological material, the fuel cell may utilise membranes allowing $H^+$ to diffuse through, but not $H_2$. It is instructive to compare the way the reactions in (3.16) proceed in biological systems such as photosystem II (considered in section 2.1.5) and in fuel cells and batteries. The proposed mechanism for water splitting in photosystem II is one of sequential stripping of four protons, one at a time, as illustrated in Fig. 3.3a. For manmade electrochemical devices, the process is by conventional chemical (ionic) theory considered to consist of two steps, as illustrated in Fig. 3.3b:

$$4H_2O \leftrightarrow 4OH^- + 4H^+ \quad \text{and} \quad 4OH^- \leftrightarrow 2H_2O + O_2. \tag{3.17}$$

The underlying thought is that the second proton is not, in the case of an electrochemical device, able to directly leave the $OH^-$ radical, but will form new water molecules plus molecular oxygen in conjunction with a second, similar process (Fig.3.3b, top). The combination of the two oxygen atoms to an oxygen molecule (or the opposite process in normal fuel cell operation) requires the presence of a catalyst, such as Ni or Pt. The first step (bottom to middle line in Fig. 3.3) is the same for organic and inorganic water splitting. The reason that the organic molecule in Fig. 3.3 (a) (middle to upper step) can get away with direct $O_2$ formation from just two $OH^-$ radicals is, if this is the right explanation, that the molecules are held in place by the amino acid chains (capable of storing and releasing any supplementary energetic requirements), in conjunction with the action of the special cluster of four Mn atoms (see Chapter 2, Fig. 2.18) acting as a Pt (or Pt compound) catalyst does in the fuel cell. Early suggestions that Mn existed in photosystem II as oxides (Hoganson and Babcock, 1997) have not been confirmed by the structural studies (Kamiya and Shen, 2003). Although not all details are revealed at the resolution achieved (0.37 nm), it is unlikely (but not impossible) that 8 oxygen atoms could have remained undetected near the 4 Mn atoms detected.

Quantum mechanical calculations are made in the following section in order to shed light on the reaction pathways. They can help determine the distribution of charges over the entire system of a metal surface and the O and H atoms. In Figure 3.3, no charges are indicated. In classical electrochemistry, these are as given in (3.17), and the specific need for a catalyst is not very clear. Looking at the fuel cell case (Fig. 3.3b), the hydrogen atoms need to be positively charged in order to move from the negative to the positive electrode. At the negative electrode, they may transfer 4 electrons to the metal catalyst. This would explain why it has to be there. For electrolysis (left-to-right reaction in (3.17)), energy has to be added by applying a voltage across the external circuit between the electrodes. This means having a surplus of electrons migrate through the external circuit to the electrode surface on the hydrogen-producing side of the membrane and away from the electrode on the oxygen side. The difference from the classical explanation is that the OH molecules may not be in the form of simple ions. The charge distribution may be more complex, which turns out to be an important observation for carrying out the quantum mechanical calculation considered in section 3.1.2.

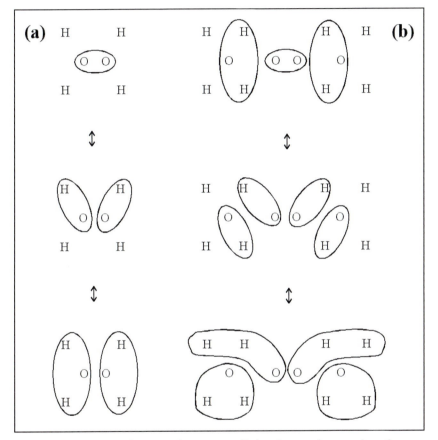

*Figure 3.3.* Reaction mechanisms for water splitting (upward arrows) or the reverse power-producing reaction (downward arrows), proposed for organic photosystem II (a) and for inorganic electrolysers and fuel cell systems (b). Charges are omitted in the figure, but discussed in the text (Sørensen, 2004e).

This type of explanation means that in the left side case, the intermediate situation in Fig. 3.3a is one where there may still be 2 charges on the metal electrode, while in both cases, the OH entities may have no overall charge. In the upper situation in both Fig. 3.3a and Fig. 3.3b, there must be 4 positive charges on the hydrogen atoms, i.e., they are 4 protons without electrons attached, ready to transfer to the negative electrode in the electrolysis case.

Returning to the thermodynamical level of description, the drop in free energy (3.7) is usually considered to be associated with the negative electrode reaction (3.15), and $\Delta G$ may be expressed in terms of the chemical potential (3.5) of the $H^+$ ions dissolved in the electrolyte. Writing the single chemical potential $\mu$ involved as Faraday's constant times a suitable potential $\phi$, the free energy for $n$ moles of hydrogen ions is (cf. (3.4) and (3.5))

$$G(H^+) = n\mu = n\mathscr{F}\phi = nN_A\,e\phi. \tag{3.18}$$

When the hydrogen ions "disappear" as a result of the reaction at the positive electrode ((3.17) in the direction towards the left), this chemical free energy is converted into the electrical energy by (3.7), and since the numbers of electrons and hydrogen ions in (3.16) are equal, $n = n_e$, the chemical potential $\mu$ is given by

$$\mu = \mathscr{F}\phi = \mathscr{F}\Delta\phi_{ext}. \tag{3.19}$$

The quantity $\phi$ is usually referred to as the electromotive force (e.m.f.) of the cell, or the "standard reversible potential" of the cell, if taken at standard atmospheric pressure and temperature. From the value of $\Delta G$ quoted above in connection with (3.14) for two molecules of water, the corresponding value of $-2.37 \times 10^5$ J is obtained for producing (or splitting) a mole of $H_2O$. The cell e.m.f. $\phi$ then becomes

$$\phi = -\Delta G / n\mathscr{F} = 1.23 \text{ V}, \tag{3.20}$$

with $n = 2$ since there are two $H^+$ ions for each molecule of $H_2O$ formed. The chemical potential (3.19) may be written in the form (3.6), and the cell e.m.f. thus expressed in terms of the properties of the reactants and the electrolyte (including the empirical activity coefficients appearing in (3.6) as a result of generalising the expression obtained from the definition of the free energy, (3.1), assuming $P$, $V$ and $T$ to be related by the ideal gas law, $PV = \mathscr{R}T$, valid for one mole of an ideal gas (cf., e.g., Angrist, 1976)).

The efficiency of a fuel cell is the ratio between the electrical power output (3.4) and the total energy lost from the fuel. However, it is possible to exchange heat with the surroundings, and the energy lost from the fuel may thus be different from $\Delta G$. For an ideal (reversible) process, the heat added to the system is

$$\Delta Q = T\,\Delta S = \Delta H - \Delta G, \tag{3.21}$$

and the efficiency of the ideal process thus becomes

$$\eta^{ideal} = -\Delta G / (-\Delta G - \Delta Q) = \Delta G / \Delta H. \tag{3.22}$$

For the hydrogen–oxygen fuel cell considered above, the enthalpy change during the two processes (3.15) and (3.16) (in the direction left to right) is $\Delta H = -9.5 \times 10^{-19}$ J or $-2.86 \times 10^5$ J per mole of $H_2O$ formed, and the ideal efficiency of power production in this case is thus

$$\eta^{ideal} = 0.83.$$

There are reactions with positive entropy change, such as $2C + O_2 \rightarrow 2CO$, which may be used to cool the surroundings and at the same time create electric power with an ideal efficiency above one (1.24 for CO formation).

In actual fuel cells, a number of factors tend to diminish the power output. They have traditionally been expressed in terms of "expenditure" of cell potential fractions on processes not contributing to the external potential,

$$\Delta \phi_{ext} = \phi - \phi_1 - \phi_2 - \phi_3 - ..., \tag{3.23}$$

where each of the terms $-\phi_i$ corresponds to a specific loss mechanism. Examples of loss mechanisms are blocking of pores in the porous electrodes, e.g. by piling up of the water formed at the positive electrode in connection with the processes (3.17), internal electric resistance of the cell (heat loss) and the building up of potential barriers at or near the electrolyte–electrode interfaces, e.g. due to impurities. Most of these mechanisms limit the reaction rates and thus tend to place a limit on the current of ions that may flow through the cell. There will be a limiting current, $I_L$, beyond which it will not be possible to draw any more ions through the electrolyte, because of either the finite diffusion constant in the electrolyte (in case the ion transport is dominated by diffusion) or the finite effective surface of the electrodes at which the ions are formed. Figure 3.4 illustrates the change in $\Delta \phi_{ext}$ as a function of current,

$$I = \Delta \phi_{ext} R_{ext} = I_- + I_+ = (\phi_- - \phi_+) R_{ext}, \tag{3.24}$$

expressed as the difference between potential functions at each of the electrodes, $\Delta \phi_{ext} = \phi_- - \phi_+ = \phi_c - \phi_a$ (the latter notation used in Fig. 3.4). This representation gives some indication of whether the loss mechanisms are connected with processes at one or the other electrode, and it is seen that the largest fraction of the losses is connected with the more complex positive electrode reactions in this example. For other types of fuel cells it may be negative ions that travel through the electrolyte, with corresponding changes in characteristics (cf. Jensen and Sørensen, 1984).

Equation (3.24) also expresses the total current as a sum of contributions from each of the electrode areas. They can be related to the potential loss terms in (3.23), which describe the loss terms at each electrode as well as Ohmic loss terms for each of the components of the cell. At each electrode these involve obstacles to transferring electrons from electrode to electrolyte (especially at low current) and at higher currents the obstacles to diffusing charges into the electrolyte from their creation near the electrodes, or to maintain sufficient amounts of reactants in the electrode area. The exponential dependence of electrode currents on these loss potentials is described by the Butler-Volmer equations (Hamann *et al.*, 1998; Bockris *et al.*, 2000),

$$I_- = I_-^0 \left(\frac{C_-}{C_-^0}\right)^{\gamma_-} \left[\exp\left(\frac{\alpha_- \mathscr{F}}{\mathscr{R}T}\phi_+\right) - \exp\left(\frac{\alpha_+ \mathscr{F}}{\mathscr{R}T}\phi_-\right)\right]$$

$$(3.25)$$

$$I_+ = I_+^0 \left(\frac{C_+}{C_+^0}\right)^{\gamma_-} \left[\exp\left(\frac{\alpha_+ \mathscr{F}}{\mathscr{R}T}\phi_-\right) - \exp\left(\frac{\alpha_- \mathscr{F}}{\mathscr{R}T}\phi_+\right)\right].$$

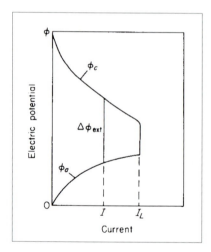

*Figure 3.4.* Fuel cell negative electrode potential $\phi_a$ and positive electrode potential $\phi_c$, as a function of current. The main cause of the diminishing potential difference $\Delta\phi_{ext}$ for increasing current is at first incomplete electrocatalysis at the electrodes, for larger currents also ohmic losses in the electrolyte solution, and finally a lack of ion transport (cf. Bockris and Shrinivasan, 1969). From B. Sørensen, *Renewable Energy*, 2004, used by permission from Elsevier.

The parameters $\alpha_-$ and $\alpha_+$ are charge transfer coefficients for each of the electrodes, $C_-$ and $C_+$ are the concentrations of the reactants (such as hydrogen and oxygen) near the electrode, while $C_-^0$ and $C_+^0$ are the corresponding concentrations in the bulk electrolyte. The exponents $\gamma_-$ and $\gamma_+$ are empirical quantities taken as 0.5 for oxygen and 0.25 for hydrogen (Nguyen *et al.*, 2004). Finally, $I_-^0$ and $I_+^0$ are reference exchange currents for each electrode. They are the currents corresponding to neglecting the loss contributions to the potentials, which is the Nernst potential depending on the activities of the reversible electrode reactions without losses (such as (3.20) for the combined oxygen and hydrogen reactions).

The Butler-Volmer equations simplify in the case where one of the exponential terms dominate, which will happen if the loss potential in question is large. In this case the electrode potential is linearly related to the logarithm of the corresponding current, and if the losses occur equally at both electrodes, then the total potential becomes linearly related to log($I$). This is called the Tafel relation, and the slope of the line is called the "Tafel slope". The following sections will give many examples of potential-current relationships, and except for the interval of smallest currents, the Tafel approximation is often a valid one.

It follows from diagrams of the type shown in Fig. 3.4 and from the equations (3.25), that there will be an optimum current, usually lower than $I_L$, for which the power output will be maximum,

$$\max(E) = I^{opt} \, \Delta\phi_{ext}^{opt}. \tag{3.26}$$

Dividing by the rate at which fuel energy $\Delta H$ is added to the system in a steady state situation maintaining the current $I^{opt}$, one obtains the actual efficiency for the conversion maximum,

$$\max(\eta) = I^{opt} \, \Delta\phi_{ext}^{opt} / (dH/dt). \tag{3.27}$$

The potential losses within the cell are fundamentally similar to the thermodynamical energy dissipation described in the beginning of this section, implying that energy cannot be extracted in a finite time without losses.

## 3.1.2 Modelling aspects

Quantum mechanical modelling can be used to describe processes involving individual molecules. Employing such models is also possible for describing surface processes, such as the adsorption and chemisorption at electrode surfaces and the action of catalysts on individual molecules, as well as describing reactions taking place between molecules moving within a solvent such as an electrolyte (e.g. Redox couples). The theoretical framework used in quantum chemical modelling is described in section 3.1.3.

Gas flows in channels leading to the electrodes, including the gas diffusion layers, are modelled, e.g., by finite element models of flows such as those described in section 3.1.5. Such models may also, with suitable modifications, be applied to the ion flow through membranes.

Based upon the results of detailed models of these two types, overall models for the performance of fuel cells may be formulated in terms of simple equivalent electric circuit models that parametrise the loss terms and allow calculations of overall efficiencies as a function of such parameters.

The electrical circuit analogues have been extensively used in connection with impedance spectroscopic measurements. These are measurements of the response of a total cell upon application of an alternating current across its outer electrodes and thus are aimed at the type of characterisation possible without dismantling the device for experiments on individual constituents (which, however, can also be done, e.g., for a half cell as in Fig. 3.5). The response to the application of an external time-dependent potential of the form $V=V_0\cos(\omega t)$ is a measurable current that may be out of phase with the applied voltage variation. In circuit theory, the phase differences are modelled by an equivalent description in terms of complex functions, which allows the use of imaginary numbers for quantities 90° out of phase with the potential as it is written above. In general, the cosine is replaced by $\exp(i\omega t) = \cos(\omega t) + i\sin(\omega t)$, where $i = (-1)^{1/2}$, and the measurements will then determine a complex impedance $Z$, which differs from the Ohmic resistance by an imaginary part describing a phase delay between voltage and current. The prototype causes for this behaviour are capacitors and coils, but other components, for example, with a dependence on the frequency $\omega$, may be invoked. Writing $Z = Re(Z) + i\, Im(Z)$, the outcome of an experiment may be expressed in a diagram such as the one for a proton exchange membrane (PEM) fuel cell shown in Fig. 3.5, with $Re(Z)$ as abscissa and $Im(Z)$ as ordinate (Ciureanu et al., 2003). The straight line at smaller frequencies may be interpreted in terms of diffusion of $H^+$ through the membrane, while the curve at high frequencies can be modelled by bulk capacitance and Ohmic resistance of the cell.

In Fig. 3.6, impedance spectroscopy has been performed on the negative electrode-side half cell of a direct methanol fuel cell (Müller et al., 1999), under conditions of (a) limited methanol supply (about two times the supply rate required by stoichiometry) and (b) generous methanol supply. The four parameters of the equivalent circuit proposed for case (b) may be simply fitted, or alternatively they may by modelling be associated with physical properties of the cell (Harrington and Conway, 1987). A similar behaviour is found for proton exchange membrane half cells (Ciureanu et al., 2003). The behaviour in Fig. 3.6b can be explained in terms of the rates of electrons received from the negative electrode, the net rate of CO adsorption on the electrode and the fractional coverage of the electrode by CO. The slow relaxation of adsorbed CO makes modelling in terms of an inductance $L$ do well in reproducing the phase delay. Figure 3.6a exhibits three arcs. The one for low frequencies varies with methanol flow rate and thus may have to do with the probability of the methanol in reaching the active sites, the middle arc is due to methanol oxidation kinetics; and the arc at high frequencies, which is independent of electrode potential, represents an Ohmic loss that is insignificant at lower frequencies (Mueller and Urban, 1998).

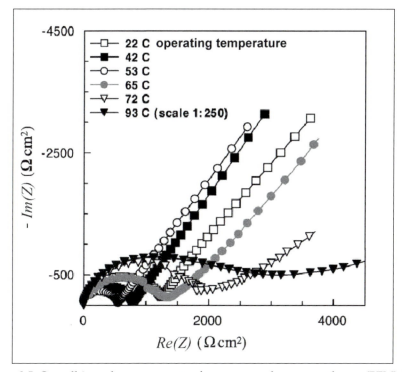

*Figure 3.5.* Overall impedance response of a proton exchange membrane (PEM) fuel cell for different cell temperatures, depicted as corresponding values of the real and imaginary parts of the complex impedance (sometimes denoted a Nyquist plot). Each sequence of points represents frequencies ranging from $10^{-1}$ to $10^7$ Hz, with the highest values corresponding to the leftmost points. From M. Ciureanu, S. Mikhailenko, S. Kaliaguine (2003). PEM fuel cells as membrane reactors: kinetic analysis by impedance spectroscopy. *Catalysis Today* **82**, 195-206. Used with permission from Elsevier).

The negative electrode reaction (3.15) involves the processes depicted in Chapter 2, Fig. 2.5: adsorption of a hydrogen molecule at $d = 0.74$ nm followed by splitting into two chemisorbed H atoms. The mechanism for these to part from the surface and move across to the opposite electrode involves for each hydrogen atom the delivery of an electron to the metal surface. The actual travelling need not be performed by $H^+$ ions or by any other ions (say of an electrolyte) in a bulk fashion, but may involve short-distance hopping from molecule to molecule, as suggested by Hamann *et al.* (1998) and illustrated in Fig. 3.7. The water molecule lends itself perfectly to this process. Due to its asymmetry (and hence dipole moment), it may easily accept the approach of a third hydrogen atom and form $H_3O^+$ as shown in the Fig. 3.7. A proton on the other side of the molecule may then leave and move the charge to a neighbouring molecule, which is now the $H_3O^+$, and so on.

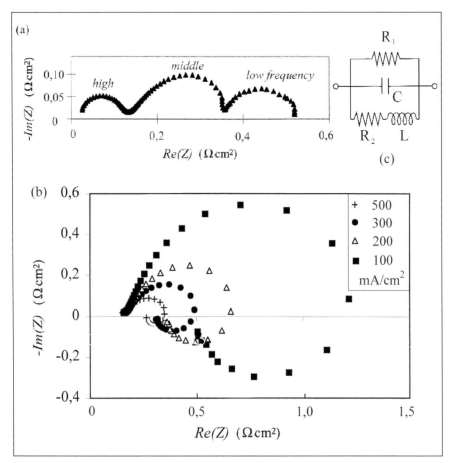

*Figure 3.6.* Impedance spectra for negative electrode half of direct methanol fuel cell. (a) Methanol mass flow limited situation (see text for explanation of the three arcs). (b) Situation for generous mass flow, with impedance behaviour shown for different current densities. (c) Equivalent circuit model for the situation in (b). Based on J. Mueller and P. Urban (1998). Characterization of direct methanol fuel cells by ac impedance spectroscopy, *J. Power Sources* **75**, 139-143; J. Müller, P. Urban, W. Hölderich (1999). Impedance studies on direct methanol fuel cell anodes. *J. Power Sources* **84**, 157-160. Used by permission from Elsevier.

There have been considerable amounts of theorising about the behaviour of ions in a solution such as an electrolyte. The layer of oppositely charged ions near a charged electrode is well established, but the possible extension of this layer into an ionic cloud diffusing far out into the electrolyte, as suggested by Gouy and Chapman (see Bockris *et al.*, 2000), is less clear. The model calculation presented in Chapter 2, Fig. 2.4 only exhibits very minor variations in charge away from the electrode, and this is probably what one has to expect due to the falling strength of the Coulomb force with distance, in a model where the electrode charge is supposed to be responsible for the rearrangement of ions in the electrolyte. Within the electrolyte, only minor variations in charge are expected, due to charge diffusion by "hopping" mechanisms such as the one depicted in Fig. 3.7.

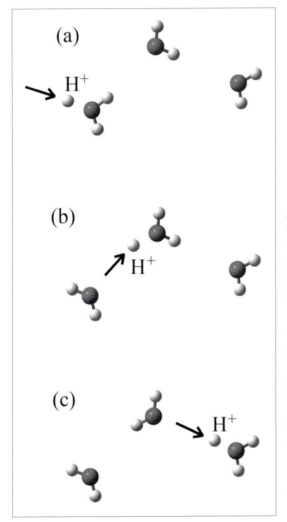

*Figure 3.7.* Transfer of a proton through water: the proton attaches to the "unoccupied" side of a water molecule, the electron density rearranges, and another proton may leave the other side of the molecule and move to another water molecule. In this way, any bulk motion of single $H^+$ atoms through the water is avoided. A potential slope is assumed to have been created by a voltage applied to the electrodes, or the displacement of the protons will create such a potential, depending on whether the electrochemical device is operated as an electrolyser or as a fuel cell.

### 3.1.3 Quantum chemistry approaches

The many-body problem of describing molecules in terms of their nuclear and electronic constituents requires some level of simplification of the general quantum mechanical approach in order to be numerically tractable. The first approximation made is usually the Born-Oppenheimer assumption, that only the dynamic behaviour of electrons has to be modelled. The atomic nuclei are much heavier and thus move less rapidly, implying that one may look at the electron motion for each configuration of the atomic centres (called the *molecular structure*). One can then repeat the calculation for different structures, in case one is interested in finding one or more equilibrium configurations of a molecular structure, and hope to identify minima in the total energy calculated, as a function of structure parameters. Often there will be more than one minimum, e.g. indicating isomerism in ring structures.

The Schrödinger equation for describing a molecular system has the form

$$H\Psi = i\ D_t\Psi, \tag{3.28}$$

where   is Planck's constant over $2\pi$ ( $= 1.05 \times 10^{-34}$ J), $D_t$ is the partial derivative, here with respect to time, $\Psi$ is the wave function of the system, and $H$ is the Hamiltonian, an energy operator which in the molecular case may be written

$$
\begin{aligned}
H \ =\ T+V \ =\ & -\tfrac{1}{2}\hbar^2 \sum_{i\in\{e,N\}} m_i^{-1}(D_x^2 + D_y^2 + D_z^2) \\
& + e^2(4\pi\varepsilon_0)^{-1}\left( \sum_{i<j\in\{e\}} \left|r_i - r_j\right|^{-1} + \sum_{i<j\in\{N\}} Z_i Z_j \left|R_i - R_j\right|^{-1} - \sum_{i\in\{e\},j\in\{N\}} Z_j \left|r_i - R_j\right|^{-1} \right)
\end{aligned}
\tag{3.29}
$$

in terms of kinetic energy $T$ (first line), potential energy $V$, with $m_i$ the mass of an electron or of a particular atomic nucleus, depending on whether $i$ belongs to the set of electrons in the system, $\{e\}$, or to the set of atomic nuclei, $\{N\}$ (each consisting of a number of protons and neutrons not modelled here), and in terms of the Coulomb interaction between all pairs of the charged particles, whether negatively charged electrons or positively charged atomic nuclei. The radius vector is denoted $r_i$ for the $i$th electron and $R_j$ for the $j$th nucleus. The elementary charge is denoted $e$ (= $1.60 \times 10^{-19}$ C) and the dielectric constant in vacuum $\varepsilon_0$ (= $8.85 \times 10^{-12}$ C$^2$ m$^{-1}$ J$^{-1}$). Thus $e^2/(4\pi\varepsilon_0) = 2.30 \times 10^{-28}$ J m. Although no special notation is used, all radius vectors $r_i$ or $R_i$ are indeed vectors, with three co-ordinates such as $x$, $y$ and $z$.

The wavefunction $\Psi = \Psi(r_i\in\{e\}, R_j\in\{N\},t)$ is assumed normalised for any fixed time $t$ by the condition

$$\int_{all\ space} \Psi^*\Psi\ dr_1...dr_{n_e} dR_1...dR_{n_N} = 1, \tag{3.30}$$

and the density (a function of all space variables as well as time) is given by

$$\rho\,(r_i\in\{e\},\,R_j\in\{N\},t) \;=\; \Psi^*\,\Psi \;=\; |\,\Psi\,|^2. \tag{3.31}$$

$\Psi^*$ is the complex conjugate of $\Psi$ (that is $\Psi^* = a-ib$ if $\Psi = a+ib$). Note that by this definition, $\rho$ is normalised to unity and not, e.g., to the total number of particles, an equally valid alternative that will be discussed below. The interpretation of this density is that it describes the probability of finding the system with its particles at the specified space vectors $\{r_i, R_j\}$ at a given time $t$.

That is the very beauty of quantum mechanics. It is a perfectly deterministic theory for calculating the development of a physical system, once you know it at a given time. But because only the absolute square of the wavefunction is accessible to experiment, it is not possible for us to know the state precisely at such an initial time, and hence all the further development of the system is to the human observer clouded with uncertainty relations and similar indeterminacy, despite the fact that the system itself knows perfectly well what it is doing and follows a uniquely given path of development in time and space.

The solutions to (3.28) have the following general properties: there are a countable number of bound, stationary state solutions, $\psi_q$, $q = 1,2,3,\ldots$, with energies $E_q$. This is the meaning of calling the spectrum of solutions *quantised* and the theory *quantum mechanics*. Additionally, there may exist unbound state solutions that may form a continuum in energy space. The set of quantised solutions are labelled by $q$. The collection of values that $q$ may have is called quantum numbers. Although time dependence is sometimes important (e.g., for chemical reactions), many problems can be reduced to finding the stationary states, and finding these is also a good starting point, if time dependence has to be subsequently studied. The form of the wavefunction for a particular stationary case $q$ is

$$\Psi_q(r_i\in\{e\},R_j\in\{N\},t) \;=\; \psi_q(r_i\in\{e\},R_j\in\{N\})\,exp(-iE_q t/\,\,). \tag{3.32}$$

Since the $H$ in (3.28) does not have an explicit time dependence, this equation reduces to what is called the *stationary Schrödinger equation*,

$$H\psi_q \;=\; E_q\psi_q. \tag{3.33}$$

The Born-Oppenheimer approximation can now be stated as follows: treat all the variables of the atomic nuclei, $R_j$, as constants and forget about the nuclear part of the wavefunction. More precisely, one may assume that the wavefunction can be written in the form

$$\psi_q(r_i\in\{e\},\,R_j\in\{N\}) \;=\; \varphi_q(r_i\in\{e\})\,\Phi_q(R_j\in\{N\}), \tag{3.34}$$

which allows the Schrödinger equation to be written, with obvious notation for the electronic and nuclear parts of the kinetic and potential energy operators in (3.29),

$$(T_e + T_N + V_e + V_N + V_{eN}) \, \varphi_q \Phi_q = E_q \varphi_q \Phi_q. \tag{3.35}$$

Although the electron wavefunction $\varphi_q(r_i; i \in \{e\})$ is primarily a function of the electron co-ordinates, it indirectly depends also on the nuclear co-ordinates because they appear in the potential terms $V_N + V_{eN}$. The Born-Oppenheimer approximation now neglects the influence of nuclear motion, i.e., the kinetic term with $T_N$, on the solutions $\varphi_q$ for the electron part of the system, by solving for the latter by the following equation:

$$(T_e + V_e + V_N + V_{eN}) \, \varphi_q = E_{q,eff} \, \varphi_q. \tag{3.36}$$

All that is done is to leave out the action of the operator $T_N$ and, as a consequence, get a modified, effective energy $E_{q,eff}$. The rest of the energy, $E_{q,rest} = E_q - E_{q,eff}$, may subsequently be found by solving the remaining part of (3.35), i.e., what was left out in (3.36). If the indirect dependence of $\varphi_q$ on the $R_j$s is disregarded, what is left is only

$$(T_N + E_{q,eff}) \, \Phi_q = E_q \Phi_q \quad \text{or} \quad T_N \, \Phi_q = E_{q,rest} \, \Phi_q \tag{3.37}$$

The formation of molecules in a model with only Coulomb forces does of course rely strongly on the presence of both positive and negative particles, but the approximations made here have shifted all the non-trivial calculations to the electron treatment in (3.36). Because all electrons are *fermion* identical particles, quantum theory makes additional formal requirements of antisymmetry on their wavefunctions, which will be discussed below.

## Hartree-Fock approximation

The splitting of a wavefunction into product form, as in (3.34), is a widely used method for seeking computationally simple solutions to the Schrödinger equation. The Hartree (1928) approximation consists in writing the overall wavefunction $\varphi_q$ as a product of wavefunctions depending only on the co-ordinate of one electron,

$$\varphi_q = \prod_{i \in e} \psi_{q,i}(r_i). \tag{3.38}$$

This wavefunction does not satisfy the antisymmetry property required for electrons, and it does not include the spin ($s = \frac{1}{2}$ for electrons, with projection $m_s = -\frac{1}{2}$ or $+\frac{1}{2}$ on the quantisation axis). To remedy these shortcomings, the wavefunction should rather be of the form

$$\varphi_q = (n!)^{-1/2} \det\left( \prod_{i \in e} \psi_{q,i}(r_i)\chi_{q,i} \right), \tag{3.39}$$

where the product over $i$ goes over all electrons $e$ of the system and the determinant ensures that the interchange of any two electrons will give an overall minus sign on the wavefunction (this is a standard property of determinants). The spin functions are denoted $\chi_{q,i}$. They must equal one of the two "universal" spin eigenfunctions mentioned above as possible for electrons. They are universal in the sense that there is only one spin-up ($m_s = \frac{1}{2}$) and only one spin-down ($m_s = -\frac{1}{2}$) spin function and no variables known to describe them further (except in superstring theory and the like). The "guess" (3.39) is now ready to be introduced into (3.36) for solution.

A well-known theorem in quantum mechanics states that the expectation value of the Hamiltonian, taken in the state defined by any wavefunction, is minimum when the wavefunction equals the exact ground state wavefunction. Starting from a trial wavefunction (that most likely is different from the exact ground state wavefunction) involving some parameters, one can then use a variational approach to find the best wavefunction within the space spanned by the trial wavefunctions considered. The energy average to be varied is (from here on dropping the index $q$ labelling the stationary energies)

$$<E_{eff}> = \frac{\int \varphi^* (T_e + V_e + V_N + V_{eN})\varphi \, d\tau_1...d\tau_e}{\int \varphi^* \varphi \, d\tau_1...d\tau_e}, \tag{3.40}$$

where $\tau_i$ is a shorthand for the space and spin variables of the $i$th particle and where in contrast to (3.36), the division by the normalising integral has to be there because variations of $\varphi$ may not preserve normalisation. This energy has a minimum,

$$\delta <E_{eff}> /\delta \varphi = 0, \tag{3.41}$$

where it is closest to (but not smaller than) the exact ground state energy, and the corresponding function $\varphi$ of space and spin variables is the corresponding ground state wavefunction, in the form implied by the approximation used. Performing the variation (3.41) with a wavefunction of the form (3.39) leads to $e$ equations of the form (see, e.g., Scharff, 1969):

$$(H_k + U^{(k)}(r_k)) \, \varphi_k = E_k \varphi_k; \quad k = 1,\dots, e, \tag{3.42}$$

where $H_k$ is the part of the Hamiltonian in (3.36) that depends only on the $k$th electron (using below a suffix $_{,k}$ to indicate that only the $k$th term in the sum over electrons in (3.29) should be included),

$$H_k = T_{e,k} + V_N - V_{eN,k} \tag{3.43}$$

and

$$U^{(k)}(r_k) = \frac{\int \varphi^{(k)} * V_e \, \varphi^{(k)} \, d\tau_1 \dots d\tau_{k-1} d\tau_{k+1} \dots d\tau_e}{\int \varphi^{(k)} * \varphi^{(k)} \, d\tau_1 \dots d\tau_{k-1} d\tau_{k+1} \dots d\tau_e}, \tag{3.44}$$

where

$$\varphi^{(k)} = (n!)^{-1/2} \det \left( \prod_{i \in e, \, i \neq k} \psi_i(r_i) \chi_i \right) \tag{3.45}$$

as distinguished from $\varphi_k = \psi_k(r_k) \, \chi_k$. These equations constitute the Hartree-Fock approximation (Fock, 1930; in Heisenberg form: Foresman et al., 1992).

The physical significance of this approximation is clear from the form of (3.42): each electron is supposed to move in a potential, which in addition to the atomic nuclei consists of the average over the Coulomb interactions with all electrons other than the one focussed upon. The construction ensures that the Pauli principle demanding antisymmetry for the total electron wave-function is satisfied in regard to other electrons with the same spin direction $m_s$ as the one focussed upon. However, it is not satisfied with respect to other electrons with opposite spin direction This can be a serious restriction in the appropriateness of the Hartree-Fock approximation. Solving for the best de-scription of a single electron moving in the average potential of all the other particles of the system (nuclei and $e-1$ electrons) will clearly not address the issue of antisymmetry, and only if the initial antisymmetric states are not mixed appreciably will the solution retain a reasonable amount of this prop-erty. Furthermore, constructing wavefunctions comprising several configu-rations of many-electron states characterised by $m_s$ equal to $+\frac{1}{2}$ or $-\frac{1}{2}$ will not ensure that the many spin-$\frac{1}{2}$ components couple correctly to one definite total spin (such as either $S = 0$ or $S = 1$ for two electrons), as they should in quantum mechanics.

## Basis sets and molecular orbitals

Most numerical algorithms solve the Hartree-Fock equations by iterative methods, implying that a good initial wavefunction guess is useful, but also that calculations can be lengthy. To cater to these problems, it is customary to expand the solutions (seeked) in terms of known basis wavefunctions, implying that only the expansion coefficients need to be modified at each iteration, while all integrals in principle can be kept on file. Use of a basis set is an exact method, provided that the basis set is complete (i.e., sufficiently large to span the whole space of solutions). However, and again for reasons of computational rationality (read: computing time required), practical calculations do not use complete, infinite sets of basis functions, but truncate them and often quite dramatically, say to just 2 to 6 basis functions per electron in the system modelled. It is then important to have "good" basis functions, with "good" meaning functions that approximate well in a truncation at low order, and functions applicable for a wide range of different molecules (rather than basis sets doing well only for selected problems). It is important to be able to go into uncharted territory and still have reasonable confidence in using the methods that worked well in problems studied previously. Such basis sets are usually taken from the electron wavefunctions of single atoms, which in many cases are known with high precision or even are exactly known in analytical form, such as the Laguerre polynomials and spherical harmonic functions entering the solution for the hydrogen atom.

The most widely used bases for molecular calculations are constructed with use of Gaussian functions, due to the ease with which such functions can be differentiated or integrated. A Gaussian is a real function of the form

$$g_{l,m,n,\alpha}(x,y,z) = A x^n y^m z^l e^{-\alpha(x^2+y^2+z^2)}, \tag{3.46}$$

where A ensures that the integral of $g^2$ over all space is unity (normalisation) and $x,y,z$ are co-ordinates relative to a particular atomic centre. The parameter $\alpha$ determines the width of the density distribution. The actual basis functions are constructed as linear combinations of the primitive Gaussian functions (3.46), $b_j = \sum_i a_{i,j} g_{li,mi,ni,\alpha i}$, and are fixed quantities for solving a given problem: one chooses which basis set to use and then starts calculating, seeking to solve the equations (3.42) and find solutions of the form (3.39) with

$$\psi_{q,k} = \sum_j c_{q,kj} b_j, \tag{3.47}$$

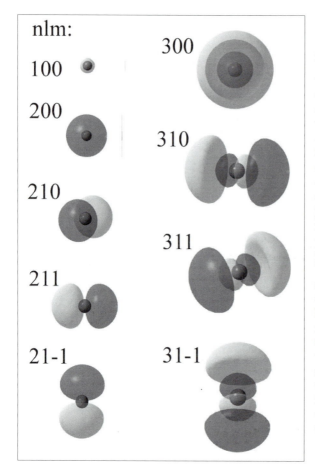

nlm:

100

200

210

211

21-1

300

310

311

31-1

*Figure 3.8.* Molecular orbitals for the oxygen atom, with indication of their quantum numbers (main, orbital angular momentum and projection along axis of quantisation). Shown is the oxygen nucleus and the electron density $\psi^*\psi$ (identical for each pair of two spin projections), but with two different shades used for positive and negative parts of the wavefunction. The calculation uses density functional theory (B3LYP) and a Gaussian basis of 9 functions formed out of 19 primitive Gaussian functions (see text for further discussion). The first four orbitals (on the left) are filled in the ground state, while the remaining ones are unoccupied.

where the set of $c_{kj}$s for the lowest energy state defines both the Hartree-Fock approximation to the actual ground state $q = 0$ of the system through (3.39) and the wavefunctions $\psi_{0,k}$ called *molecular orbitals*. It is because the solutions $c_{kj}$ also appear in the integrand of (3.44) that the solution must be found through iteration. The lowest $e$ levels (some possibly degenerate, i.e., with same energy) will be filled by electrons in the ground state, but the Hartree-Fock solution through (3.42) gives a full set of orbital solutions comprising not only occupied orbitals, i.e., those filled when electrons are added from the lowest one, but also the higher energy solutions corresponding to virtual orbitals that may be excited to form states other than the ground state.

Figures 3.8 and 3.9 show[a] the molecular orbitals for an oxygen atom and

---

[a] The calculations presented in these and the following figures are made using the software Gaussian (2003) and displayed by GaussView or WebLab ViewerLight.

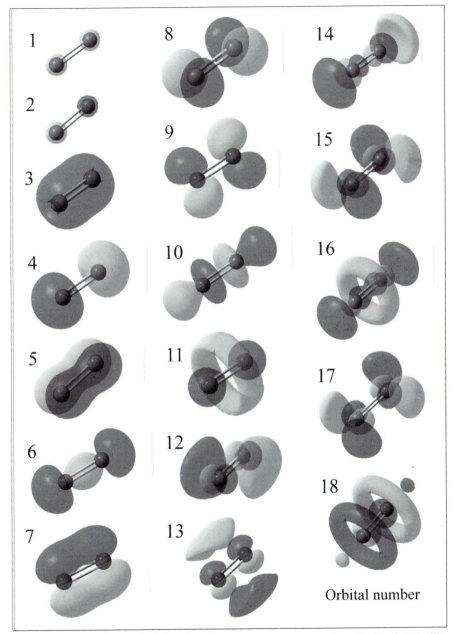

*Figure 3.9.* Molecular orbitals for the diatomic oxygen molecule, calculated as in Fig. 3.8 using a Gaussian basis of 18 functions formed from 38 primitive Gaussian functions. The basis states numbered 1 to 8 are occupied in the $O_2$ ground state.

an oxygen molecule at a proper atomic distance ($R = 0.12$ nm). The atomic orbitals are good indications of the spatial shape of the Gaussian functions and may be compared with the Laguerre polynomial/spherical harmonics solutions often depicted in elementary chemistry textbooks. The molecular orbitals of Fig. 3.9, on the other hand, give some insight into the types of solutions involved in calculations for more complicated systems. The accuracy of the Hartree-Fock approximations for the $O_2$ energy will be discussed in the next section. The SV basis used in the figures (Schaefer *et al.*, 1992; 1994) contains 9 basis functions for O, formed out of 19 primitive Gaussian functions, and twice as many for $O_2$. This is only about 2 basis functions per electron in the system, but using instead the EPR-III basis (Barone, 1996) with about 8 basis functions per electron does not change either the shapes shown or the energy of the ground state significantly.

### Higher interactions and excited states: Møller-Plesset perturbation theory or density function phenomenological approach?

Perturbation theory is a standard method in quantum physics (Griffith, 1995). It consists in writing the Hamiltonian as $H_0 + \lambda H_1$, where $H_0$ may be the Hartree-Fock Hamiltonian and $H_1$ the part containing the remaining correlations, and then expanding the overall wavefunction solution as a power series of $\lambda$. The introduction of $\lambda$ is only for identifying the order of approximation: in the end, $\lambda$ is put equal to unity, meaning that it is actually $H_1$ that is assumed "small". The application of perturbation theory for calculating corrections to Hartree-Fock wavefunctions and energies arising from the interaction between electrons was first proposed by Møller and Plesset (1934).

As one progresses through second, third, fourth or higher order corrections the calculations become increasingly lengthy, despite refinements (Schültz *et al.*, 2004), and efforts have been made to find other methods. As it is seen from (3.44) and (3.45), the integrals in the Hartree-Fock approximations can be cast into a form depending only on matrix elements of $V_e$ and of the unity matrix between two electron wavefunctions $\varphi_{k1}{}^*\varphi_{k2}$. Around 1930, it was recognised that the indistinguishability of electrons meant that instead of keeping track of the co-ordinates of each particle, one might introduce an overall density function, $\rho(r)$, normalised to the actual numbers of particles and a complete set of basis states $b_j(\tau) = b_j(r,\sigma)$, where $\tau$ as in (3.40) is a shorthand for $r$, the position vector of a point in space, and $\sigma$, a general spin-related co-ordinate, assumed discrete so that integrals over d$r$ must be accompanied by sums over $\sigma$. In the Heisenberg matrix formalism, all properties of the system can therefore be related to the density matrix, as already noted by Dirac (1930) and used in Hartree-Fock related theories such as that of Møller and Plesset in 1934 for both stationary and time-dependent situations. The density matrix may be constructed in the space of the real wavefunctions of different states of the system, or it may be formed for an arbitrary complete basis such as the $b_j$s considered in the previous subsection.

This suggests a method for introducing electron correlations based on the density matrix, which was taken up by Hohenberg and Kohn (1964) and Kohn and Sham (1965), but which had its roots in much earlier work (Thomas, 1927; Fermi, 1928; Slater, 1928, 1951 and several papers in between). The method is now called *density functional theory*.

The first observation of density functional theory is that there is a one-to-one correspondence between the ground state total electron density and the ground state energy, under the assumption that there is no degeneracy of states. In chemistry, in the past the interest was often confined to the ground state, and using a real density rather than the complex wavefunction appealed to many people working in chemistry[a]. The ground state energy for a given density is of the form (Kohn and Sham, 1965)

$$E_0 = \int (V_N + V_{eN}(r))\rho(r)dr + \int V_e(r_1, r_2)\rho(r_1)\rho(r_2)dr_1 dr_2 + T_e[\rho] + E_{xc}[\rho] \quad (3.48)$$

$$\rho(\mathbf{r}) = \sum_k \varphi^* \varphi_k = \sum_k \chi_k^* \chi_k \psi_k(r)^* \psi_k(r) \qquad k=1,\ldots,e, \qquad (3.49)$$

Here, the square brackets in (3.48) denote a possibly complicated and indirect dependence on the density $\rho$. The second observation is that a minimum found by varying this energy with respect to the density will describe the ground state. Unfortunately, the terms describing antisymmetry and deviations from independent electron motion beyond those described by the previous term cannot be written in simple terms as a function of the density and therefore must be approximated and/or parametrised. Zero and first order gradients of the density are used, under the names of *local spin density* and *generalised gradient approximation* (Perdew et al., 1996, 1999). Effectively, these methods may be viewed as replacing the antisymmetrisation by a short-range repulsive potential, which in the local spin approximation is just a function of a radial co-ordinate.

Iterative solution of the Kohn-Sham equations is similar to that of the Hartree-Fock equations, and a number of popular methods use a parametrisation with both Hartree-Fock and density function bits, each with a phenomenological weight factor. The density functional bit may consist of the local spin density part plus a phenomenological parametrised part representing exchange and other correlations, such as the popular three-

---

[a] Of course, use of complex numbers is essential to quantum mechanics and necessary for explaining the limitations to observability. Indeed, the matrix elements in any "density functional theory" aimed at explaining more than just one state of the system must have this essential feature.

parameter B3LYP method (Becke, 1993; with use of Vosko *et al.*, 1980; Lee *et al.*, 1988). The *ad hoc* parametrisations are made such that on average they fit selected properties for a range of simple molecules, but without necessarily doing well for a particular molecule, even in the sample used for parameter choice. There is continued effort to find new approximations capable of addressing the known problems and preferably reducing the use of non-exact treatment of the basic Schrödinger equation (see, e.g., Adamo and Barone, 2002; Kudin *et al.*, 2002; Kümmel and Perdew, 2003; Staroverov *et al.*, 2004).

Treatment of excited states in Hartree-Fock or in a density functional-type approach is not possible using simple variational methods. Considering the time-dependent Schrödinger equation (3.28), one may instead describe those excited states that can be reached by a one-step transition from the ground state, by adding a time-dependent potential and looking at the resulting density distributions of the states that can be formed. Giving the transition matrix a simple form, e.g. by using a Taylor expansion, it can be inverted to yield a correspondence between the excited state densities and the potentials responsible for them (Runge and Gross, 1984). Switching on these potentials adiabatically allows the calculation of the energy dependence of the density function by solving a matrix diagonalisation eigenvalue problem of twice the physical dimension (Bauernschmitt and Ahlrich, 1996). This is called time-dependent Hartree-Fock or time-dependent density functional theory depending on the starting point.

| Method | $E_1(O_2)$ (eV) | $E_2(O\text{-}O)$ (eV) | $E_3(2O)$ (eV) | $E_1\text{-}E_2$ (eV) | $E_1\text{-}E_3$ (eV) |
|--------|-----------------|------------------------|----------------|-----------------------|-----------------------|
| HF     | −4 064          | −4 044                 | −4 059         | −19.5                 | −4.4                  |
| MP2    | −4 071          | −4 401                 | −4 063         | +319                  | −11.7                 |
| MP3    | −4 071          | +4 083                 | −4 064         | −8 154                | −6.8                  |
| MP4    | −4 071          | −163 400               | −4 064         | +159 279              | −7.3                  |
| B3LYP  | −4 084          | −4 074                 | −4 076         | −10.1                 | −8.1                  |
| PBE    | −4 080          | −4 073                 | −4 071         | −7.8                  | −9.4                  |
| PBE1   | −4 080          | −4 069                 | −4 071         | −11.1                 | −8.4                  |

*Table 3.1.* Calculations of ground state energies for the oxygen molecule, $E_1(O_2)$, for the oxygen atoms moved 2.0 nm apart, $E_2(O\text{—}O)$, and twice the energy of a single oxygen atom, $E_3(2O)$, plus $O_2$ binding energies estimated as the difference between $E_1$ and each of the two calculations for separated atoms. The value determined experimentally is −5.2 eV. All calculations use the SV basis set. HF = Hartree-Fock, MP = Møller-Plesset. For the remaining rows see discussion in text.

As an example of the accuracy of the higher order corrections to Hartree-Fock solutions obtained by perturbation methods and by parametrised density functional methods, Table 3.1 looks at the binding energy (or minus the atomisation energy) for an oxygen molecule, calculated in a number of different approximations. All calculations use the same basis set SV as Figs. 3.7 and 3.8 (Schaefer *et al.*, 1992), and adding polarisation to the basis functions as in Schaefer *et al.* (1994) only increases the binding energy by some 10%. The binding energy is already too large in all calculations except HF, as the measured value is –5.2 eV (CRC, 1973). It is seen that the Møller-Plesset perturbation expansion is well behaved for the oxygen atom or molecule, but when the atoms of the molecule are pulled apart, the expansion breaks down. The pure Hartree-Fock also does poorly with the two atoms far apart, while the HF difference $E_1$–$E_3$ = –4.4 eV is closer to the measured value than any of the higher approximations, of which the MP3 comes next, but overshooting. The parametrised calculations are the B3LYP (Becke, 1993) mentioned above and two variants using PBE, based on Perdew *et al.* (1996), with PBE1 and PBE having different weightings between the exchange and correlation parts of the functional $E_{xc}$ in (3.48) (0.25:0.75 and 0.5:0.5).

The conclusion drawn from this small exercise is that the perturbational but otherwise exact methods can be dangerous, especially for large molecules with parts bound loosely together, where the expansion may diverge as it does in the O--O case studied in Table 3.1. The density functional methods seem computationally much more stable, but are not necessarily correct in predicting absolute energies. Yet relative energies may be all right, as the measured $O_2$ bond distance of 0.12 nm is very well reproduced by the potential surface charting by most of the methods used in Table 3.1. This can be seen from Chapter 2, Fig. 2.6, using HF for the Ni surface and B3LYP for the oxygen molecule. The same is true for the bond distance of $H_2$, shown in Chapter 2, Fig. 2.5, and generally, the structure determination obtained even with the simpler quantum chemical methods seems acceptable, at least in cases without multiple, close-lying minima in the potential surface. However, the problem in reproducing binding energies encountered with oxygen suggests that one should exercise much care in attributing too much confidence to the results of energy calculations, although the general features of excitation spectra, such as succession of low-lying states, would in many cases come out in fair agreement with data, for cases where such are available. In other words, the quantum chemical tools available at the moment give a reasonable qualitative picture, without being necessarily reliable at a quantitative level.

### 3.1.4 Application to water splitting or fuel cell performance at a metal surface

In Fig. 3.3, mechanisms were presented for water splitting in an electrolyte or the reverse reaction of electric power production in a fuel cell. Here I shall attempt to use quantum chemical calculations to assess if such proposed schemes are possible and in which way they need the assistance of a metal surface ("catalysis"). Is the metal a template or catalyst satisfying the older textbook definition of "not participating in the reaction", or does it play a much more essential role in the processes, as the calculations shown earlier in Chapter 2, Figs. 2.5 and 2.6 have already suggested?

The dissociation of a hydrogen molecule at a distance of just over 0.12 nm from the metal surface shown in Chapter 2, Figs. 2.5 and 2.58 lead to two hydrogen atoms separated by around 0.12 nm, but rather insensitive to the precise distance. Similarly, an oxygen molecule would dissociate at a distance from the metal surface only slightly larger than for hydrogen (about 0.16 nm, reflecting the larger "radius" characterising the oxygen electron density distribution), and the distance between the two oxygen atoms is around 0.15 nm, again slightly larger than for hydrogen, as expected (Chapter 2, Fig. 2.6).

The dissociation of molecular hydrogen is the main story of what happens at the negative electrode of a fuel cell. The "liberated" hydrogen atoms can move through the (liquid or solid membrane) electrolyte towards the positive electrode, by the mechanism described in Fig. 3.7, and the electrons will enter the electrode metal and take part in an external current, if the cell is connected to a load, or will just build up a potential across the cell, creating the conditions for the transfer of positive hydrogen ions.

At the positive electrode of a fuel cell, a more complex set of reactions must be at work in order for the hydrogen ions to join with oxygen and form water (as illustrated by the sequences in Fig. 3.3, read from top to bottom). Figure 3.3 does not indicate ionic charges, because part of the investigation is to find out at which stage molecules are ionised or electrons are captured to form neutral molecules. The electrons are taken from or donated to the electrode metal, which then becomes positively charged in the fuel cell case (or already is due to the external potential in the electrolyser case). It follows that the electrons likely react with ions quite near to the electrode surface.

In order to determine more precisely what is happening, the quantum calculations must be made in the presence of either water molecules or hydrogen atoms and OH molecules capable of making use of the effective positive charge caused by the deficiency of electrons on the positive metal electrode surface. The types of calculations described in the previous subsection in principle predict the distribution of charges over the entire system, but in case of divided systems such as the metal surface and the molecules in the electrolyte, this may not emerge correctly (due to the deficiencies of the approximate quantum mechanical treatment, which as illustrated by the ionisation energy calculations above does not do well for separated subsystems). Indeed, straight calculations using the density functional theory fail to describe the water formation and subsequent movement away from the metal surface in a fuel cell, and inspection of the wave functions determined show that electrons are invariably moved from the metal surface to the protons, even when they are far from the surface. This failure to distribute charges correctly is due to poor initial guesses and to optimisation routines being un-successful due to multiple minima in the potential energy surface.

The solution used here to avoid such problems is to divide the system into parts, with the metal surface being one part and the hydrogen and oxygen atoms included in the calculation forming one or two other parts. This type of calculation allows each part of the system to have a prescribed charge. It was initially proposed to allow lower approximations to be used for parts of the system allowing this, and it uses a prescription called ONIOM to sum the energy contributions from different levels of theory (Dapprich *et al.*, 1999).

Consider the process (a) in Fig. 3.3, involving two water molecules or six atoms for which to vary the position co-ordinates over a metal surface. To limit the variations to a manageable set[°], I first look at a case where the water molecules are both in a plane at a distance of $z$ above the metal surface, with the centre of mass above a low-density location in the surface layer of the metal lattice (Fig. 3.10). For reasons of display, it is convenient to illustrate variations of two parameters at a time. Assuming first, that the hydrogen at-oms are at the water equilibrium distance $d = 0.096$ nm from the relevant oxygen atom, while the distance $2e$ between the oxygen atoms is varied, the potential energy surface as a function of $z$ and $e$ takes the form shown in Fig. 3.10. It is seen that with the four hydrogen atoms close, the two oxygen at-oms cannot form a molecule at $2e \approx 0.12$ nm as in Chapter 2, Fig. 2.6, but they willingly separate further when they are at a distance of $z \approx 0.4$ nm from the second Ni surface (like in the figures of Chapter 2, I use the lower of the two Nickel layers to measure distances from, purely out of convenience based on the way I set up the calculations). Nickel is a commonly used catalyst, in the same transition group as Pd and Pt.

Figure 3.11 shows that once the oxygen atoms have moved away from each other (0.24 nm), the two water molecules can drift away from the Ni surface, but soon without gain in energy. In Fig. 3.12, the dependence on the distance $g$ from the hydrogen atoms to the oxygen atom of the water mole-cule to which they belong is explored. All four $g$s are still the same, and the H-O-H angles are kept at 104.4°. It is seen that if the hydrogen atoms start far away from the oxygen atoms, they will move towards them until they get below the distance (0.096 nm) characterising their equilibrium distance in a water molecule and get repelled if they try to move closer.

The next step is to explore the energy behaviour when each oxygen atom has the two H atoms at different distances from the oxygen atom. This is il-lustrated by repeating the calculation with one H atom in each water mole-cule staying at $g = 0.096$ nm, but letting the other ones increase their dis-

---

[°] To draw a potential energy surface, you need, say, 10 values for each of the $6 \times 3$ position co-ordinates. In total that is $10^{18}$ point energy calculations. Even with a fast computer at your disposal, assuming each point to require an hour of CPU time, you see that this cannot be done. Thus, some restricted variation strategy must be found.

tance, which I shall denote $h$. Figure 3.13 shows the potential energy for $h$ larger than 0.096 nm. If the height over the Ni surface is at the same time kept equal to the value $z = 0.4$ nm that lead to a minimum potential energy in Figs. 3.10 and 3.11, then one hydrogen atom from each molecule will slide away for $h$ larger than 0.15 nm, but approach the oxygen atoms if $h$ is below that value. If the water molecules are closer to the Ni surface ($z \approx 0.32$ nm), then one hydrogen atom from each water molecule could theoretically move away, but as the overall energies are higher, it will not do that. Instead, it will fall into the higher $z$ minimum (for $h > 0.14$ nm, the energy is lower for $z = 0.32$ than for $z = 0.4$, and here the H atom will run away in any case).

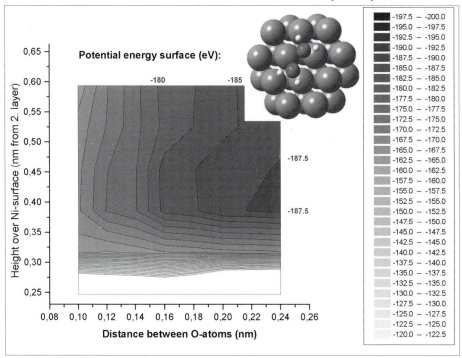

*Figure 3.10.* Potential energy surface of two water molecules over an Ni surface, as a function of height over the surface ($z$ measured from the second layer of Ni atoms) and $2e$, the distance between the oxygen atoms. The quantum chemical calculation of the potential energies employs Hartree-Fock for the two layers of 24 Ni atoms and density functional theory (B3LYP) for the rest, the set of basis function called SV (cf. section 3.2). The position of the water molecules for one set of parameter values is shown in the insert. The zero of the energy scale is chosen arbitrarily (Sørensen, 2004b).

The implication of Fig. 3.13 is that capturing the second H atom to form a water molecule (by the process going from the middle to the lower line in the left-hand side of Fig. 3.3a) cannot proceed spontaneously. What about attaching the first H atom to an O atom (top-to-middle line in Fig. 3.3a)?

Figure 3.14 shows the results of the quantum chemical calculation for this situation. The potential energy is calculated as a function of the distance between the oxygen atoms ($2e$) and the H-O distance ($g$) for the single hydrogen atoms (the other remaining far away). It is seen that the two already dissociated oxygen atoms will willingly move further away from each other, while for all distances above $2e = 0.12$ nm, single hydrogen atoms from a larger distance will move towards the O atom until they reach the equilibrium distance $g = 0.096$ nm characterising H-O distances in water molecules. In other words, getting the first H atom to join with an oxygen atom is energetically favoured (top-to-middle line in Fig. 3.3a).

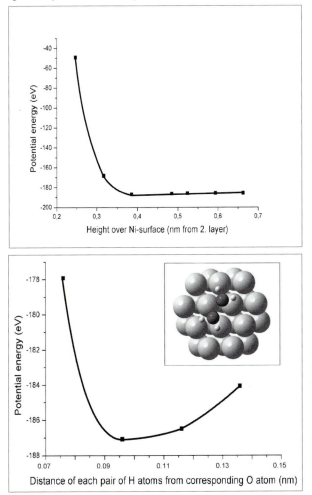

*Figure 3.11.* Detail of Fig. 3.10 showing the potential energy as a function of $z$, the height over the Ni surface, for a fixed distance between the O atoms of 0.24 nm (Sørensen, 2004b).

*Figure 3.12.* Potential energy for the Ni plus two water molecule system as in Fig. 3.10, with the distance of the H atoms from the relevant O atom varied from the value $g = 0.096$ nm valid for an isolated water molecule. The O atoms are $2e = 0.24$ nm apart, and the distance over the Ni surface is kept at the value $z = 0.4$ nm, leading to the minimum potential energy for all values of $g$ (Sørensen, 2004b).

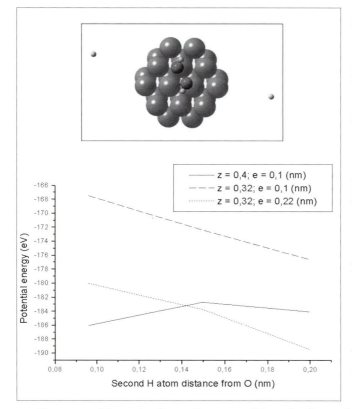

*Figure 3.13.* From a second set of calculations for two water molecules over a nickel surface, now allowing one hydrogen atom in each water molecule to move, while the other is fixed at $g = 0.096$ nm. The dependence of the total potential energy on the distance $h$ of the two H atoms allowed to move is shown for two heights over the Ni surface and for the lower height $z$ for two distances between the oxygen atoms (the distance being $2e$ or 0.2 and 0.44 nm) (Sørensen, 2004b).

To proceed to water formation something else is needed, such as another two water molecules (Fig. 3.3b) or a catalyst, which could be the metal surface or a nearby organic molecule such as the *tyr* in the photosynthetic water splitting illustrated in Chapter 2, Fig. 2.18b. The calculations above have been done for neutral molecules. It is now time to look in more detail at the distribution of charges, starting with the four positive charges brought along by the four H atoms from the negative electrode processes.

Figure 3.15a shows the results of a series of calculations all with an overall charge of +4 elementary charge units on the system. The first ones have the 4 charges on the $H^+$ (protons) entering above the electrode surface area from the electrolyte membrane. One proton is transferred to each O atom to form OH (initially positively charged) and two charges remain at the distant H atoms. This is like in the neutral case. Now, electrons may be transferred from the Ni surface to the $OH^+$ or $H^+$ at little energy change. The charge of the Ni surface now increases towards +4. However, like in the case discussed in connection with Fig. 3.13, pulling the second $H^+$ in to form a water molecule still would increase the potential energy.

Fig. 3.15b shows how this obstacle can be overcome. While charges accumulate on the metal surface in the case of an open external circuit, an electricity-producing fuel cell will receive electrons from the external circuit to the Ni-surface, and the calculation is thus repeated with a reduced positive charge on the surface. This creates the condition for water molecules to form and move away from the surface, in a succession of downwards energetic steps. The possible further reduction of the charge on the Ni surface to near zero (in the condition of a high external current), actually makes any water molecules present stick to the surface, and only those already moved away from the surface can fully escape. These peculiarities at different levels of external current and closed/open circuit should in principle be observable.

In other words, the presence of a metal such as Ni as a catalyst-electrode is essential both for the charge transfer constituting the electric current of the fuel cell and also for making it energetically favourable to form water molecules and move them away from the reaction site. The set of calculations described here is the first to successfully explain the mechanism of an operating fuel cell on a quantum mechanical level.

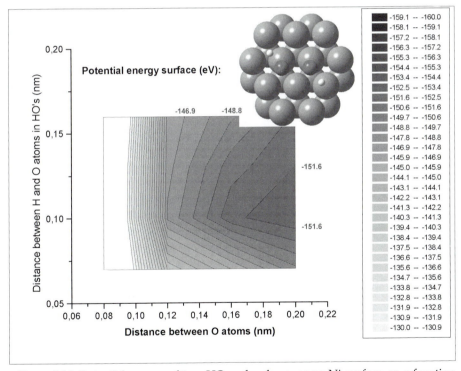

*Figure 3.14.* Potential energy of two HO molecules over an Ni surface, as a function of the O—O distance $2e$ and the O—H distance $g$, calculated with the same methods as those used for Figs. 3.10-3.13 (Sørensen, 2004b).

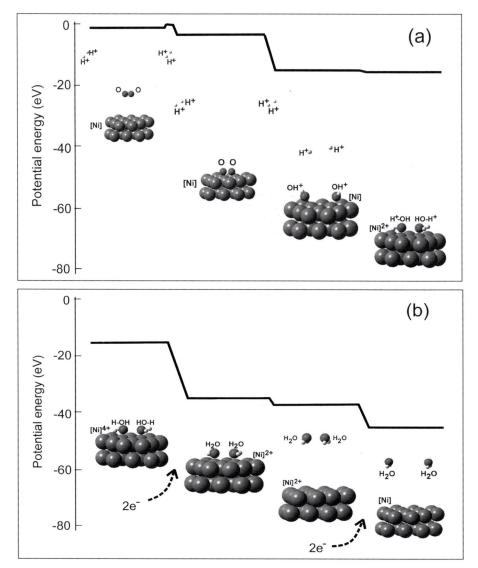

*Figure 3.15a,b.* Series of restricted potential energy minima for a sequence of configurations of an Ni surface, two O atoms and four H atoms. The calculation uses an ONIOM method with the Ni surface as the low layer, the two Os or O-H pairs as the intermediate layer and the last four or two H atoms as the high layer. The intermediate and high layers both use the same B3LYP density functional method. The overall charge is in the cases shown at the top (a) +4. In the first case (left), these charges sit on the 4 H atoms, and proceeding to the right and to (b) below, first 2 and then all 4 charges are moved to the Ni layer. In the bottom sequence (b), electrons from the external circuit reduce the overall charge in the calculations to +2, allowing water to be formed and to escape from the Ni surface, and finally to zero (Sørensen, 2004b).

## 3.1.5 Flow and diffusion modelling

While an understanding of the molecular processes at the fuel cell electrodes requires a quantum mechanical description, the flows through the inlet channels, the gas diffusion layer and across the electrolyte can be described by classical physical theories such as fluid mechanics and diffusion theory. The equivalent of Newton's equations for continuous media is an Eulerian transport equation of the form

$$\frac{\partial}{\partial t}(\rho A) + \mathrm{div}(\rho v A) + \mathrm{div}(s_A) = E_A, \tag{3.50}$$

where $t$ is time, $\rho$ is the density, $v$ is the velocity at a given point $(x,y,z)$ in space, $s_A$ is a vector describing microscopic transport due to molecular processes, and $E_A$ describers external sources of the quantity $A$ looked at. For fluid flow, $A$ would be a component of the velocity $v$, while for $A = 1$, (3.50) is an equation of continuity,

$$\frac{\partial}{\partial t}(\rho) + \mathrm{div}(\rho v) = 0. \tag{3.51}$$

When $A = v$, the molecular transfer of velocity is related to the stress tensor $\tau_{ij}$ by (see, e.g., Sørensen, 2004a)

$$-(s_{v_j})_i = \tau_{ij} = \left( -P + \left( \eta' - \frac{2}{3}\eta \right) \rho \, \mathrm{div}(v) \right) \delta_{ij} + \eta \rho \left( \frac{\partial v_i}{\partial x_j} + \frac{\partial v_j}{\partial x_i} \right), \tag{3.52}$$

where $P$ is the pressure, $\eta$ is the kinematic viscosity and $\eta'$ is the volume viscosity. If the latter is neglected and the fluid is treated as incompressible, the simpler version of equation (3.50) for $A = v$ with (3.52) inserted is called the Navier-Stokes equation.

Numerical solution of equations such as (3.50)-(3.52) can be done in various ways, each suitable for certain classes of problems. Flow modelling is used in many problem areas, including climate modelling, weather forecasting, aircraft and vehicle air resistance and flow over wings and similarly wind turbine blades, flow in pipes and combustion boilers and turbines, and so on. Flow in the gas channels of fuel cells is just one example. The numerical methods used comprise two types. In one, the integration space is discretised, i.e., the functions considered are given only on a mesh of points in space and differential equations become difference equations. Each point is assigned average values of the quantities studied, and the solutions may be sought in terms of an initial value problem (levels and velocities, corresponding to variables and their derivatives in the continuous case, known at each point at one time $t_0$) or of a boundary problem (levels and velocities specified at the boundaries that define the volume for which calculations are performed).

The other method of calculation defines a discrete set of (generally) three-dimensional regions spanning the volume of interest in space and expands the solutions on a polynomial basis set, typically just piecewise linear functions, for each of the elemental regions (called *finite elements*). Each of the basis polynomials are zero, except for a few elemental regions, and the solution to, e.g., the flow field $v$ is a linear combination of the basis functions. The coefficients are determined either from applying the differential equations at a finite number of points spanning the elements or by minimising a suitable aggregate function such as the total energy, in case the solution sought corresponds to a stationary point in energy space. The finite elements, or elemental regions, selected would typically be large in regions where little variation in $v$ is expected, but smaller in regions where interesting variations are expected. For some problems, symmetry in design may allow the transport equations to be solved in less than three dimensions.

Examples of employing the fluid dynamics or finite element method will be shown in the following sections. The methods can obviously be used also for variables other than the velocity, by a suitable selection of $A$ in (3.50), such as, e.g., temperature, once the basic velocity field has been determined (as it enters in the equations for all other quantities). In the case of temperature $T$, the source term $E_T$ in (3.50) includes external sources of heat and condensation, and the ideal gas law may be used to connect temperature and pressure.

Neither the finite element method nor a discretised integration will "catch" eddy or other motion below a certain scale determined by the choice of mesh. Small-scale motion in many cases may be better described as random, in which case the transport of the quantity $A$ is called *diffusion*. Diffusion can be described by Fick's law, assuming that the flux density $f$, i.e., the number of "particles" (here a small parcel of a gas), passing a unit square in a given direction, is proportional to the negative gradient of particle concentration $n$ (Bockris and Despic, 2004),

$$f = -D \operatorname{grad} n. \tag{3.53}$$

The proportionality factor $D$ characterising the medium is called the *diffusion constant*. It is assumed that each particle, i.e. each parcel of the fluid, is moving under the influence of the average field from the rest of the fluid. Fick's law is then combined with the continuity equation similar to (3.51)

$$\text{div} f + \frac{\partial n}{\partial t} = 0, \tag{3.54}$$

expressing that changes in concentration are equal to minus the sum of fluxes out of the parcel volume in all directions. Combining (3.53) and (3.54), one gets the diffusion equation

$$D \Delta n = \frac{\partial n}{\partial t}, \tag{3.55}$$

where the notation $\Delta = \nabla^2 = (\text{div grad})$ is used. Equation (3.55) describes diffusion of a single substance. In practical cases such as the fuel cell types considered in the following sections, there will often be several chemical components characterised by their mole fractions $x_i$ (cf. (3.6)). The equations corresponding to Fick's law are now called the Stefan-Maxwell equation, and may, with use of the ideal gas law in the form valid for partial pressures $p_i$ and molar fractions of the $i$th substance, $n_i$, be written (Bird $et\ al.$, 2001)

$$\text{grad } x_i = \frac{\mathscr{R}T}{P} \sum_{j \neq i} \frac{x_i f_j - x_j f_i}{D_{ij}}, \tag{3.56}$$

$$p_i = n_i \mathscr{R}T/V, \tag{3.57}$$

$$P = \sum p_i; \quad n = \sum n_i, \quad n_i = x_i n, \tag{3.58}$$

where $\mathscr{R}$ is the gas constant (cf. (3.6)), $T$ is the temperature and $P$ is the total pressure. The flux density of the $i$th component is denoted $f_i$ and the total volume is denoted $V$. The diffusion constant $D_{ij}$ describes the inter-diffusion of components $i$ and $j$ and is proportional to the mean free path for collisions between the substances $i$ and $j$ and to their relative speeds $((v_i^2 + v_j^2)/2)^{1/2}$. The flux density of the $i$th substance is related to the velocity of this substance at a given point by

$$f_i = \varepsilon c_i v_i, \tag{3.59}$$

where $\varepsilon$ is the porosity and $c_i$ is the molar concentration of substance $i$ ($\sum c_i = c_{total}$). For the small pore sizes found in many gas diffusion layers, collisions with pore walls become as important as collisions between the particles in the substances, and the right-hand side of (3.56) receives an additional contribution (Knudsen, 1934),

$$-\frac{\mathcal{R}T f_i}{P D'_i},\tag{3.60}$$

where $D'_i$ is a pressure-independent diffusion coefficient depending on pore structure, and where the particle collision-type diffusion parameters $D_{ij}$ in (3.56) may have to be somewhat adjusted by the presence of pore walls (Bruggeman, 1935).

Many numerical models make additional assumptions, valid if only some specific questions are being asked. For example, if one is not interested in the start-up phase or in changing the operation of a fuel cell, one may apply the steady state condition that time-independent solutions are requested. In certain problems, one may disregard temperature variations, and in the free gas ducts, laminar flow may be imposed. The diffusion in porous media is often approximated by an assumption of isotropy for the gas diffusion or membrane layer, and the coupling to chemical reactions is often simplified or omitted. Water evaporation and condensation, on the other hand, are often a key determinant for the behaviour of a fuel cell and thus have to be modelled at some level.

The management of water is an integral part of operating many fuel cell types. For those operating at high temperatures, water vapour can be modelled as just one additional component of gas flow. However, for cells operating at lower temperatures, a two-phase flow model is required, and in general, there will be both liquid and gas forms of water present in both gas diffusion layer and gas channels (plus in the liquid electrolyte or membrane). It is thus necessary to consider both transport of both phases and also evaporation and condensation processes taking place in various parts of the cell. The water generated by the cell from hydrogen and oxygen may at its origin be considered as gaseous and generated in proportion to the current density. Subsequently, the water vapour condenses to liquid form, likely to appear predominantly at the positive electrode. The motion of the liquid water is then proceeding through the gas diffusion pores, with capillary pressure gradients driving the flow being the most probable mechanism of transport (Wang *et al.*, 2001; Nguyen *et al.*, 2004). The relation between the fluid velocity and the pressure gradient may be expressed by Darcy's law

$$v = -\frac{K(s)}{\varepsilon\eta}\,\text{grad}\,P,\tag{3.61}$$

where $K$ is the permeability which may be taken to depend on the level of water saturation in the pore, $s$, i.e. the liquid volume fraction. The pressure $P$ is for this unsaturated flow proportional to a capillary pressure function that also depends on $s$ (Stockie, 2003; Stockie *et al.*, 2003). The transport of water in pores may be enhanced by impregnating the gas diffusion layer with, e.g., teflon to reduce adhesion to pore walls.

## 3.1.6 The temperature factor

The following sections will describe a number of fuel cell concepts. The reason there is so many in part has to do with the historical development of the technology, but also to a large extent the temperature regime for which they are designed. In many cases, high operation temperature will entail a higher efficiency of electricity production, but of course maintenance of a high temperature will also influence efficiency and in some cases, such as cold-start of a vehicle, be inconvenient.

Generally, large stationary fuel cell installations aim at the highest possible efficiency, which points in the direction of molten carbonate fuel cells (operating around 670°C) or solid oxide fuel cells (currently operating at around 800°C or higher). For decentralised stationary uses, lower temperatures may be contemplated, although presently used boiler units for heating operate at temperatures similar to the high-temperature fuel cell concepts. For automotive applications, the absence of combustion implies that heat would have to be supplied initially and sustained during driving with insulation and top-up heat supply, e.g. from batteries. This is generally considered inconvenient, because fuel cells are often seen as an alternative to electrically powered vehicles, and even in hybrid concepts, the energy use for heating would substantially decrease the overall efficiency of operation. This statement can be made more precise by considering the loss of efficiency implied by using low-temperature fuel cells and comparing it with the energy cost of having to heat the fuel cell to a higher operating temperature.

The bottom line is that at the present conception of the various technologies, proton exchange membrane (PEM) fuel cells operating near ambient temperature are considered the best choice for (road) vehicles, while the high-temperature fuel cells are considered as suitable for centralised stationary applications. The market development may interfere with this line of thought, because the automotive market is seen as a key entry point for fuel cell technology, and if this large market succeeds in bringing the cost of PEM fuel cell technology down, such cells may out-compete the other options. Signs of this are already evident in the current development phase, because several manufacturers take advantage of the strong effort made to ready PEM cells for the automotive market to develop small building-integrated PEM fuel cell units for replacement of natural gas boilers. Because low-efficiency and high-pollution problems are much more severe in the transportation sector than in the fossil fuel-based power production sector, the initial market for high-temperature fuel cells for general power production may be very small and thereby prevent such fuel cells from taking advantage of a production volume-related cost decline.

# 3.2 Molten carbonate cells

A fuel cell cycle employing carbonate ions penetrating a solid matrix electrolyte at high temperatures is schematically illustrated in Fig. 3.16. It is aimed at stationary applications and promises high efficiency. The electrode reactions for this electricity-producing molten carbonate fuel cell (MCFC) are

$$H_2 + CO_3^{2-} \rightarrow CO_2 + H_2O + 2e^-, \tag{3.62}$$

$$\tfrac{1}{2} O_2 + CO_2 + 2e^- \rightarrow CO_3^{2-}. \tag{3.63}$$

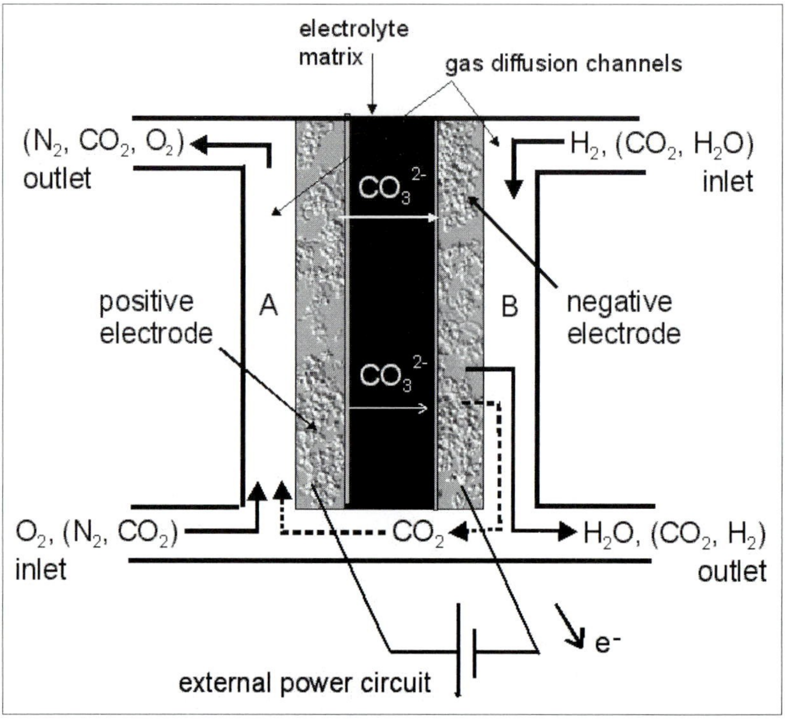

*Figure 3.16.* Schematic picture of a molten carbonate fuel cell. The gases in parentheses are not necessarily taking part in the electrochemical processes, but they are required either for ensuring flow or for carrying excess gases away. The arrangement of $CO_2$ recycling from negative to positive electrode can be made in different ways.

The carbon dioxide from (3.63) is in most designs recycled as input to (3.62) together with the hydrogen fuel. It has been suggested that $CO_2$ emissions from fossil power plants could be used as input as a way of reducing greenhouse gas emissions from the current system. However, then the output $CO_2$

from (3.62) would have to be collected and disposed of, e,g, by transforming it once more to the form of carbonates for deposition as waste (Lusardi *et al.*, 2004). As indicated in Fig. 3.16, all the gases may have to be present in the appropriate channels in order to provide the required flow and to recycle fuel gases not used in a particular pass. This signals a need for fuel processing loops separated from the fuel cell stack. The cell electrolyte consists of Li-K (for systems operating near atmospheric pressure) or Li-Na compounds (for elevated pressure systems) able to carry the molten alkaline carbonate mixture, while for stability and strength the assembly is being supported within a porous aluminate (e.g., $LiAlO_2$) matrix.

Electrode materials are for the negative electrode Ni with Cr or Al additives for providing material strength and for the positive electrode NiO with additives of Mg or Fe, mainly serving the purpose of avoiding short-circuit phenomena (conductivity across the electrolyte from positive to negative electrode). The carbonate ions formed at the positive electrode travel through the matrix and combine with hydrogen at the negative electrode at an operating temperature of 660°C or higher. The process was originally aimed at hydrogen supplied from coal gasification or natural gas conversion. The first research installations (being built in the 1980s, after a few European precursors) suffered severe degradation problems in the electrolyte, including cracking of the matrix (Mugikura, 2003). Also the prototype units put into operation around the year 2000 are considered vulnerable to possible structural change effects after longer periods of operation, notably affecting seal coatings and the integrity of the alumina matrices (Frangini and Masci, 2004; Mendoza *et al.*, 2004; Jun *et al.*, 2002). The first prototype tests of MCFC power plants (250 kW or larger) were conducted during the period 1996-2000 in the USA, Japan, Italy and Germany (Farooque and Ghezel-Ayagh, 2003). Conversion efficiencies in the range of 55-60% were reached, but, theoretically, the efficiency could become higher (near 70%). Despite the relatively short periods of operating the demonstration plants, degradation of key components affecting plant lifetime were observed, and efforts to improve the situation are ongoing. It should be kept in mind that degradation of electrolyte and electrodes is inherent in all electrochemical systems and is the main reason for the limited life of conventional or advanced batteries. The challenge is whether it is possible to extend the lifetime for fuel cells beyond the under 5 year level characterising most battery types, which by some are seen as necessary for reaching economic viability.

The electrolyte constitutes a separator for the gas flows shown in Fig. 3.16, in addition to providing ionic transport. Physically, it consists of a compressed powder behaving as a soft paste at the operating temperatures, thus substantiating the name "molten carbonate" for the liquid Li-Na or Li-K carbonates inhabiting the electrolyte substrate. The $LiAlO_2$ material used for the support structure can exist in three different forms (phase $\alpha$, $\beta$ and $\gamma$), and transitions have been observed (e.g., $\gamma$ to $\alpha$) after extended fuel cell operation times, leading to enhanced degradation of the electrolyte integrity. Once the degradation processes have reached a certain level, the paste pore structure opens up and shorting by precipitation of Ni between the electrodes ends the electrolyte life. The Li-K electrolyte suffers a higher rate of acidic dissolution than Li-Na, but the latter shows a more pronounced sensitivity to high temperatures (Hoffmann *et al.*, 2003). Using $\alpha$-type $LiAlO_2$ rather than the conventional $\gamma$-type may reduce degradation (Batra *et al.*, 2002).

The stability of the negative electrode is fairly good, but the gas flowing past the positive electrode is highly corrosive and responsible for a significant dissolution of the NiO material. The Ni erosion increases with pressure and is very high at the temperatures of 660-700°C prevailing in MCFCs. It declines when the temperature is raised above 1000 K. NiO dissociation rates in the range of 0.01-0.02 g m$^{-2}$ h$^{-1}$ have been measured at ambient pressure, rising by an order of magnitude for higher pressures (0.7 MPa; Freni *et al.*, 1998). The high-pressure value implies a service life of only a few months, and even the lower value for ambient pressure systems leads to a lifetime (under 3 years) lower than what is considered necessary. Impedance measurements of the catalytic activity of Li-Ni-O (used to enhance the conductivity of the positive NiO electrode) under the influence of an eutectic Li-K melt and a corrosive $CO_2$-$O_2$ atmosphere have showed a transfer of lithium and related morphological change (expansion) of the NiO electrode material (Escudero *et al.*, 2003). Also, the reduction of electrode area has a negative influence on cell performance (Freni *et al.*, 1998). It has been proposed that separating the $CO_2$ and $O_2$ inlet flows (Fig. 3.16) into two separate streams, with the $CO_2$ flow entering the electrolyte matrix directly, on the opposite side of the positive electrode relative to the $O_2$ flow, may reduce the corrosion problem significantly, at a modest loss of efficiency (Au *et al.*, 2003).

In actual designs of MCFCs, flows are controlled by separator plates with corrugated patterns, and wet seals are used to separate corrosive gas flows from the sensitive electrolyte matrix and still allow the molten carbonate to pass through. Adding carbonates of Ca, Sr and Ba may diminish the NiO dissociation (Tanimoto *et al.*, 2004). The cell voltage at zero current density is about 1 V and diminishes by around 30% for a current density of 2 kA m$^{-2}$ (Freni *et al.*, 1997). It is hoped that a lifetime exceeding 5 years can be achieved by implementing improvements as suggested above.

Simple models for performance simulation have been developed for stationary flow situations, with the options of verifying the voltage-current relations and calculating the overall efficiency under consideration of total plant auxiliary power and heat inputs. The efficiency calculated with such models exhibits a few percent efficiency improvement from raising the pressure above ambient by the factors considered above and, as expected, a decline with increased gas flow rates (Simon *et al.*, 2003).

# 3.3 Solid oxide cells

The high-temperature fuel cell to which most attention is paid at present is the solid electrolyte cell. Solid oxide fuel cells (SOFC) use zirconia compounds as the electrolyte layer to conduct oxygen ions formed at the positive electrode. The electrode reactions involve oxygen ion transport (in contrast to the hydrogen ion transport in the basic scheme given by (3.15) and (3.16)), whence the appearance of "oxygen" in the cell name,

$$H_2 + O^{2-} \rightarrow H_2O + 2e^-, \tag{3.64}$$

$$\tfrac{1}{2} O_2 + 2e^- \rightarrow O^{2-}. \tag{3.65}$$

These reactions take place at the surface of a solid state electrolyte at a temperature of 600-1000°C. The lower temperatures are desirable, due to a wider choice of materials capable of maintaining integrity. A number of materials have been contemplated for use as electrodes and electrolyte.

The electrolyte should be able to conduct oxygen ions, but must be impermeable to the hydrogen and oxygen gases flowing along its sides. For this purpose, a solid membrane structure is used as electrolyte. It may consist of a thin layer of yttrium-stabilised zirconia, $ZrO_2$, doped with 3-8 mol% $Y_2O_3$ and possibly adding other dopants such as $Sc_2O_3$. A molecular structure is shown in Fig. 3.17. This electrolyte material may be sprayed onto the negative electrode as a ceramic powder with a thickness of around 10 μm (Kahn, 1996) or may be self-supporting at a thickness above 100 μm (Weber and Ivers-Tiffée, 2004). Generally, electrolyte losses increase with electrolyte thickness, so when possible the thinnest sheets not giving rise to short circuiting or gas penetration are preferable. For the lowest temperature cells, alternative electrolytes are needed in order to achieve sufficient conductivity. Among these are (Y-)doped ceria ($CeO_2$), used, e.g., for direct hydrocarbon SOFCs or in two-layer electrolytes. The second layer may be a lanthanum gallate electrolyte ($La_{1-x}Sr_xGa_{1-y}Mg_yO_3$), which exhibits high oxide ion conductivity but may be discouraged by stability problems and gallium cost.

The electrodes should be able to catalyse the relevant reactions (i.e., (3.64), (3.65) and necessary fuel transformation reactions, in case the initial fuel is not hydrogen). They need to have a large effective surface, but still possess long-term stability at the operating temperatures and with considerations of the temperature cycling taking place in practical applications. For designs such as the one shown in Fig. 3.20, where the gas flow is on the outer side of the electrodes, the electrodes must also be permeable for the gases involved in the cell reactions. The electrodes suitable for the temperature regime relevant to SOFCs are typically made of oxides of metals in or near the rare earth sequence in the periodic system. For the positive electrodes, the materials used include compounds such as $La_{1-x}Sr_xMn_{1-y}Co_yO_3$, possibly with Fe instead of Mn and Ca instead of Sr. The molecular structure of $LaCoO_3$ is shown in Fig. 3.18. Typically, some 20% of the La atoms are replaced by Sr ($x$ = 0.2), and the amount of Co ($y$) controls the electric conductivity and the thermal expansion rate of the electrode, both increasing with $y$. In particular, the positive electrode requires the dopants to be stable in a high-temperature oxygen atmosphere.

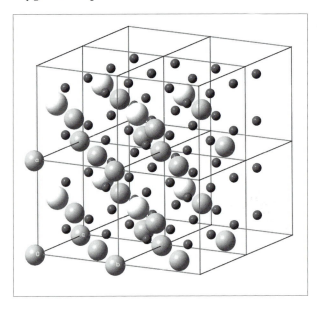

*Figure 3.17.* Structure of eight unit cells of an yttrium-stabilised zirconium oxide, $Zr_{0.75}Y_{0.25}O_2$. The smaller (dark) atoms are oxygen, and the larger ones are Y (lightest grey shade) or Zr, sitting at cube corners or face centred. The oxygen atoms have been placed in the most symmetric configuration. For independent $ZrO_2$ molecules, the angle between the O atoms is only 21° and the O-Zr distance is 0.2 nm, a little larger than in the lattice.

*Figure 3.18.* Structure of eight unit cells of $LaCoO_3$, with body-centred La atoms (large, light shaded) and Co (medium, darker) at the unit cell corners and O atoms (smallest, darkest) halfway between them, in the same arrangement as for perovskite (e.g., $CaTiO_3$ or $MgSiO_3$). Perovskites can exist in another structure, but only at the very high pressure (125 GPa) and temperature (2700 K) found at the Earth's core–mantle interface (Murakami *et al.*, 2004).

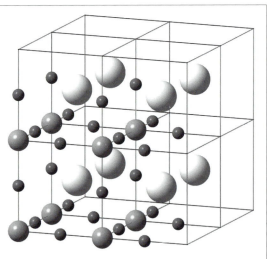

The negative electrode may consist of an Ni-based compound, using, e.g., a mixture of NiO with a gadolinium-doped cermet, $Ce_{1-x}Gd_xO_{1.95}$ ($x$ around 0.1), and $RuO_2$ as a catalyst for use at around 600°C (Hibino et al., 2003). Temperatures around 600°C are required for the direct use of hydrocarbon fuels (e.g. methane, ethane or propane) instead of hydrogen, which is then formed in a second step by internal reformation. Also in use are negative electrodes combining the same yttrium-stabilized zirconia as used for the electrolyte with Ni and Ce oxide (Weber and Ivers-Tiffée, 2004). Techniques used in metal–organic photoelectrochemistry (cf. Sørensen, 2004a; Chapter 4, section 4.2.3) may be used to create a large surface area of contact between the electrolyte and the electrode material.

Figure 3.19 summarises data on conductivity of some materials considered for SOFC electrolytes, as a function of temperature (values along top, inverse temperatures along bottom abscissa). The uniform decline in conductivity at lower temperatures illustrates the compromise on efficiency associated with designing SOFCs to operate at lower temperatures.

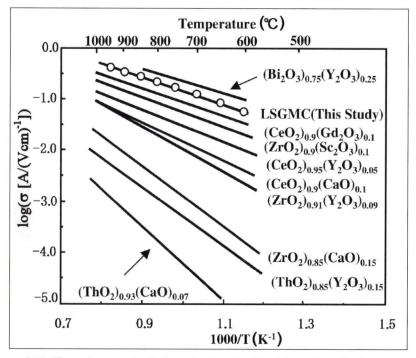

*Figure 3.19.* Electrolyte conductivity $\sigma$ for $La_{0.8}Sr_{0.2}Ga_{0.8}Mg_{0.115}Co_{0.085}O_3$ (open circles) and other materials (the unit $AV^{-1}$ is sometimes referred to as a *siemens*), as a function of temperature. (From Ishihara et al. (2004). Novel fast oxide ion conductor and application for the electrolyte of solid oxide fuel cell. *J. European Ceramic Soc.* **24**, 1329-1335. Used by permission from Elsevier.)

Figure 3.20 shows a cylindrical layout often used for high-temperature SOFCs. Alternatives are a stack of planar cells or a disk concept with feed tubes in the centre. A consideration of efficient heat exchange is the common design strategy for the high-temperature fuel cell geometry.

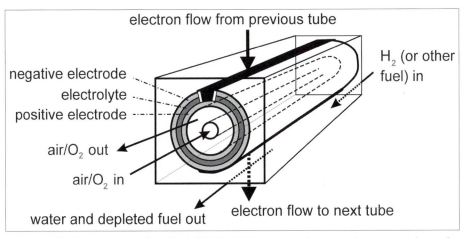

*Figure 3.20.* Layout of a cylindrical SOFC. Air (or oxygen) enters the centre tube and is deflected back through the next cylinder ring after being heated to operating temperature. The hydrogen (or other fuel) flow is external to the cylinder, and cells may be assembled in packages with electric interconnects as indicated.

Modelling the behaviour of solid oxide fuel cells employs a combination of an electrochemical model and a heat and materials flow model. The electrochemical model starts from calculating the internal cell potential by combination of the equations (3.5)-(3.7) with (3.20) and (3.23),

$$\phi = \sum_i n_{e,i}\,\Delta\mu_i - \phi_{losses} = \sum_i n_{e,i}\mu_i^0 + \mathscr{R}T\sum_i n_{e,i}\log(f_i x_i) - \phi_{losses}. \quad (3.66)$$

For the SOFCs, the sums in (3.66) have three terms, two positive ones for $H_2$ and $O_2$ and one negative one for $H_2O$. The constant part of the first terms on the right hand side, $\phi^0$, deviates from the one given in (3.20) for an ideal gas at ambient pressure and temperature, as it must take into account the high temperature in the SOFC. It is estimated as (Campanari and Iora, 2004),

$$\phi^0 = 1.2723 - 2.7645 \times 10^{-4}\,T_{cell}. \qquad (3.67)$$

The second sum on the right-hand side of (3.66) is called the Nernst potential, and finally, the loss term is as in (3.23) made up of contributions from bulk resistive losses and losses occurring at each of the two electrodes. These terms can be modelled in terms of measured Ohmic resistance and parametrised activation losses at electrodes (Costamagna *et al.*, 2004; Campanari and Iora, 2004).

For describing heat flow and mass transport, models based on the finite element method or some volume averaging scheme for handling general Eulerian transport are used, as described in section 3.1.5. Several programs are available for calculating the time-dependent behaviour of such systems, in terms of fields of velocity, matter concentration and temperature, and considering couplings to the electrochemistry and other chemical reactions that may take place. Examples of the results of such calculations for SOFCs are given below. Also simulations of entire SOFC stacks are available (Roos *et al.*, 2003).

Figure 3.21 shows an example of the air velocity field in an SOFC of the shape given in Fig. 3.20. The segment shown goes from the centre of the air inlet to the wall of the second tube used for return air (vertical axis) and horizontally covers the innermost part of the tube, where the air (with its oxygen) is forced to deflect back into the outer of the two tubes. This is evidently the most difficult part to model, and most fluid dynamics models allow the number of computation elements to be increased in such regions.

The temperature field from a similar simulation is shown in Fig. 3.22, comprising both the two air/$O_2$ carrying tubes and their internal wall and also the cell area (electrodes and electrolyte, cf. Fig. 3.20), which in the model cell is at radii between 0.5 and 0.72 cm, and finally, a piece of the channel for fuel flow. The temperature distribution over the length of the cell is not quite uniform, but shows 100-150°C lower values near the ends of the tube cylinder for the cell and fuel compartments. It is seen that the air/oxygen entering the inlet tube at lower temperature does not acquire a uniform temperature over the radial dimension of the tube, but only reaches the cell operating temperature in the part of the cylinder closest to the electrode assembly (and only in the middle of the cylinder). The assembly of positive electrode, solid electrolyte and negative electrode, as well as the fuel flow region adjacent to the negative electrode, on the other hand exhibits a temperature dependent only on the position $x$ along the tube length. Figure 3.23 shows similar temperature profiles as a function of $x$ only.

The two calculations underlying Figs. 3.22 and 3.23 agree on the behaviour of the cell assembly and fuel temperatures, but the calculation shown in Fig. 3.23 assumes pre-heating of the air/$O_2$ flow to a higher inlet temperature than the one appearing from Fig. 3.22. The return air/$O_2$ flow in Fig. 3.23 is therefore losing temperature when it is slowed down at the internal bend, under influence of the lower temperature in the cell assembly. It indicates that, for this particular case, less heating of the inlet air might be better and that a higher inlet fuel temperature could be desirable, although possibly problematic for safety reasons. In any case, the figures illustrate the usefulness of being able to model the dynamic behaviour of the cells.

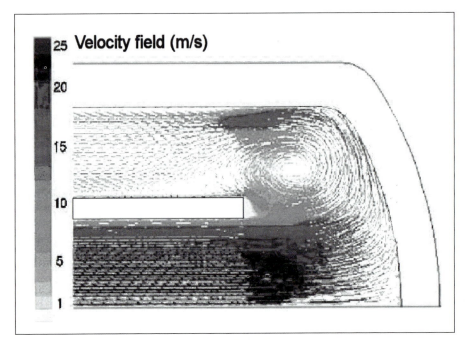

*Figure 3.21.* Velocity field calculated by a fluid dynamics model for the turning area of the air/oxygen flow in a solid oxide fuel cell at operating temperature. (Reprinted from S. Campanari and P. Iora (2004). Definition and sensitivity analysis of a finite volume SOFC model for a tubular cell geometry. *J. Power Sources* **132**, 113-126. Used by permission from Elsevier.)

The fuel input streams for the models illustrated in Figs. 3.22 and 3.23 are considered as the mixture of gases typically arising from previous steam reforming of natural gas. The calculation shown in Fig. 3.23 and the following assumes an inlet fuel composition of 26% $H_2$ (molar fraction), 11% $CH_4$, 23% $CO_2$, 6% CO, 6% $N_2$ and 28% water (Campanari and Iora, 2004). The calculation shown in Fig. 3.22 assumes more methane and no nitrogen. The option of internal reforming of fossil fuels within the SOFC will be discussed below. The inclusion of the possible reactions between fuel and gas flow components, in addition to those at the electrodes, must be considered in models for combined reforming and fuel cell reactions.

The temperature in an SOFC bundle is high enough for reforming of many hydrocarbons to take place within the reactor, so that the hydrogen required will be created internally, and the fuel to handle externally could be natural gas (methane), gasified coal or liquid hydrocarbons. For methane, the reactions (2.1), (2.2), (2.10) and (2.14) in Chapter 2 may take place near the hot fuel cell electrode, and for higher hydrocarbons the reactions (2.9) and (2.11) take place. Figure 3.24 shows a process diagram for a proposed natural gas-based SOFC, showing some of the additional problems introduced with the use of fuels other than pure hydrogen. The natural gas has to be de-sulphurised and eventually pre-reformed in a more efficient reformer than the fuel cell. The fuel cell faces additional corrosion problems due to carbon depositing on electrodes, formed by partial cracking of the types

$$C_nH_m \rightarrow nC + (m/2) H_2. \tag{3.68}$$

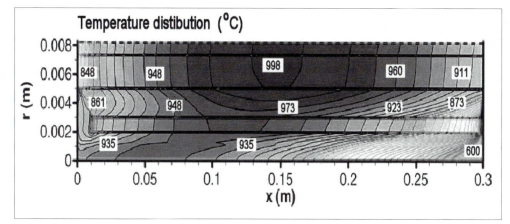

*Figure 3.22.* Temperature field along the length $x$ of a tubular SOFC cylinder (air/$O_2$ inlet to the right, tube end with flow bending at left, radially from $r = 0$ to 0.5 cm), as simulated by a dynamic flow model. The middle horizontal band is the electrode-electrolyte assembly, and the top part (above $r = 0.72$ cm) is the fuel channel. (Reprinted from P. Li and M. Chyu (2003). Simulation of the chemical/electrochemical reactions and heat/mass transfer for a tubular SOFC in a stack. *J. Power Sources* **124**, 487-498. Used by permission from Elsevier.)

For high-$n$ hydrocarbons, cracking is abundant at the temperatures prevailing in an SOFC, while for methane, it is only a problem in the presence of certain catalysts, particularly Ni-based ones. A further problem is associated with the gases leaving the fuel cell area. Figure 3.25 shows the composition of gases along the flow path in the SOFC of Fig. 3.20 and used in Fig. 3.23, the initial gas composition being described above. It is seen that the $CH_4$ reformation happens during the early part of the flow path. The carbon is seen to be mainly continuing in the form of CO. The hydrogen formed plus the amount already present in the input fuel serves as fuel for the SOFC during the middle flow path, and at the end, some 30% of the hydrogen leaves the fuel cell unprocessed, along with the quantities of water initially present or formed by the fuel cell reactions. In Fig. 3.24, it is proposed to add a catalytic afterburner to burn the remaining hydrogen and possibly remove more pollutants. The scheme does not sequester the $CO_2$ in the output stream and considers a fuel with 88% methane, with the rest made up by higher hydrocarbons. As it stands, SOFCs operating on fuels other than hydrogen do not live up to the fuel cell ideal of emitting "just water".

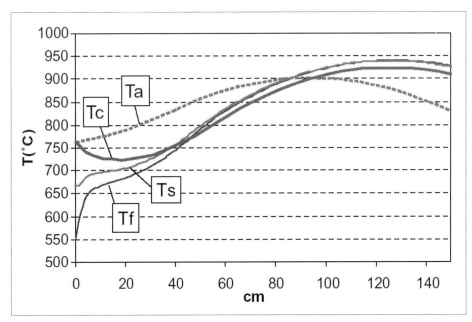

*Figure 3.23.* Temperature profiles from a flow model of SOFC, as a function of distance from the $H_2$/fuel inlet (cf. Fig. 3.20). The four temperature curves are *Ta:* inlet $O_2$/air centre tube; *Tc:* return $O_2$/air tube along positive electrode; *Ts:* electrode-electrolyte assembly; and *Tf:* $H_2$/fuel flow. (Reprinted from S. Campanari and P. Iora (2004). Definition and sensitivity analysis of a finite volume SOFC model for a tubular cell geometry. *J. Power Sources* **132**, 113-126. Used by permission from Elsevier.)

The cell losses discussed in connection with (3.66) include the three components depicted in Fig. 3.26 as a function of the SOFC flow path for the mixed gas fuel cell underlying Figs. 3.21 and 3.23. They are the bulk Ohmic losses and losses associated with overcoming the potential barrier for each of the electrodes. Further bulk loss terms associated with diffusion polarisation are much smaller (Campanari and Iora, 2004; cf. also the dissociation barrier discussion in Chapter 2, section 2.1.1 and 2.1.3). Figure 3.26 shows that the voltage losses mainly take place in the part of the electrolyte layer residing near the fuel inlet side of the cell cylinder, due to the lower temperatures prevailing in this area. At the maximum temperature locations along the electrodes (cf. Fig. 3.23), the loss terms are substantially smaller.

In principle, reverse operation of SOFCs is possible, although little data on efficiency obtained in praxis are available (European Platform for Hydrogen and Fuel Cell Technologies, 2004).

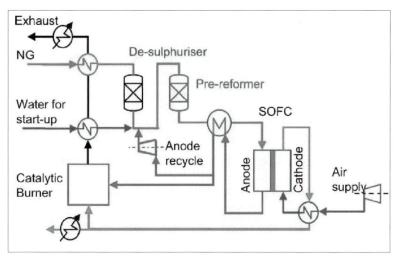

*Figure 3.24.* System layout for SOFC plant based on natural gas fuel (NG), with gas cleaning except for $CO_2$ and with the possibility of combined power and heat output. "Anode" and "cathode" are obsolete names for the negative and positive electrodes. (From Fontell *et al.* (2004). Conceptual study of a 250 kW planar SOFC system for CHP application. *J. Power Sources* **131**, 49-56. Used by permission from Elsevier.)

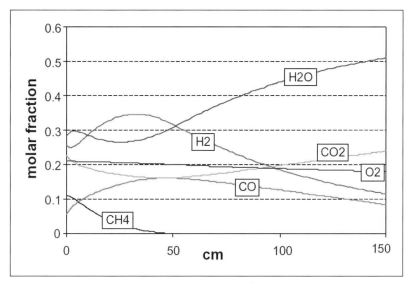

*Figure 3.25.* Molar composition profiles along the flow path in the fuel channel, calculated using a flow model of the tubular SOFC, with the abscissa being the distance from the $H_2$/fuel inlet (cf. Fig. 3.20). The six curves reflect the six components in the fuel used in the cell modelled (see text). (Reprinted from S. Campanari and P. Iora (2004). Definition and sensitivity analysis of a finite volume SOFC model for a tubular cell geometry. *J. Power Sources* **132**, 113-126. Used by permission from Elsevier.)

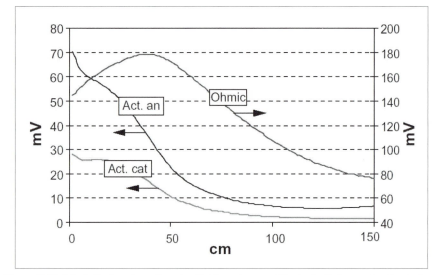

*Figure 3.26.* Modelled main cell-potential loss terms. *Ohmic:* bulk losses in electrolyte; *Act. an:* activation losses at negative electrode; *Act. cat:* activation losses at positive electrode. (Reprinted from S. Campanari and P. Iora (2004). Definition and sensitivity analysis of a finite volume SOFC model for a tubular cell geometry. *J. Power Sources* **132**, 113-126. Used by permission from Elsevier.)

Overall efficiency of SOFCs depends on fuels, materials and operating temperature and on the total device capability of supplying and recycling heat required for maintaining the operating temperature, beyond what can be generated as waste heat of the fuel cell reactions themselves. Figure 3.27 shows the cell potential and current density for the modelling exercise pursued through Figs. 3.23, 3.25 and 3.26, as a function of the path through the length of the cell cylinder. The Nernst part of the cell potential $\phi$ (3.66) declines after the initial stretch, at least as fast as the decline in hydrogen content in the fuel stream (Fig. 3.25). The current density $i$ exhibits a maximum in the early middle of the path, when the fuel cell reactions are most active and the fuel is not yet much depleted. Figure 3.28 gives a number of measured values of $\phi$, now as a function of current density $i$ for a planar single cell SOFC device, at different operating temperatures. The right-hand axis gives the corresponding power density, $e = \phi i$. The voltage drops to zero at higher $i$s, and the power density thus declines to zero after having passed the maximum (approximately where the curves in the figure stop).

Efforts to optimise design are not only concerned with the selection of materials for electrodes and electrolyte, but also with the microstructure of the surfaces between them. Activity areas must be large, causing as mentioned above interest in the surface-increasing deposition techniques developed for metal-organic solar cells. The surface reactions to control comprise bulk and surface diffusion, adsorption, dissociation, charge transfer and chemical reactions, with consideration of the dynamics of motion ("kinetics") for each constituent, whether a molecule, an atom or a charged entity (ion or electron) (Kawada and Mizusaki, 2003). Figure 3.29 shows a cut through the electrode–electrolyte interface of a laboratory manufactured SOFC.

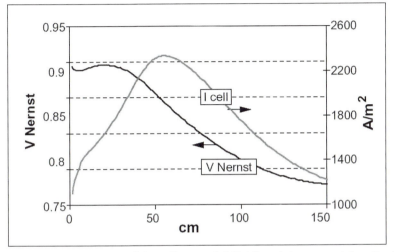

*Figure 3.27.* The Nernst part of the cell potential (3.66) and the current density, from an SOFC model calculation and like the previous figures given as a function of distance from the fuel inlet. (Reprinted from S. Campanari and P. Iora (2004). Definition and sensitivity analysis of a finite volume SOFC model for a tubular cell geometry. *J. Power Sources* **132**, 113-126. Used by permission from Elsevier.)

The bipolar plates (Fig. 3.20) providing electric contact between cells and forming impermeable channels for gas transport are in most SOFC designs made of ferritic stainless steel (low thermal expansion coefficient), although at the highest operating temperatures, more advanced (and expensive) metal structures must be used.

Target design lifetime is of the order of 40 000 h for stationary installations (Tu and Stimming, 2004; a 5000 h target for mobile systems is less relevant for SOFC). The planar designs seem limited to the lower temperature regimes, and high-temperature systems all use the tubular design shown in Fig. 3.20 or disk array designs of a similar complexity. Measured degradation rates for current planar cell prototypes are about 1.7% per 1000 h of operation (Borglum, 2003), which is an order of magnitude too high.

A number of SOFC prototype plants (in the 100-kW size range) are in operation. The current direct conversion efficiency (i.e., neglecting system auxiliaries) is about 55%, but could reach 70–80% in the future, at least for the high temperature range. System power efficiencies are currently around 45% and around 75% for SOFCs producing both power and heat.

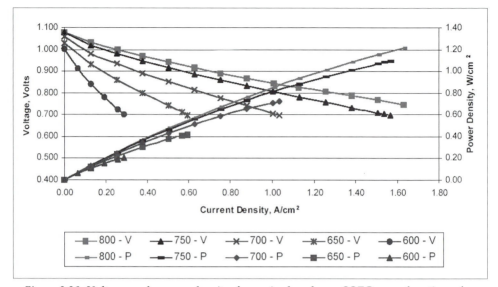

*Figure 3.28.* Voltage and power density for a single, planar SOFC, as a function of current density in the range from zero to approximately the value giving maximum power. The different curves are for different operating temperatures, as indicated in the box. (From D. Ghosh (2003). Development of stationary solid oxide fuel cells at Global Thermoelectric Inc. Presented at "14th World Hydrogen Energy Conf., Montréal 2002". Used by permission from Canadian Hydrogen Society.)

The efficiencies quoted are based on the thermodynamical efficiency (3.22), corrected for voltage losses (see (3.23) and the discussion of (3.66)) and for fuel passing to exhaust without being reacted (Fig. 3.25).

SOFCs share with other fuel cell types a very low tolerance to fuel impurities of $H_2S$ (< 1 ppm), $NH_3$ (< ½%), HCl and other halogens (< 1 ppm), but in contrast to the low-temperature cells accept fuel impurities of CO in addition to the methane and hydrocarbons reformed internally (Dayton *et al.*, 2001).

Table 3.2 shows typical compositions of potential fuels other than pure hydrogen. It is seen that gasified coal and biomass both fail to meet the criteria regarding poisoning by the mentioned poorly tolerated substances. The same is the case for MCFCs and phosphoric acid cells, while proton exchange membrane fuel cells do not tolerate CO in the fuel stream. The implication is that if solid oxide fuel cells are not used with hydrogen, there has to be additional cleaning operations such as the sulphur removal (cf. Fig. 3.24) in the case of natural gas or higher hydrocarbons, and in the case of coal or biomass based fuels removal by some cleaning process of both sulphur and halogens, plus possibly ammonia. Simplification by mixing $CH_4$ fuel and $O_2$ in a single gas channel has been investigated (Hibino *et al.*, 2002).

| (%) | Natural gas (pipeline) | Coal ($O_2$ blown) | Biomass (steam/air blown) |
|---|---|---|---|
| $H_2O$ | | 27–62 | 0–40 |
| $H_2$ | | 38–42 | 15–21 |
| $N_2$ | 0.06 | 0.3–0.8 | 0–40 |
| $CO_2$ | 1.3 | 23–31 | 13–22 |
| CO | | 15–37 | 11–43 |
| $CH_4$ | 88.1 | 0.1–9.0 | 11–16 |
| $C_nH_m$ | 10.4 | ~0.8 | 0.1–5.0 |
| $H_2S$ | 10 ppm | 0.2–1.3 | 0.01–0.1 |
| $NH_x$ | 0.3 | 0.3–0.8 | 0.1–0.4 |
| Tars | | ~0.24 | 0.3–0.4 |
| HCl | | 200 ppm | |

*Table 3.2.* Composition of potential SOFC fuels derived from North Sea natural gas (vol%) and from gasification of coal and biomass (mol%). Except for water, the mol% given is on a dry basis. Steam gasified biomass is at low pressure, and the air blown is at high pressure. Based on Fontell *et al.* (2004) and data collected by Dayton *et al.* (2001).

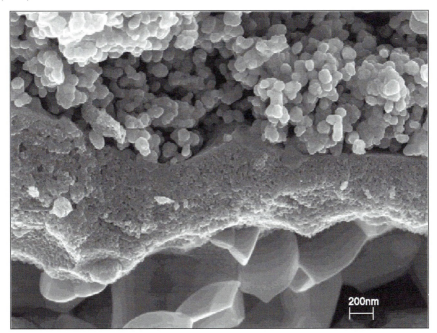

*Figure 3.29.* Scanning electron microscope picture of the electrode-electrolyte structure along a perpendicular cut. Top: screen-printed $La_{0.6}Sr_{0.4}Co_{0.2}Fe_{0.8}O_3$ positive electrode. Middle: spray-deposited electrolyte, $YSZ$ = 8 mol% $Y_2O_3$ stabilised $ZrO_3$. Bottom: negative electrode, NiO and $YSZ$ in ratio 7:3, in cermet $CeO_2$. (From D. Perednis and L. Gauckler (2004). Solid oxide fuel cells with electrolytes prepared via spray pyrolysis. *Solid State Ionics* **166**, 229-239. Reprinted by permission from Elsevier.)

# 3.4 Acid and alkaline cells

Phosphoric acid cells have been developed for stationary uses. They employ porous carbon electrodes with a platinum catalyst and phosphoric acid as electrolyte and feed hydrogen to the negative electrode and oxygen (or air) to the positive electrode, with the basic electrode reactions given by (3.15) and (3.16). The operating temperature is in the range of 175-200°C, and water is continuously removed. Several 200-kW installations have been operating for a number of years for emergency power at hospitals and military facilities.

Fluid acid electrolytes are generally good conductors with reasonable levels of stability (as witnessed by the 40 000 h operation between major overhauls achieved by some operating plants according to King and McDonald, 2003). Corrosion is a problem at the electrodes and is the reason for having to use catalysts based on noble metals such as Pt on electrodes made of porous graphite paper. The amount of graphite is reduced to 20% of the original after 40 000 h of operation (Kordesch and Simader, 1996). The use of $H_3PO_4$ as the electrolyte, rather than other conventional fluid acids, is due to considerations of low evaporation and stability at the temperatures of 150-200°C suitable for operation of the phosphoric acid fuel cell (PAFC). If the fuel is based on reformed natural gas, it would contain $CO_2$ at a level of typically 20%, which is acceptable for the PAFC reactions to proceed at an overall efficiency of around 40%. The cell voltage versus current density relationship is similar to the ones shown in Fig. 3.28 for SOFCs, but maximum voltages are slightly lower and drop as fast as the SOFC curve for 650°C (Kordesh and Simader, 1996). Like the high-temperature fuel cells, PAFCs need hours to start up and are thus not suited for automotive applications (Spakovsky and Olsommer, 2002). Despite sales of several hundred units, the price has remained over US$ 3000/kW, and it is generally held that radical breakthroughs are required for this technology to compete with the other fuel cell types.

In general, because PAFCs are proton conductors like the proton exchange membrane (PEM) and the subcategory direct methanol fuel cells (sections 3.5 and 3.6), there is a continuous conceptual transition between them. The polymers used in PEM cells usually contain weakly acidic components such as $HSO_3$, but may be reinforced with a stronger acid in order to increase conductivity or allow operation at higher temperatures.

Several suggestions have been made for adding acids to a polymer material, while maintaining the solid structure as much as possible. One proposal adds phosphor-tungsten acid ($H_3PW_{12}O_{40}$) to a mixed organic–inorganic polymer network based on an organic hydrocarbon plus inorganic zirconium structure (Kim and Honma, 2004). This material has a conductivity of around

$10^{-3}$ A V$^{-1}$ cm$^{-1}$ (slightly higher when humidity-saturated) at temperatures in the interval 100-160°C (compare, e.g., with Fig. 3.19) and good temperature stability below 200°C. A development along similar lines is the replacement of the perfluorinated membranes common in PEM cells by polybenzimidazole (PBI), e.g. as a polymer chain with two benzimidazole molecules followed by an extra benzene ring (a unit segment seen at the top and bottom of Fig. 3.30). This polymer is stable between 100 and 200°C and may be doped with H$_3$PO$_4$ to get a strongly increased proton conductivity (Li *et al.*, 2004). The mechanism of proton transfer is illustrated in Fig. 3.30. Laboratory fuel cells made with this technology offer high CO tolerance and need no water management, but may be operated using dry gas inputs (Jensen *et al.*, 2004). Loss of acid molecules from the membrane may be a problem (Wang, 2003).

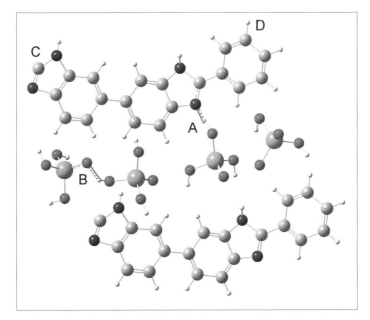

*Figure 3.30.* Proton conduction in acid polymer PBI membrane based upon benzimidazole. The smallest molecules are hydrogen, and the polymer backbone chains at the top and bottom are mainly carbon in three benzene rings, but with two attached pentagonal rings with two nitrogen atoms each. In the middle are four phosphoric acid molecules (the 4:2 ratio of doping acid to PBI gives maximum protonation, the largest atoms being phosphor and the middle ones being oxygen). The proton transport is illustrated by transfer from a PBI nitrogen atom (A) and an acid-to-acid molecule transfer (B). If water is present, further proton transport is facilitated by reactions such as those illustrated in Fig. 3.7. The next polymer chain segments are attached at points C and D (eventually twisted by 180°).

Another possibility is to introduce a solid acid directly as electrolyte in a cell similar to the PAFC (i.e., using a Pt catalyst and carbon materials for electrodes). Materials tried include $CsHSO_4$ (Haile *et al.*, 2001), $CsH_2PO_4$ (Boysen *et al.*, 2004) and $Tl_3H(SO_4)_2$ (Matsuo *et al.*, 2004). These solid electrolytes exhibit rapidly rising conductivities, from some $10^{-8}$ A $V^{-1}$ cm$^{-1}$ at ambient temperature to over $10^{-2}$ A $V^{-1}$ cm$^{-1}$ at temperatures in the range of 150-200°C. This leads to cell power levels of about 40 mW cm$^{-2}$ or at least 5 times lower than those of conventional direct methanol fuel cells (see section 3.6), but 25 times lower than some of the SOFC power densities in Fig. 3.28, with methanol cells being one of the competing technologies over which the solid acid cells have the advantage of not requiring careful water management. Stability problems have been reduced by initial water pressure cycling.

Alkaline fuel cells (AFCs) use an aqueous potassium hydroxide (KOH) solution (around 30%) as electrolyte and have electrode reactions of the form

$$H_2 + 2OH^- \rightarrow 2H_2O + 2e^-, \tag{3.69}$$

$$\tfrac{1}{2} O_2 + H_2O + 2e^- \rightarrow 2OH^-. \tag{3.70}$$

These cells operate in the temperature range of 70-100°C, but the specific choice of catalysts (e.g. Pt or Ni) requires maintenance of a fairly narrow temperature regime. Also, the hydrogen fuel must have a high purity and notably not contain $CO_2$, due to the alkaline pH prevailing. Carbon dioxide reacts with $OH^-$ to form carbonate ($K_2CO_3$), which reduces ion generation and transport within the electrolyte. There is some dispute over the importance of this effect, and a recent paper has suggested that alkaline fuel cells degrade equally rapidly with and without $CO_2$ impurities in the fuel (Gülzow and Schulze, 2004). The likely explanation is that $CO_2$ poisoning is very important with the small amounts of KOH present in fixed matrix configurations, while the currently preferred design (Fig. 3.31) with KOH being circulated through the cell (in order to be able extract the extra water generated by the reactions (3.69) and (3.70) and evaporate it) diminishes the problem (Gouérec *et al.*, 2004). However, using large quantities of circulating electrolyte also has implications for the physical size of the system, which is important, e.g. for automotive uses. The alternative of cleaning the electrolyte for carbonates has not been put in practical use and would add to bulkiness and cost.

Alkaline fuel cells have been used extensively on early spacecraft until they were superseded by more reliable solar cells. The high cost of the space cells and the use of corrosive compounds requiring special care in handling have been held against AFCs. Current AFC development employs multi-component electrodes using Ni for structural stability and as catalyst, carbon black as electron conductor and polytetrafluoroethylene (PTFE) pore-forming powder for gas diffusion and water repulsion (hydrophobic quality, Fig. 3.31).

General objections to liquid electrolytes include corrosion and difficulty in minimising physical volume requirements for applications such as vehicle systems. The lifetime of AFCs are of the order of 5000 h with a large spread in experience (McLean *et al.*, 2002). The electrolyte penetrates into the electrode pores and degrades functionality by increasing diffusion paths (Cifrain and Kordesch, 2004). Extending the lifetime requires that KOH be drained out of the system when not in use.

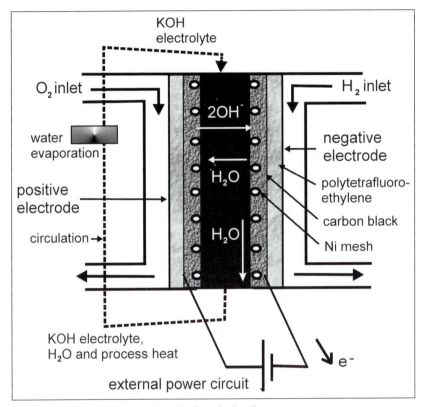

*Figure 3.31.* Schematic layout of an alkaline fuel cell.

The individual components used in an AFC are not necessarily expensive compared to those of other fuel cell types under development. Use of Pt catalysts can be avoided, while the bipolar plates collecting the electron flows typically have to be made of fairly expensive black carbon to avoid corrosion. The peripherals needed for water management and electrolyte draining add to the cost, but do not necessarily lead to drawbacks such as long start-up times if proper controls are part of the system design. Process heat is used to evaporate water in the electrolyte circulation cycle and would have to be recovered by condensation for energetic efficiency.

The energy efficiency of the AFC itself is similar to or slightly better than that of other low-temperature cells, lying in the range of 45-60%, with an open circuit cell voltage around 0.9 V and a current density of 0.2-1.0 A cm$^{-2}$, highest for space cells (Jo and Yi, 1999; Spakovsky and Olsommer, 2002; McLean *et al.*, 2002).

Mass production cost of AFCs is quoted as 400-500 US$/kW (Gülzov, 1996) and in another study as 155-643 US$/kW (McLean *et al.*, 2002), which compares favourably with the estimated cost of proton exchange membrane (PEM) cells (60-1220 US$/kW), except for a lower expectation of options for cost reduction by mass production. The difference in lower bounds may be decreased if peripheral costs are added, for reasons of, e.g., water management costs for the PEM cells. However, as mentioned above, similar or even more elaborate water and electrolyte management costs might have to be added to AFC modules for durability and efficiency in practical applications. The lifetimes of prototype AFCs are fairly low for stationary uses, but are considered acceptable to mobile applications, where the bulkiness of the system is a more likely limitation. Cost comparisons are difficult to make because all systems have too high costs at present, and the decisive parameters have to do with the possible advantages obtained by mass production and reductions in material use, which by many observers is judged to be less probable for AFC technology than for PEM fuel cell technology.

Variations on the alkaline reaction scheme (3.70) at the positive electrode are the possible creation of hydroperoxide,

$$O_2 + H_2O + 2e^- \rightarrow OH^- + HO_2^-. \tag{3.71}$$

This is an example of the use of alkaline fuel cells for production of industrially interesting compounds rather than for electricity production (Alcaide *et al.*, 2004).

# 3.5 Proton exchange membrane cells

The type of fuel cell currently undergoing the most rapid development is the proton exchange membrane (PEM) fuel cell[a]. It has been developed over a relatively short period of time, and it is hoped for economic application in

---

[a] In recent years, some public research programmes have used the term "polymer electrolyte membrane", which happens also to be abbreviated as "PEM".

the transportation sector. It contains a solid polymer membrane sandwiched between two gas diffusion layers and electrodes. The membrane material is typically polyperfluorosulphonic acid. A platinum or Pt–Ru alloy catalyst is used to break hydrogen molecules into atoms at the negative electrode, and the hydrogen atoms are then capable of penetrating the membrane and reaching the positive electrode, where they combine with oxygen to form water, again with the help of a platinum catalyst. The all-solid state design makes the cells compact and suitable for stacking (Fig. 3.32). Fig. 3.33 gives a schematic view of the assembly.

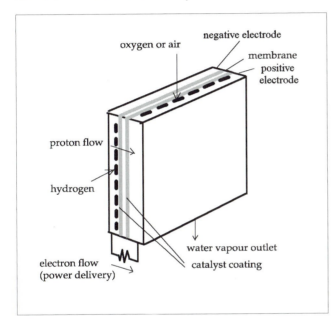

*Figure 3.32.* Layout of a PEM fuel cell layer, several of which may be stacked. Electrode areas include gas diffusion and electrode components in a grid-type structure (cf. Fig. 3.33) (from B. Sørensen, *Renewable Energy,* 2004, used by permission from Elsevier).

The electrode reactions are (3.15) and (3.16), and operating temperatures are in the range of 50-100°C. Protons ($H^+$) travel through the membrane material. Figure 3.33 shows a typical layout for an individual cell. Due to a low operating temperature and flexibility in design, proton exchange membrane (PEM) fuel cells can be used for a range of applications from portable over vehicular to general power supply. PEM cell stacks are dominating the current wealth of demonstration projects in road transportation and dispersed building-integrated areas of use. The efficiency of conversion for these systems is between 40 and 50%, and lifetimes are under 5000 h with intense efforts to improve stability. As indicated in Fig. 3.34, an advantage of particular importance for automotive applications is the high efficiency at part loads, which alone gives a factor of two improvement over current internal combustion engines.

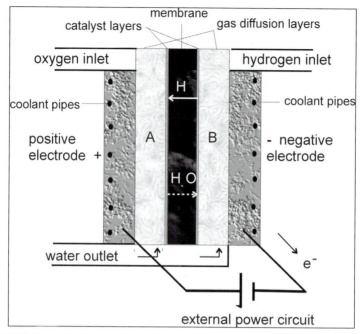

*Figure 3.33.* Schematic picture of a proton exchange membrane fuel cell. Modelling of reactions at the gas diffusion layer/catalyst/membrane interfaces A and B is discussed in section 3.5.2. Details of design are discussed in the following subsections.

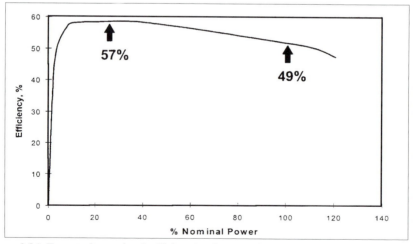

*Figure 3.34.* Expected part-load efficiencies for a 50-kW PEM fuel cell system, projected from measurements involving 10-20 cell test stacks (Patil, 1998).

In the following subsections, each component of the PEM fuel cell will be described, and modelling and empirical experience of the overall system will be presented, as well as stability and durability evaluations.

### 3.5.1 Current collectors and gas delivery system

The mechanical stability of a PEM fuel cell is usually provided by a set of bipolar plates, which also act as collecting terminals for the electric current generated and which are further forming the walls of the access flow channels for hydrogen and oxygen (air), as well as the outlet channels for water and excess gases. Additionally, in connection with the plate configuration, coolant flow (e.g. water) has to be considered, as indicated in Figs. 3.33 and 3.35. The plates may be made of suitable metals or graphite. In case metal is selected, a corrugated plate design (Fig. 3.35A) is a natural choice, while for graphite, a machined structure as shown in Fig. 3.35B may be considered (Wilkinson and Vanderleeden, 2003). Metals are subject to corrosion in the fuel cell chemical environment, and many metals must be coated to protect them from decay. Titanium has been considered, but cost speaks against it. While most metals have both adequate conductivity and high mechanical strength, the graphite has only fair conductivity and poor mechanical strength. It is therefore used in composites, e.g., with a polymer resin as support, which also allows carbon powder to be used instead of solid graphite.

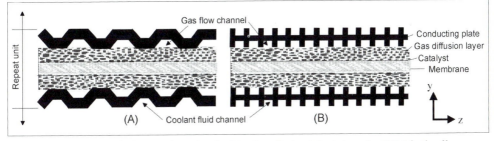

*Figure 3.35.* Two design options for the bipolar electrode[a] plates of a PEM fuel cell.

The gas diffusion layer (GDL) must be capable of transporting gases (hydrogen or oxygen) from the gas input channels to the active area at the catalyst-membrane interface. At the same time, it must be able to transport electrons to or from the active area and deliver them to or take them from the

---

[a] I use the term "electrode" for the electron conductors. A different terminology is also in use, where the name "electrode" is reserved for the catalyst and gas diffusion layer assembly.

bipolar plates connected to an external circuit. In other words, the pore structure should be such that there are contiguous gas channels through the material, as well as contiguous electron conducting wall elements.

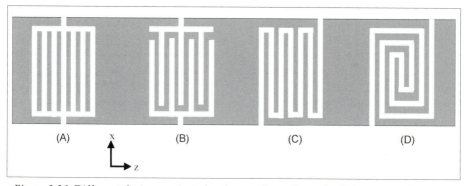

*Figure 3.36.* Different design options for the gas flow channels: (A) straight, (B) interdigitated, (C) serpentine and (D) spiral. For the interdigitated design, the incoming flow must proceed through the gas diffusion layer to reach the outlet channel.

The design of gas inlet and outlet channels offers some potential to control cell behaviour. Figure 3.36 illustrates some of the possibilities. The serpentine (C) and spiral (D) designs used with a higher pressure than the conventional straight channel design (A) are capable of reducing the amount of liquid water accumulating in the gas channels, and the interdigitated design (B) goes one step further by forcing the flow to pass through the gas diffusion layer to the adjacent channel, thereby helping to clean the gas diffusion layer for accumulating water from the reaction (3.16) or from the water initially soaking the membrane (as a condition for its proton conducting property).

The co-ordinates indicated in Figs. 3.35 and 3.36 and used in Figs. 3.37-3.40 and Figs. 3.47-3.48 are defined such that the $y$-axis is in the perpendicular direction (to the paper plane in Fig. 3.36) through the sequence hydrogen inlet channel → gas diffusion layer → negative electrode catalyst → membrane → positive electrode catalyst → gas diffusion layer → oxygen (air) channel. The $x$-axis is along the main gas channel flow directions, and the $z$-axis is across the design as shown in both figures.

Figure 3.37 shows the flow field in two adjacent oxygen channels and a bridging piece of the gas diffusion layer, from simulation of an interdigitated design with use of a three-dimensional fluid dynamics model (as discussed in section 3.1.5) including modelling of the electrochemical reactions. The left part (A) of Fig. 3.37 shows that the gas flow (as expected) is slowed to zero at the dead end of the channel. The middle part (B) of Fig. 3.37 shows the behaviour in the $y$-$z$ plane, highlighting a strong and asymmetric flow around the bridge caused by forced convention (in contrast to the straight channel design in Fig. 3.36A, for which penetration into the gas diffusion layer is by diffusion and where the flow field therefore is symmetric around the mid-point current collector). The right-hand side (C) of Fig. 3.37 shows the flow in the uptake channel, which is building up along the $x$-axis and is largest at the end of the channel.

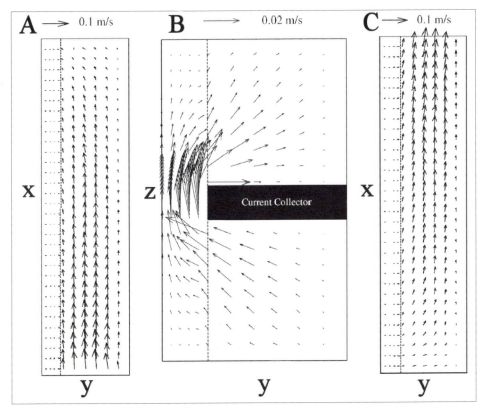

*Figure 3.37.* Computed velocity fields (m/s) for flows in adjacent interdigitated oxygen channels (with gas diffusion layer on the left side) of a PEM fuel cell (A: inlet, C: outlet, in $x$-$y$ plane). In the middle (B), the flow from one gas channel through the gas diffusion layer to the adjacent gas channel is shown in the $z$-$y$ plane, for the midpoint of the cell extension in the $z$-direction. The flows in A and C are for the midpoint value of $z$. (From S. Um and C. Wang (2004). Three-dimensional analysis of transport and electrochemical reactions in polymer electrolyte fuel cells. *J. Power Sources* **125**, 40-51. Used with permission from Elsevier.)

In addition to improving water management, the interdigitated design causes higher currents at lower voltages, because the limiting factor is mass transport in the positive electrode gas diffusion layer (Um and Wang, 2004).

Figure 3.38 shows the variations in oxygen concentration from a similar fluid dynamics calculation, also for two adjacent oxygen channels of an interdigitated design and the bridging gas diffusion layer, where a substantial reduction in oxygen concentration is seen to take place. At the exit of the outgoing channel, 34% of the incoming oxygen has been removed to the electrochemical reaction at the catalyst. Adding more channel pairs in the flow path design of Fig. 3.36 allows larger fractions of the oxygen to become processed.

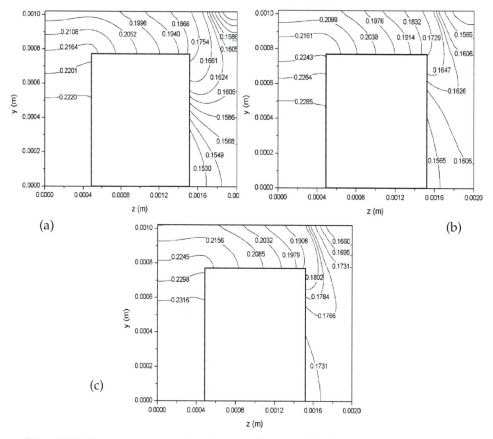

*Figure 3.38.* Oxygen concentrations in two adjacent outlet flow channels of an inter-digitated PEM fuel cell structure, with gas diffusion layer on top. The three planes are at $x$-values corresponding to the flow channel exit (a: large $x$), the middle (b) and the entrance (c: $x = 0$). The cell current is at its maximum (about 0.8 A cm$^{-2}$). (From M. Hu *et al.* (2004). Three dimensional, two phase flow mathematical model for PEM fuel cell: Part II. Analysis and discussion of the internal transport mechanism. *Energy Conversion & Management.* **45**, 1883-1916. Used with permission from Elsevier.)

## 3.5.2 Gas diffusion layers

In order for the gas diffusion layers (GDLs) to handle the hydrogen and oxygen flows, respectively, water from the recombination (3.16) must be led to the exit flow channels. This water is created at the oxygen side membrane to the gas diffusion layer interface, but it may find it easier to move through the membrane to the hydrogen side than to penetrate into the oxygen gas diffusion layer. This means that the membrane will get soaked even if it was dry at first (although in that situation, the $H^+$ penetration would be reduced) and hence that the functionality of the membrane sheet will build up. Whether the reaction water will leave through the oxygen or hydrogen side channels is therefore difficult to predict intuitively. Figure 3.39 present the results of the model calculation also underlying Fig. 3.38, showing the calculated water concentrations (for three $x$-values) in the two adjacent hydrogen channels and the corresponding gas diffusion layer (top of each pair) and for the two oxygen channels on the other side of the cell, with their diffusion layer. In the hydrogen channels, water concentrations are high, indicating that much water has crossed the membrane after its creation at the oxygen catalyst by (3.16), back to the gas diffusion layer on the hydrogen side and reaching the hydrogen outlet channel.

For the straight channel geometry (Fig. 3.36A), water concentration declines uniformly when progressing along the channel length x, both in the hydrogen and in the oxygen channels, with liquid water forming in the gas diffusion layer on the oxygen side and eventually (for larger x) showing up in one of the gas channels (Wang *et al.*, 2001; Yu and Liu, 2002; Hu *et al.*, 2004). For the interdigitated design, Hu *et al.* (2004) find condensed water only on the hydrogen side when the current density is low ($\approx 0.17$ A cm$^{-2}$), but on the oxygen side when the current density is high ($\approx 0.8$ A cm$^{-2}$), as shown in Fig. 3.40. In the first case, water is transported across the membrane to the hydrogen side, while, in the second case, the high reaction rate creates a need to replenish water in the membrane, thus removing water from the hydrogen side, while at the oxygen side, the high level of water generation by (3.16) exceeds what can be absorbed by the membrane, so that a surplus appears in the oxygen channel.

The water profiles show a dependence on pressure (Futerko and Hsing, 2000), and most PEM cells have been found capable of recovering from water flooding (Nguyen and Knobbe, 2003). Recently, direct views of the water distribution have been obtained by neutron imaging techniques, taking advantage of the different neutron cross-sections of different atoms (Satija *et al.*, 2004). Like the differences between straight and interdigitated cell designs seen here (cf. Fig. 3.47), the water handling patterns are also different for serpentine designs (Nguyen *et al.*, 2004).

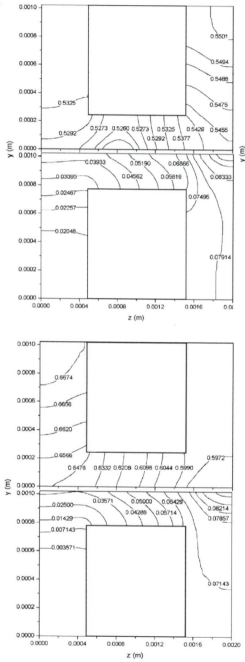

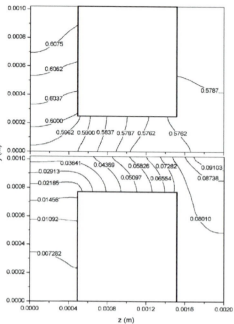

*Figure 3.39.* Water concentration in an interdigitated PEM fuel cell structure, for three planes at *x*-values corresponding to the flow channel exit (a: top left), the middle (b: top right) and the entrance (c: bottom left). In each pair of pictures, the *y-z* plots depict the hydrogen side at the top (GDL is lower bar) and the oxygen side at the bottom (GDL is upper bar). The cell current is at its maximum (about 0.8 A cm$^{-2}$). (From M. Hu *et al.*, (2004). Three dimensional, two phase flow mathematical model for PEM fuel cell: Part II. Analysis and discussion of the internal transport mechanism. *Energy Conversion & Management.* **45**, 1883-1916. Used with permission from Elsevier.)

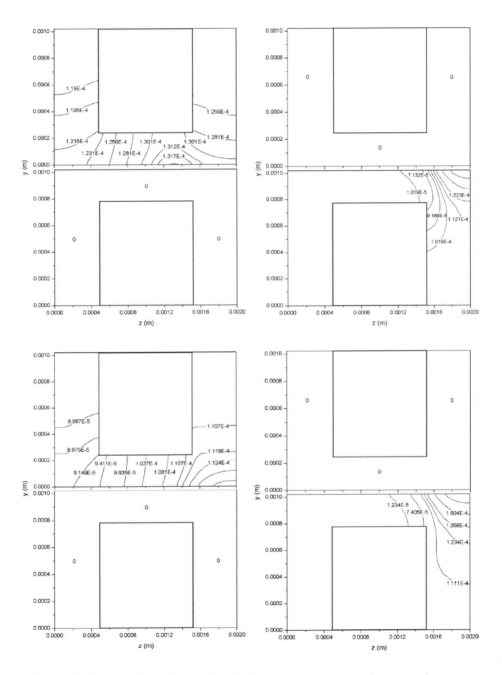

*Figure 3.40.* Saturated liquid water distributions (caption continued next page).

*Figure 3.40 (previous page).* Saturated liquid water distributions for cell with inter-digitated channel design, at low current density (≈0.17 A cm⁻², left column: a, c) and at high current density (≈0.8 A cm⁻², right column: b, d). Two planes are depicted, at *x*-values corresponding to the flow channel exits (a: top left, b: top right) and to the middle (c: bottom left, d: buttom right). Each of the four pairs of pictures depicts *y-z* plots for the hydrogen side at the top (GDL is lower bar) and the oxygen side at the bottom (GDL is upper bar). (From M. Hu *et al.,* (2004). Three dimensional, two phase flow mathematical model for PEM fuel cell: Part II. Analysis and discussion of the internal transport mechanism. *Energy Conversion & Management.* **45**, 1883-1916. Used with permission from Elsevier.)

Materials commonly used for the gas diffusion layers are carbon paper or woven carbon mats (examples of which are shown in Fig. 3.41). They combine the connectivity allowing electron transport with a pore structure suitable for hydrogen or oxygen gas access to the catalyst layer. In cell manufacture, the catalysts may be deposited either on the gas diffusion layer or on the membrane.

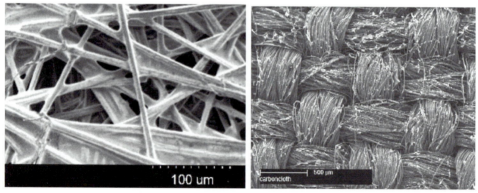

*Figure 3.41.* Structure of carbon paper (left) and carbon cloth (right) used for gas diffusion layers in PEM fuel cells. A coating of 20% (by weight) fluorinated ethylene propylene has been applied. (From C. Lim and C-Y. Wang (2004). Effects of hydrophobic polymer content in GDL on power performance of a PEM fuel cell. *Electrochimica Acta* **49**, 4149-4156; G. Lu and C-Y. Wang (2004). Electrochemical and flow characterization of a direct methanol fuel cell. *J. Power Sources* **134**, 33-40. Used with permission from Elsevier.)

### 3.5.3   Membrane layer

The membrane layer consists of a polymer structure capable of transporting hydrogen ions with high conductivity, inspiring the name "proton exchange membrane". The membrane is thus a solid electrolyte. For the current emphasis on PEM fuel cells operating below 100°C, the Nafion® (trademark of

Dupont de Nemours) or analogous perfluorosulphonic acid membranes have been the dominating choice. The structure of the repeat structure of the polymer fluorocarbon backbone and a side chain with sulphonic acid ends upon which Nafion is based is shown in Fig. 3.42 (the commercial product is sold with various thicknesses and dimensions denoted by a number code such as "Nafion-117", related to non-SI units). The membranes should have high protonic conductivity, low gas permeability and, of course, a suitable mechanical strength and low temperature sensitivity.

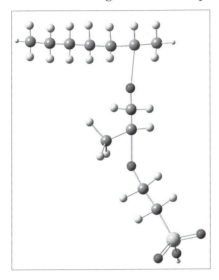

*Figure 3.42.* Segment of a perfluorinated sulphonate ion exchange membrane of Nafion® type. The top polymer chain is made from C atoms as backbone and pairs of F atoms on the sides (end hydrogens are replaced when the chain is enlarged). The acid $HSO_3$ molecule promoting H transport is attached at the end of an arm of 4 $CF_2$ molecules interspersed with two O atoms, and with one F replaced by a $CF_3$ group.

*Figure 3.43.* Structure of Nafion-115 at ambient humidity derived from small-angle X-ray spectroscopy. The lighter areas are cluster structures in the material. (Reprinted with permission from J. Elliott, S. Hanna, A. Elliott, G. Cooley (2000). Interpretation of the small-angle X-ray scattering from swollen and oriented perfluorinated ionomer membranes, *Macromolecules* **33**, 4161-4171. Copyright American Chemical Society.)

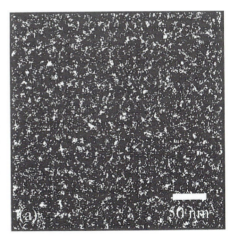

Spectroscopic studies of the structure of Nafion membranes suggest the presence of nodules of diameter 1-10 nm (the lighter parts in Fig. 3.43). The nodules are fairly regularly distributed and have been explained as spherical or oblong extrusions (with radius around 2 nm) of a pair of $(CF_2)_n$ sheets, i.e. a double layer. Some models have the acidic side chains on the outside of the membrane bi-layer (Gierke and Hsu, 1982), while other models place the $SO_3^-$ ions on the inside of the spheres (Vankelecom, 2002). It has not so far been possible to carry out full quantum chemical calculations confirming the tendency of such structures to "fold up" into a nodular structure. However, molecular dynamics calculations (cf. Chapter 2, Fig. 2.3) have been made (Jang et al., 2004) for systems containing four Nafion chains of 70 $CF_2$s and 10 side chains, together with 560 water molecules and 40 hydronium molecules ($H_3O^+$). Figure 3.44 shows results from this study, indicating a structure with clusters of curled-up Nafion backbone structures with the side chain sulphur atoms close to both water and hydronium molecules. The figure assumes a Nafion structure similar to Fig. 3.42 with side-chains every 7 units. If the side chains are lumped at fewer places, the backbone nodules become larger than in Fig. 3.44. The average distance between pairs of S atoms in the case shown in Fig. 3.44 is 0.68 nm, with a rather narrow distribution (spread about ± 0.2 nm).

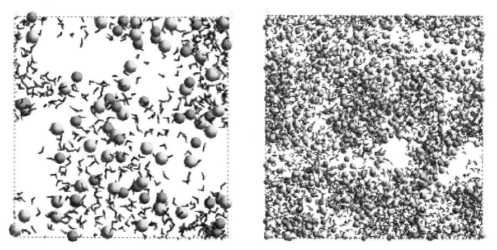

*Figure 3.44.* Molecular dynamics model calculation of Nafion-117 (for small system, left, and large system, right). The Nafion backbone structure is omitted and therefore shows up as white areas. Large, light grey atoms are S, and large dark ones are the O's of hydronium molecules. (Reprinted with permission from S. Jang, V. Molinero, T. Cagin, W. Goddard III (2004). Nanophase-segregation and transport in Nafion 117 from molecular dynamics simulations: Effect of monomeric sequence. *J. Phys. Chem.* **B108**, 3149-3157. Copyright American Chemical Society.)

The molecular dynamics simulation behind Fig. 3.44 does not consider double polymer layers of the kind postulated earlier for the nodule structure, and it does not include the hopping process of proton transfer between the water molecules (as in Fig. 3.7), because the water and hydronium molecules are treated as fixed entities. The stability and regularity of the polymer membranes is also poorly explained by the "loose" structure of Fig. 3.44. An alternative modelling approach is to depart from a double Nafion membrane structure such as the one shown in Fig. 3.45. The distance between the two membrane layers is initially taken as 1.8 nm, corresponding to the thin channels identified experimentally (Barbi *et al.*, 2003). The two sheets would have to be pulled away from each other to roughly the double distance in order to form the observed nodules of diameter around 4 nm. The side chains have in Fig. 3.45 been assumed to face the inter-membrane volume, but could alternatively face outward. The sulphuric acid terminals are already in the double-membrane structure facing "empty" areas of the structure, where water molecules might be placed to help the $H^+$ transport, but pulling the side chains apart would furnish a more continuous route for transfer. Proton transport has been studied in isolation, using a model including one acid side chain and a number of water molecules (Paddison *et al.*, 2001).

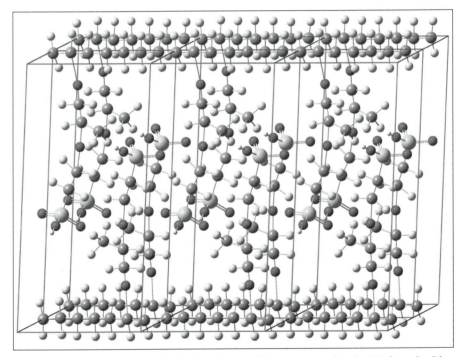

*Figure 3.45.* A $2 \times 3$ unit cell chunk of a possible structure for the Nafion double-membrane channels in the absence of nodule enlargement. The structure has been loosely optimised by 50 iterations of a Hartree-Fock calculation (Sørensen, 2004b).

The reason for attempting to explore the structure by theoretical means is that the small-angle X-ray and nuclear magnetic resonance scattering experiments are not very selective techniques for structure determination, and interpretations of results such as those shown in Fig. 3.43 involve a choice between a multitude of possibilities. Alternative guesses were made earlier by M. Ise (quoted in Kreuer, 2001), suggesting like Jang *et al.* (2004) that the membrane bundles observed are disorganised chunks of $(CH_2)_n$ and that all acid and water molecules are placed outside the clusters. I find the acid-inside and acid-outside structures to have identical energies. The inconsistency of some literature findings is pointed out by Barbi *et al.* (2003). They use their own X-ray experiments to confirm the magnitude of the nodule radius found earlier (they find 1.9 nm) and to determine an average distance between domains (3.6 nm), but refrain from further guessing.

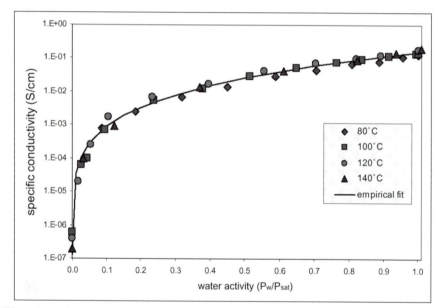

*Figure 3.46.* Conductivity of Nafion-115 as a function of humidity (partial pressure of water relative to saturation pressure, the latter corresponding to 18% water by weight or 11 water molecules per sulphonate molecule). (From C. Yang, S. Srinivasan, A. Bocarsly, S. Tulyani, J. Benziger (2004). A comparison of physical properties and fuel cell performance of Nafion and zirconium phosphate/Nafion composite membranes. *J. Membrane Sci.* **237**, 145-161. Used with permission from Elsevier.)

Figure 3.46 shows that dry Nafion is a poor conductor and that operating the fuel cell membrane close to full water saturation is essential. Proton conductivity increases modestly with temperature in the range of 20-80°C, i.e., below the values shown in Fig. 3.46 (Gil *et al.*, 2004).

The membrane water balance and conductivity was also studied by the model behind Figs. 3.38-3.40. The water model included electro-osmotic drag due to the $H_3O^+$ ion transport (Springer *et al.*, 1991) and diffusive processes. Figure 3.47 shows the results for the two cell designs A and B in Fig. 3.36, at high current densities. For the straight gas channel design (left), the water content is high on the oxygen side, but low on the hydrogen side, due to the fact that the calculation was started with a dry membrane and water creation is at the oxygen side, combined with the water transport to the oxygen side caused by the drag from the positive $H^+$ ion movements. At low currents, the water content would be highest on the hydrogen side. For the interdigitated design (Fig. 3.47, right), the more complex water distribution varies across the membrane plates (z-direction) and the oxygen side content is high on the outflow side of the gas channel, whereas the content on the hydrogen side is higher at the inflow side of the gas channel. The cell processes will start even with low initial water content and will saturate the membrane at high power output, causing membrane resistance to increase in the straight design but less so in the interdigitated design.

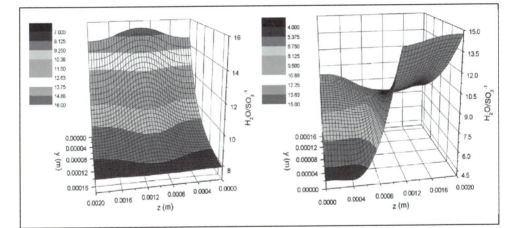

*Figure 3.47.* Distributions of water content across the membrane of a PEM cell with either straight (left) or interdigitated (right) channel design, at maximum current density ($\approx$0.8 A cm$^{-2}$). The membrane thickness is 0.16 mm. The figures correspond to an x-value at the middle of the gas flow channels. The y-z co-ordinates at the floor of each picture have the y-axes in opposite directions in order to better display the different behaviour in the two cases: the left-hand picture has the oxygen side (y = 0) at the back, while the right-hand picture has the oxygen side in front. Also, the z co-ordinate across the cell is opposite in the two pictures, corresponding to them representing views from either the front or the back side of the cell. (From M. Hu *et al.* (2004). Three dimensional, two phase flow mathematical model for PEM fuel cell: Part II. Analysis and discussion of the internal transport mechanism. *Energy Conversion & Management.* **45**, 1883-1916. Used with permission from Elsevier.)

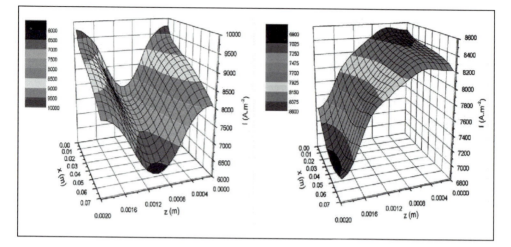

*Figure 3.48.* Distributions of local current density along the membrane of a PEM cell with either straight (left) or interdigitated (right) channel design, at maximum average current density ($\approx$0.8 A cm$^{-2}$). The membrane thickness is 0.16 mm. The figure co-ordinate system is $x$-$z$ (cf. Fig. 3.36) rather than $y$-$z$ as used in several previous figures. (From M. Hu *et al.* (2004). Three dimensional, two phase flow mathematical model for PEM fuel cell: Part II. Analysis and discussion of the internal transport mechanism. *Energy Conversion & Management.* **45**, 1883-1916. Used with permission from Elsevier.)

The current density as a function of position along the gas flow direction ($x$) and the $z$ co-ordinate spanning the membrane surface is shown in Fig. 3.48, for the cell designs A and B in Fig. 3.36 and for a high average current density. In both cases the current density decreases somewhat along the $x$-axis, as fuel becomes depleted, and as a function of $z$, the distribution is symmetric for the straight design, but highly favouring the inlet side of the gas channel ($z = 0$) for the interdigitated design, and as a whole showing less variation with the $x$ co-ordinate.

Alternatives to Nafion membrane materials are investigated, partly in the hope to reduce cost and partly in consideration of the advantages of increasing operating temperature by modest amounts (50-100 K) in order to obtain more efficient operation. Key objectives are to produce a thin and flat membrane without compromising on gas crossover. One such membrane is Gore-select (trademark of W. L. Gore & Assoc. Inc.), which is only 20-30 $\mu$m thick (as compared to 160 $\mu$m for the Nafion membrane studied in Fig. 3.47) and yet has high tensile strength and a potentially very small H$_2$ crossover (Nakao and Yoshitake, 2003). It is based on a porous, helical-structure polytetrafluoroethane (PTFE) backbone and a perfluorinated ionomer side chain.

Polyether ketones (PEEKs) are studied, but generally do not give performance advantages over Nafion, although cost may be lower. Polyphenylene sulphonic acid membranes may improve performance at low humidity, and fluorinated polyarylene ethers may be produced at lower cost, but so far have low conductivity. Also, membranes based on bacterial cellulose with added precious metals have been investigated. Generally, hydrocarbon-based materials have lesser mechanical strength than the C-F based materials (Gil et al., 2004; Lee et al., 2004; Evans et al., 2003).

For use at higher temperatures, the polybenzimidazole (PBI)-type membranes (cf. Fig. 3.30 in section 3.4) are considered. The have to be "doped" with acid at the order of around 10 mol per kg of solvent, and although the proton conductivity increases with temperature up to about 450 K , it is still slightly lower than that of Nafion membranes. Tensile strength diminishes with doping level. Among the advantages of PBI are high CO tolerance and simpler water management, because all water is in gas form at operating temperatures of 150-200°C (Schuster et al., 2004; Li et al., 2004).

Membranes are subject to degradation from peroxide ($H_2O_2$) that may be formed within the $H_2SO_4$-$H_2O$-$H_3O^+$ environment, from temperatures above the normal operating range and from a number of trace metal ions such as $Fe_2^+$, $Fe_3^+$, $Cr_3^+$ or $Ni_2^+$, possibly coming from metallic end-plates used in the cell construction. With time the membrane loses sulphonic acid, particularly at the negative electrode side, and small amounts of fluoride ions and $CO_2$ may be found in the product water. Several of the mechanisms are poorly understood. Some studies suggest that degradation is largest during low power production and small at full and steady power output. A mixture of enhanced ageing experiments and model estimates has been used to establish a basis for further work (Fowler et al., 2002; LaConti et al., 2003; Okada, 2003; Kulinovsky et al., 2004).

### 3.5.4   Catalyst action

The central part of a PEM fuel cell is the catalysts that help the basic reactions (3.15) and (3.16) to proceed at sufficiently high rates. Due to the design of the PEM cells (see Figs. 3.33 and 3.35), the catalysts are the real electrochemical electrodes[a], receiving gases from the gas diffusion pores and delivering split molecules for ionic transport through the membrane. The conducting parts of the gas diffusion layer just carry the electrons made available to the outer circuit and its loads.

---

[a] For this reason the name "electrode" is sometimes used to denote only the catalyst, while in other sources, it denotes the gas diffusion layer plus catalysts, and in still other treatments it denotes the bipolar conductors to which the gas diffusion layer delivers its electrons.

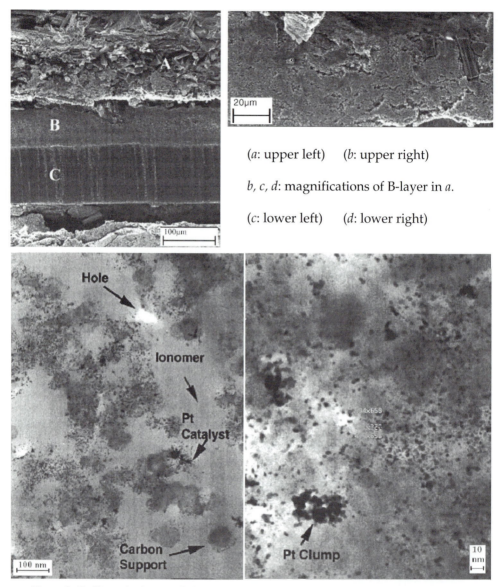

*Figure 3.49.* Slice of a PEM cell showing gas diffusion layer (A), catalyst layer (B) and membrane layer (C), at a magnification factor of 200 (*a*). Tunnelling electron microscope pictures of catalyst layer at a magnification factor of 500 (*b*), 18 400 (*c*) and in (*d*) 485 500. (From N. Siegel, M. Ellis, D. Nelson, M.v.Spakovsky (2003). Single domain PEMFC model based on agglomerate catalyst geometry. *J. Power Sources* **115**, 81-89. Used with permission from Elsevier.)

Figure 3.49 shows scanning and tunnelling electron microscope (SEM, TEM) pictures of the catalyst layer sitting between the gas diffusion layer and the membrane and at increasing magnification. One notes a pore structure (1-10 μm, see Fig. 3.49b) and the presence of irregular lumps (≈3 nm, see Fig. 3.49c,d) of the Pt catalyst, when it is in a mixed environment including carbon structures offering support for the Pt as well as electron conduction, and ionomer intrusions (presumably Nafion). The white "hole" in Fig. 3.49b is from a carbon support that has been withdrawn during manufacturing. No macroscopic pores are seen on the magnified portions of Fig. 3.49c and d, indicating that all pores are at a scale above 1 μm. It is therefore inferred that hydrogen and oxygen gases initially flow through the macroscopic pores, but then have to become dissolved and to diffuse to the reaction sites at the Pt catalyst particles (Siegel *et al.*, 2003). Since the ionomer channels indicated in Fig. 3.49c seem to fully enclose the Pt lumps, while being capable of transporting $H^+$ but not $H_2$ or $O_2$, this transport mechanism is difficult to understand. The membrane-electrode assembly (MEA) used in the experiment is manufactured by deposition of Pt on carbon paper followed by hot pressing onto a Nafion membrane. Possibly the observed polymer blocking of small pores in the catalyst layer is due to the sample preparation technique, or to application of an epoxy resin to fixate the sample before slicing. More remarks on the nearness of access and exit channels from the catalyst follow below.

If the catalyst particles during manufacturing are deposited on the membrane, they should remain chiefly on its surface (although some mixing is caused by heat application), but if they are alternatively deposited on the carbon material of the gas diffusion layer, an amount of penetration into the gas channels is acceptable, as long as the dissociated gases can reach the membrane. All three layers must be in close sequential contact to enable transport (in opposite directions) of both electrons and protons. Early designs of the catalyst layer used polytetrafluoroethene to bind the platinum and subsequently applied a Nafion impregnation by spray techniques (some 2 mg cm$^{-2}$). This method required rather high percentages of Pt (some 20% by weight or over 400 μg cm$^{-2}$). Performance increased up to a certain Nafion loading and then decreased, presumably due to pore blocking as discussed above (Lee *et al.*, 1998; Qi and Kaufman, 2003; review by Lister and McLean, 2004). Current manufacturing uses Nafion directly to bind platinum in a thin-film process. This ensures easy proton transport from the reaction site, but as mentioned requires care in retaining gas access.

The choice of catalyst is influenced by considerations of tolerance of contaminants in the gas flows. A pure Pt catalyst works well for extremely pure hydrogen feed, but even small amounts of CO in the $H_2$ fuel will deteriorate performance. This problem can be reduced by using a Pt-Ru alloy as catalyst (Liu and Nørskov, 2001).

Molecular level modelling of the catalyst reactions are made in the manner described in section 3.1.4. Such quantum chemical modelling can also be performed for CO adsorbing to the Pt or Pt-Ru catalyst surface at the negative electrode, where a number of reactions may take place in addition to (3.15), describing the competition between splitting of hydrogen molecules and the conversion of CO and any water present to $CO_2$ and protons,

$$H_2 + \text{catalyst surface} \rightarrow 2H \text{ at catalyst surface (cf. Fig. 2.5),}$$
$$2H \text{ at catalyst surface} \rightarrow 2H^+ + 2e^-. \tag{3.72}$$

$$H_2O \text{ at catalyst surface} \rightarrow OH \text{ at catalyst surface} + H^+ + e^-,$$
$$OH \text{ and CO both at catalyst surface} \rightarrow CO_2 + H^+ + e^-, \tag{3.73}$$

with the two latter equations adding to

$$H_2O + CO \text{ both at catalyst surface} \rightarrow CO_2 + 2H^+ + 2\,e^-. \tag{3.74}$$

The reaction (3.74) would remove CO, but quantum calculations find it energetically unfavourable (cf., e.g., the calculation with just two Pt catalyst atoms performed by Narayanasamy and Anderson, 2003). However, Liu and Nørskov (2001) were able to prove that although unfavourable at either Pt or Ru surfaces, the combined presence of Pt and Ru allows OH to attach to Ru faster than to Pt, while the presence of Ru reduces the binding of CO to Pt sites and thus furthers the second reaction in (3.73), even though CO is not willing to adsorb to Ru itself.

Other alloys such as Pt with Cr or Ni have been investigated, but without any definite advantages as catalysts emerging. Karmazyn *et al.* (2003) looked at the behaviour of CO contamination at Pt and Ni catalyst surfaces, taking into account the quantum modelling experience of catalytic reactions often being favoured on steps in the catalyst surface structure. Steps are layer discontinuities deviating from simple surface structures with Miller indices such as (1,1,1) (cf. the discussion of Ni catalytic reactions with $H_2$ in Chapter 2, section 2.1.3, with the crystal structures of Ni and Pt being basically identical), as first described by Hammer and Nørskov (1997).

Macroscopic modelling of the catalyst electrode–electrolyte interface behaviour is made by use of the Butler-Volmer equations (3.25), and this model is then incorporated into the gas flow and ion diffusion models described in sections 3.5.2 and 3.5.3.

The loading of catalyst particles influences cost, and values have been decreased over time, with the lowest ones about 14 µg/cm² achieved by using a sputtering process (O'Hayre *et al.*, 2002). Occasional flooding by water should be tolerated by the catalysts as well as by the other cell layers.

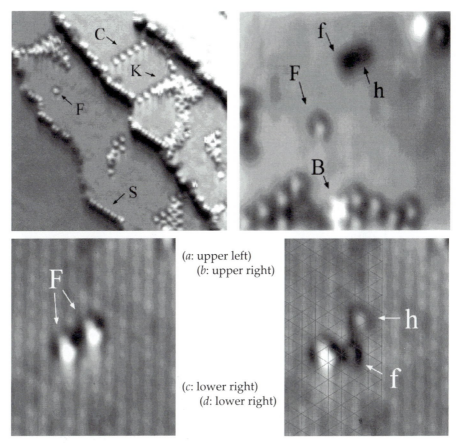

*Figure 3.50.* Oxygen dissociation at catalyst surface. *a:* Pt(111) surface with steps (S) and various clusters. *b:* $O_2$ at face-centred site (F) and at bridge site (B), and an O-O pair under dissociation (f, h). *c:* two $O_2$ molecules at face-centred sites (F). *d:* dissociated O atoms (f, h) with neighbouring $O_2$ molecule. Triangles indicate Pt atoms (distance 0.277 nm). (From B. Stipe, M. Rezaei, W. Ho, S. Gao, M. Persson, B. Lundqvist (1997). Single-molecule dissociation by tunnelling electrons. *Phys. Rev. Lett.* **78**, 4410-4413. Used with permission from the American Physical Society.)

A new angle on the proximity needed between catalyst and the adjacent layer is offered by an alternative thin-film catalyst process requiring neither carbon fibre backup nor channels of Nafion or other ionomer proton conductors (Debe, 2003). Pt-Ru catalysts are deposited by sputtering on oriented crystalline organic whiskers (length about 1 μm), and this assembly is then transferred to a Nafion membrane surface. The voltage-current behaviour is decent and fairly independent on whisker orientation, which is taken to suggest that ionomer intrusions into the catalyst layer are not necessary for proton transport and that carbon conductors within the catalyst layer are not required for efficient electron transport. That proton transport through the void structure if possible is plausible, while the unaided electron transfer over distances up to 1 μm is more difficult to understand. Most likely, the catalyst particles sit so densely on the whiskers that they are in mutual contact, so that electrons can move within the catalyst to the surface facing the electric conductors of the gas diffusion layer.

Quantum chemical description of the catalyst action includes the dissociation processes depicted in Chapter 2, Figs. 2.5 and 2.6, and the water recombination processes discussed in section 3.1.4. Among the first realistic calculations of the $H_2$ dissociation at the negative electrode is the work of Hammer and Nørskov (1995), followed by calculations of increasing sophistication (e.g., Penev et al., 1999; Horch et al., 1999) The behaviour at the positive electrode involves the well-studied dissociation of $O_2$ over a catalyst surface, including the exploration of the importance of steps on the catalyst surface, mentioned above (e.g., Gambardella et al., 2001). Figure 3.50 shows a scanning tunnelling microscope picture with identification of steps as well as a specific $O_2$ splitting process. The more complex reactions involved in the splitting (or formation) of water are under study by density functional methods, as discussed in section 3.1.4. Preliminary investigations using the simpler molecular dynamics methods have also been undertaken (Wang and Balbuena, 2004).

### 3.5.5   Overall performance and reversible operation

The overall performance of PEM cells may be evaluated on the basis of the current-voltage diagram, in the same way as for other electrochemical devices (see, e.g., Fig. 3.28). Figure 3.51 gives an example of the temperature dependence of IV-curves for a PEM cell with very low Pt loading, a possibility mentioned in section 3.5.4. The power density superimposed on the figure shows that densities as high as 0.7 W cm$^{-2}$ can be achieved in this way. Similar curves published over the last decade show that it has been possible to improve the maximum power density performance gradually over the last decade, from under 0.5 W cm$^{-2}$ (Starz et al., 1999) to the present level.

Relative to the ideal thermodynamical efficiency (3.22), the practical efficiency is diminished by the electrochemical losses through the system (3.23), discussed above for each step in the process. Furthermore, as also mentioned in the preceding subsections, not all the hydrogen fuel is utilised and there is a hydrogen content in the outflow from the fuel channel. When cells are combined to form a fuel cell stack, a fuller utilisation may be achieved by passing unused fuel from one cell to the next, but on the other hand, if the amount of fuel reaching the last part of the stack is small, less power is produced than could be accomplished with adequate hydrogen supply. Thus, the overall efficiency may by written in the form

$$\eta = \eta_{ideal}\,\eta_{voltage}\,\eta_{fuel}\,\eta_{stack}, \qquad (3.75)$$

with $\eta_{ideal} = \Delta G / \Delta H$, $\eta_{voltage}$ expressing the correction for "voltage loss" by cell processes not contributing to the electric current, $\eta_{fuel}$ the fraction of fuel used and finally $\eta_{stack}$ the stack correction to efficiency caused by non-optimal flow to each cell or other losses that are not counted for the individual cell.

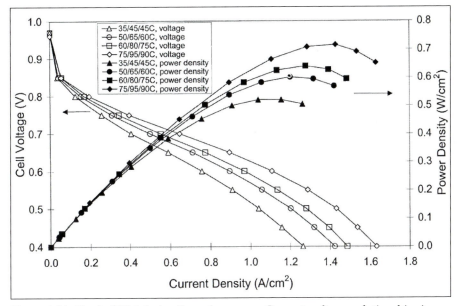

*Figure 3.51.* Single PEM fuel cell performance. Current-voltage relationship (open symbols) and implied power density (filled symbols), are shown for different operational temperatures in the range of 45-90°C, for a cell with a catalyst layer with incorporated PTFE (polytetrafluoroethene) to reduce water flooding, a low loading of a carbon-supported Pt catalyst layer (120 µg Pt cm$^{-2}$) and finally Nafion intrusions. (From Z. Qi and A. Kaufman (2003). Low Pt loading high performance cathodes for PEM fuel cells. *J. Power Sources* **113**, 37-43. Used with permission from Elsevier.)

In Fig. 3.52, the efficiency of a reversible PEM fuel cell is depicted. Any fuel cell can in principle be operated in either direction, to produce electricity from hydrogen or to produce hydrogen from electricity. However, in most cases, the cell design is optimised for one intended use and not very efficient for the reverse mode of operation. Electricity-producing PEM cells usually have a stack of many, fairly small cell assemblies, converting hydrogen with an efficiency typically below 60%, whereas a hydrogen-producing PEM electrolyser will typically have a few, large-area cells in order to produce hydrogen at an efficiency as high as 95% (Yamaguchi *et al.*, 2001). Reversible fuel cells would have an important role in dispersed, building-integrated applications, but typical electricity-producing PEM cells will produce hydrogen with an efficiency of only around 50% (Proton Energy Systems, 2003). Efforts are in progress for finding a better compromise between the efficiencies of the two modes of operation, or preferably to improve the performance of both. One recent suggestion has been the use of new catalyst where Pt is mixed with $IrO_2$ (Ioroi *et al.*, 2002; 2004). Figure 3.52 shows the power and hydrogen efficiencies together with the implied round trip efficiency that would be relevant if the cell is used to produce hydrogen from electricity and to store it for later regeneration of electricity (e.g., in connection with intermittent primary energy sources such as solar or wind energy).

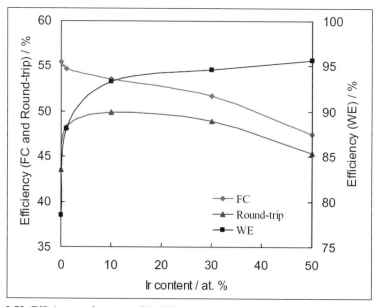

*Figure 3.52.* Efficiency of a reversible PEM fuel cell as a function of the amount (at. % or mol %) of Ir in the form of $IrO_2$ relative to Pt in the positive electrode catalyst, for fuel cell electricity production (FC) or for water electrolysis (WE). Also the product of the two efficiencies relevant for storage cycles is shown. The catalyst is otherwise similar to that of Fig. 3.51, with PTFE and Nafion channels. (From T. Ioroi, K. Yasuda, Z. Siroma, N. Fujiwara, Y. Miyazaki (2002). Thin film electrocatalyst layer for unitized regenerative polymer electrolyte fuel cell. *J. Power Sources* **112**, 583-587. Used by permission from Elsevier. See also Ioroi *et al.* (2004).

The efficiency of electrolysis takes a strong dip when the $IrO_2$ content drops and the catalyst approaches pure Pt, in accordance with earlier experience. However, already at 10% Ir content, the electrolysis efficiency is near 95%, rising only a little with further addition of Ir. At 10% Ir, the drop in the fuel cell efficiency for electricity production has been reduced only from 55 to 53%, so the technology for acceptable operation in both modes seems finally to be established. This will prove important for some of the hydrogen introduction scenarios described in Chapter 5.

Commercial PEM fuel cell stacks currently have electricity production efficiencies in the range of 40-60% at temperatures from 50 to 100°C (highest efficiency at the highest temperatures, as shown for a single cell in Fig. 3.51). An example of stack efficiency was given in Fig. 3.34. Upstart times are of the order of seconds, making the technology suitable for vehicle and other applications requiring quick upstart on demand.

Micro fuel cells intended for use, e.g., with portable electronics, will be mentioned below in section 3.6, as they are often based on direct methanol fuel. Direct methanol fuel cells are also PEM fuel cells, as they are based on the transport of hydrogen ions through a solid polymer electrolyte.

## 3.5.6 Degradation and lifetime

As mentioned in the individual subsections above, degradation of PEM cells may occur as a result of structural change in electrodes (as for batteries), possibly precipitated by the presence of foreign substances such as CO or by water. Removal of CO requires a second reactor, such as a reformer or a catalytic reduction compartment in combination with a fuel cell using e.g. liquid polyoxometalates in a positive electrode circuit (Kim *et al.*, 2004). The overall lifetime of the current "first generation" of industrially produced PEM cells is expected to be of the order of 4000 h of continuous operation in contrast to the 1000 h lifetime of most prototype cells. For automotive applications, target lifetimes of 5000 h or more are desirable, under conditions of typical driving cycles, while for stationary uses, 5 years (43 800 h) is the minimum acceptable lifetime. It is hoped that such targets can be met within the next 5 years. Other concerns include, e.g., survivability at extreme temperatures. Current cells have problems if subjected to temperatures under about –25°C, e.g., due to failure of seals preventing gas transport through the diffusion layers by mechanisms other than diffusion (Schulze *et al.*, 2004). Nafion membranes may be damaged at temperatures over 100°C, and for thin membranes, the crossover of hydrogen or oxygen remains a problem that must be reduced, just as the management of water in the membrane-electrode assembly (MEA) needs to be improved. Models for estimating the magnitude of MEA resistance increase over time has been formulated (Fowler *et al.*, 2002), with the purpose of identifying items needing improvement.

# 3.6 Direct methanol and other non-hydrogen cells

The similarity of methanol (and ethanol) to conventional fuels used for transportation has spurred an interest in using it directly as a fuel for a proton exchange membrane fuel cell, thereby retaining the advantage of the simpler methanol storage and infrastructure, relative to hydrogen, while avoiding the additional component of a methanol-to-hydrogen reformer for accommodation in passenger vehicles where size and weight is at a premium.

The electrochemical reactions at the electrodes are now

$$2CH_3OH + 2H_2O \rightarrow 2CO_2 + 12H^+ + 12e^-, \tag{3.76}$$

$$3O_2 + 12H^+ + 12e^- \rightarrow 6H_2O, \tag{3.77}$$

and the thermodynamical ideal cell voltage is $\phi = 1.20$ V, similar to the one for hydrogen (3.20). The membrane must let some of the produced water pass through to the methanol side, and the byproduct besides water is carbon dioxide, which is a problem in a world needing to reduce emissions of greenhouse gases, except if the methanol is originally derived from (woody) biomass, so that both $CO_2$ from methanol production and from methanol fuel cell conversion can be considered as balanced by $CO_2$ assimilated earlier from the atmosphere (Sørensen, 2004a).

It is possible to use methanol in all acid fuel cells, but the key research area is in PEM cells. A major difference between hydrogen PEM and direct methanol PEM cells (DMFCs) is that the latter produces exhaust gases on both sides of the membrane-electrode assembly. Both $CO_2$ and $N_2$, if air is used at the positive electrode, can block pores of the diffusion layers and catalysts. The reaction (3.76) is relatively slow, and power densities obtained so far are at least ten times lower than for hydrogen-fuelled PEM fuel cells, as illustrated in Fig. 3.53. Better catalysts than Pt are under investigation, taking into account evidence that the Pt attracts a CO subunit of the methanol molecule in a way depending on the presence of steps in the catalyst surface (cf. Fig. 3.50a). Catalyst candidates include Ru-Pt alloys and Pt-Sn alloys on amorphous Ni-Nb substrates (Sistiaga and Pierna, 2003). A problem is that of stability, both of more complex catalysts (such as Mo-Ru-S alloys) and also of the membrane, where thin membranes give higher performance but also promote methanol crossover and degradation (Hamnett, 2003).

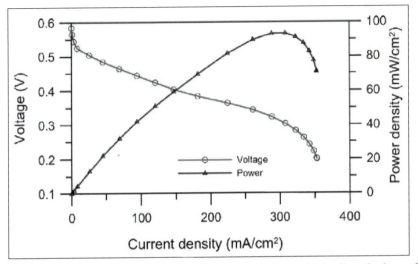

*Figure 3.53.* IV-curve and power density for direct methanol fuel cell with electrodes of carbon black coated on a carbon paper substrate, with Pt-Ru (ratio 1:1) on the negative electrode side and Pt alone on the positive electrode side, both with Nafion intrusions and hot-pressed on a Nafion-112 membrane. The 2-mol methanol solution was fed at a rate of 21 ml min$^{-1}$ and at the other side non-humidified air at a rate of 700 ml/min. The temperature was 85°C. (From G. Lu and C. Wang (2004). Electrochemical and flow characterization of a direct methanol fuel cell. *J. Power Sources*, in press. Used with permission from Elsevier.)

The maximum power density of 0.093 W cm$^{-2}$ shown in Fig. 3.53 is one of the highest currently demonstrated, but still nearly ten times lower than that of hydrogen PEM cells (see Fig. 3.51), which also achieves more than twice the current density. Typical single cell overall efficiencies are of the order of 40% (Müller *et al.*, 2003).

The crossover of methanol has caused problems in finding a suitable membrane material. On the positive electrode side, methanol combines with oxygen to form $CO_2$. Among the alternatives to pure Nafion are Nafion filled with zirconium phosphate or grafted with styrene to inhibit methanol transport (Bauer and Willert-Porada, 2003; Sauk *et al.*, 2004), as well as non-Nafion membrane materials such as sulfonated polyimide (Woo *et al.*, 2003). None have achieved performance as good as the one shown in Fig. 3.53, which, however, has a substantial methanol crossover rate.

Cell modelling techniques for flow and electrochemistry are basically the same for DMFC as for other PEM fuel cells, i.e., the techniques described in section 3.1.5 and 3.5.1-3.5.3 have been employed for DMFCs by Fuhrman and Gärtner (2003).

Fuels other than methanol have been contemplated, such as formic acid, which exhibits lower crossover fluxes than methanol and thus may be an alternative for use e.g., in small systems contemplated for portable applications (Ha *et al.*, 2004; Zhu *et al.*, 2004).

Due to the inferior performance, DMFCs are not the ideal choice for automotive applications, although the possibility is being investigated (e.g., by DaimlerChrysler-Ballard). However, for small portable devices, the convenience in carrying fuel and minimising the number of components may weigh more than efficiency in determining acceptability. The proper alternative to compare with for the small portable applications is lithium ion batteries, which currently have a power density of around 130 Wh/kg but with further development may reach 200 Wh kg$^{-1}$ (Sørensen, 2004a).

The theoretical power density of a DMFC at 0.5 V is about 1600 Wh per kg of methanol fuel, but in practice, small DMFCs for portable applications have achieved much less. If small DMFCs are designed like conventional PEM cells, including a membrane-electrode assembly (MEA), two gas diffusion layers, fuel and air channels with forced flows and current collectors, they may achieve power densities of about 0.015-0.050 W cm$^{-2}$ at temperatures in the range of 23-60°C (Lu *et al.*, 2004), consistent with the value found at 85°C in Fig. 3.53.

However, a new approach to design is to simplify the number of components and to dispense from forced flows, as these require mechanical parts not always desirable for portable applications. A resulting design may look as shown in Fig. 3.54. There are no gas diffusion layers, no forced fuel flow (the area between two MEAs constitutes the fuel container) and no air channels, but there is free exposure to air on the outside of the membrane assemblies, supported by a meshed conductor (Kubo, 2004). In order to compensate for the decline in power density implied by not using forced flow and by operating at ambient temperatures, Kubo used an improved Pt catalyst structure with high surface area and finer Pt particles (about 2 nm diameter) than conventional "lumpy" structures (such as those shown in Fig. 3.49c,d). The structure is one of "nanohorns" produced from graphene, with a sheet structure (similar to single-wall nanotubes) shown in Fig. 3.55. They assemble in the "sea urchin" landscapes shown in the lower right part of the figure, and Fig. 3.56 shows the tunnelling electron microscope pictures after application of Pt particles (a) and for comparison in (b) the same for a conventional Pt catalyst of the type used in PEM cells such as those of Fig. 3.49. With the improved catalyst structure, the measured power density was about 0.045 W cm$^{-2}$ at ambient temperature, implying that the catalyst improvement has brought the performance up to the same level other DMFCs reach at 60°C with forced air and methanol flows and without the simplifications in design such as the absence of gas diffusion layers and air channel.

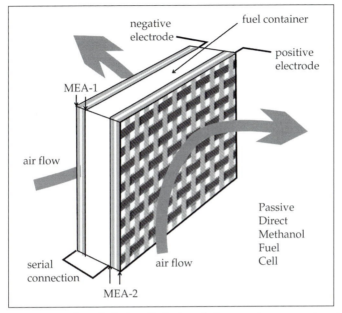

*Figure 3.54.* Direct methanol fuel cell designed for passive operation (no forced flows), with two membrane-electrode assemblies (MEA) and a central fuel container.

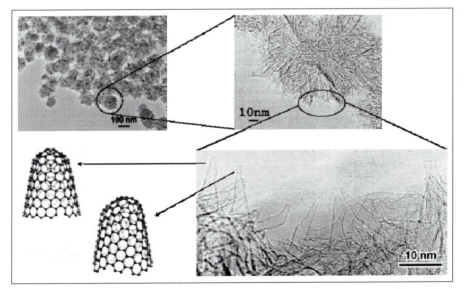

*Figure 3.55.* Carbon nanohorn substrates used for catalyst backing in NEC's passive DMFC design. (From Y. Kubo (2004). Micro fuel cells for portable electronics. In Proc. 15th World Hydrogen Energy Conference, Yokohama. Used with permission.)

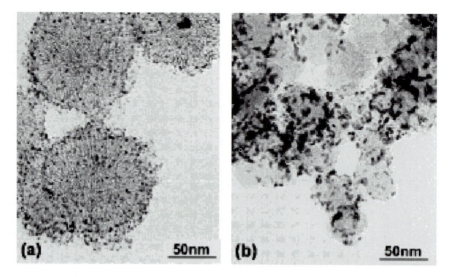

*Figure 3.56.* Platinum catalyst (dark spots) deposited on carbon nanohorn substrates in NEC's passive DMFC design (a), compared to Pt deposited on a conventional carbon substate (b). (From Y. Kubo (2004). Micro fuel cells for portable electronics. In Proc. 15th World Hydrogen Energy Conference, Yokohama. Used with permission.)

Micro fuel cells for portable applications could also be conventional hydrogen PEM cells, or a micro-reformer could be included in the concept, again due to the convenience of carrying hydrocarbon or methanol fuels as compared with carrying hydrogen around (e.g., Holladay *et al.*, 2004). Small planar PEM cells (size range 0.01-1.0 cm²) have been developed for hydrogen fuel in the same manner as the passive DMFC discussed above (Hahn *et al.*, 2004). Again, there is no gas diffusion layer and no forced flow, but in contrast to the device in Fig. 3.54, there are flow channels for both air and hydrogen, delimited by a patterned current collector with a gold layer deposited at a thickness of around 0.01 mm. This was required in order to get reliable long-term performance, resulting in a power density peaking at 0.09 W cm⁻² for a current density of 0.20-0.25 A cm⁻². Planned change to methanol fuel will show, if this slightly different design can match the performance of the one shown in Fig. 3.54.

# 3.7 Biofuel cells

The concept of biofuel cells was briefly mentioned in Chapter 2, section 2.1.6, where microbial sensitisers were employed in photoelectrochemical devices.

The production of hydrogen by photosynthesis or dark fermentation (section 2.1.5) in principle allows that hydrogen to be used as a fuel independent of the source. However, a number of integrated systems have been contemplated, where the biological reactor is adjacent to or integrated into a fuel cell in order to save transport. The problem with this approach is that the hydrogen production from biological systems is variable, particularly of course for photosynthetic systems. This means that hydrogen cannot always be fed into the fuel cell at an optimum rate, with lower efficiency as a likely result.

In a recent experiment, using *Chlamydomonas reinhardtii* to produce hydrogen, the CO-free gas was fed into a PEM fuel cell, using the accompanying gases ($CO_2$, $N_2$) to generate a flow (Dante, 2004). With 100 $m^3$ of algae culture, a power production of 475 W appeared over a period of 25 h, with rather sharp drops before and after. Of course, the production could be smoothed by adding a hydrogen store. Earlier experiments used *Anabena variabilis*, with even more modest results (Yagishita *et al.*, 1996). In all cases, the difficulty in handling the low and intermittent photosynthetic fuel production makes integration with a fuel cell unattractive. Fermentation could conceivably produce a more steady flow of hydrogen. However, if these avenues of hydrogen production should ever become viable, the hydrogen would probably be better utilised by storage and use in the same fashion as hydrogen from other sources.

Other schemes use biological molecules as catalysts in order to obtain direct enzymatic oxidation of a fuel such as glucose (Katz *et al.*, 2003; Chaudhuri and Lovley, 2003). The advantage is that a natural biological material could be used in a special fuel cell without having to first convert it to hydrogen, but there are still many problems to overcome, related to the poor ability of the biological substances to conduct, i.e., to transfer electrons to the external electrodes. Attempts are being made to integrate the biological material in nano-structured carbon materials or to find biological organisms capable of transferring electrons without special mediators.

One such organism is *Rhodoferax ferrireducens,* which is claimed to be able to convert over 80% of electrons from glucose to cell current (Chaudhuri and Lovley, 2003). Whether this concept can be translated into a practical biofuel cell remains to be seen. It does gain all the 24 electrons from glucose and seems to attach to a carbonic electrode in a way that allows electron transfer without the help of enzymes or catalysts. The methods for this are not understood but seem to involve the *R. ferrireducens* outer boundary ("biofilm") having an affinity for attaching to the carbon surface (Fig. 3.57). The energy transfer from glucose involves reduction of Fe(III),

$$C_6H_{12}O_6 + 6H_2O + 24 \text{ Fe(III)} \rightarrow 6CO_2 + 24 \text{ Fe(II)} + 24H^+, \qquad (3.78)$$

as suggested by Chaudhuri and Lovley (2003) and Finneran *et al.* (2003). Placing the bacterium in the negative electrode compartment of a biofuel cell, separated from the positive electrode chamber by a membrane, a current of up to 0.6 mA, or 0.3 mA m$^{-2}$, was observed.

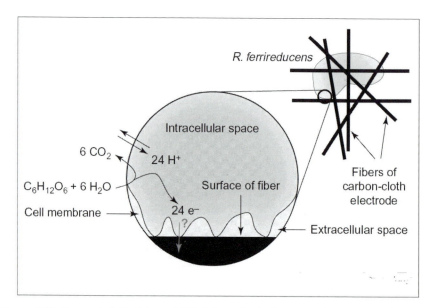

*Figure 3.57.* Suggested docking of *Rhodoferax ferrireducens* surface to a carbon electrode for efficient transfer of electrons derived from oxidation of glucose. (From G. Tayhas and R. Palmore (2004). Bioelectric power generation. *Trends Biotechnology* **22**, 99-100. Used with permission from Elsevier.)

# 3.8 Problems and discussion topics

### 3.8.1
Write down expressions in analogy to (3.12) for the efficiency of fuel cells in reverse operation, used to produce either hydrogen or both hydrogen and heat from electric power,

### 3.8.2
Using fossil fuels in solid oxide fuel cells, it is possible in cars to reform the fuels to hydrogen within the device. $CO_2$ will be a side product, unlikely to be collectable as in stationary installations. What are the implications for greenhouse gas emissions? For natural gas as fuels, what is the global gas re-

quirement and $CO_2$ emissions? Same question if coal is used, although here the direct carrying of coal in cars is improbable, and it would rather be a "producer gas" in compressed form.

### 3.8.3

Think of a fuel cell and its fuels as an energy store and compare it with other storage options. For example, estimate the energy density on a weight basis for a conventional lead battery, an advanced lithium ion battery and for several of the low-temperature fuel cells (such as PEM and DMFC). In contrast to the storage point of view in Chapter 2, section 2.4 (or in Chapter 5 of Sørensen, 2004a), you will need not only the weight of the fuel, but also of all the equipment making up the fuel cell and fuel handling system, including the fuel container. Some data on this may be found in Chapter 6. Energy and power densities have a particular relevance for portable devices, so make sure that you include some portable concepts such as the passive DMFC.

Try to make a similar comparison, not on a mass, but on a volume basis. For fuels, this may be taken from tables such as Chapter 2, Table 2.4, but for equipment, you may have to look at actual equipment (e.g., on the manufacturer's internet pages) to get an idea of just how tiny or bulky the devices are.

### 3.8.4

Try to design a fuel cell with all-solid components (except possibly for water), having the benefits of a PEM fuel cell but without its drawbacks. Would it look like a 200°C acid cell?

### 3.8.5

Could a direct carbon fuel cell work without employing, as the MCFC, melting temperatures?

# SYSTEMS

The distinction between systems and components is not always meaningful, and systems are mentioned in many places in this book. However, in this chapter, a more systematic overview of the types of systems using fuel cells is given, with system being defined as a comprehensive aggregate of components serving a particular demand, such as providing personal or freight transportation or providing heat and power to a building. These individual systems may be combined into nation-wide or global networks of interconnected energy supply systems (by convention using the term "system" also for these, rather than the equally ambiguous terms "economies" or "societies" found in such constellations as "hydrogen economy" or "hydrogen society"). National and global hydrogen systems will be discussed in Chapter 5.

# 4.1 Passenger cars

### 4.1.1 Overall system options for passenger cars

The simplest system for using hydrogen in a passenger car, not considering straight combustion (see Chapter 2, section 2.3.3), has a fuel storage tank, a fuel cell and an electric motor. The electric motor is rated at the maximum power required by the vehicle, and the fuel cell should be able to deliver the corresponding power input to the motor, while the store must be large enough to give the vehicle the desired range.

In case other fuels are used, the fuel cell must be capable of accepting them (direct methanol fuel cell, etc.), or they must be converted to hydrogen, typically using a reformer (natural gas, gasoline, methanol, etc., cf. Chapter 2, Fig. 2.44). The store is now accommodating the fuel of choice.

A control system administers the flows of fuel and timing of the components' functions. There should in most cases also be a water management system, capable of keeping the fuel cell (if it is of PEM type) at the proper humidity. In many cases, it is not convenient to use hydrogen for cold starts, and a battery is used to furnish start-up power. This could be a conventional car battery (of lead-acid type) with modest storage capacity, but generally, the functionality will be improved by using a larger, high-voltage battery.

When substantial amounts of traction are delivered both by the battery and by the fuel cell, the system is called a *hybrid* system. The most advantageous set-up in this case should allow recharging of the battery when the vehicle is operated on the fuel cell at less than full power. For this configuration, there may be an additional option to recharge the battery from external sources when the car is parked (Bitsche and Gutmann, 2004; Suppes *et al.*, 2004). Figure 4.1 shows some possible hybrid layouts.

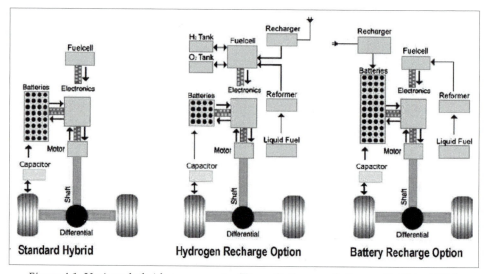

*Figure 4.1.* Various hybrid car concepts. (From G. Suppes, S. Lopes, C. Chiu (2004). Plug-in fuel cell hybrids as transition technology to hydrogen infrastructure. *Int. J. Hydrogen Energy* **29**, 369-374. Used by permission from Elsevier.)

For direct combustion hydrogen vehicles, the components include the engine and usually a liquid hydrogen store (to get sufficient range). The control equipment must therefore also comprise an exhaust system capable of safely handling the hydrogen boil-off from the store (Ochmann *et al.*, 2004).

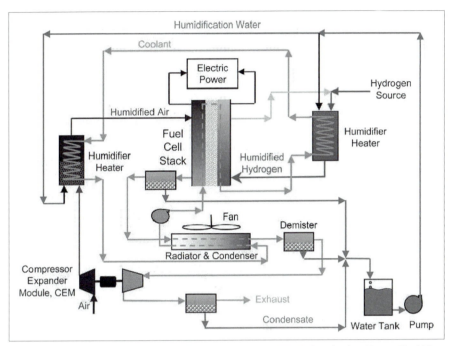

*Figure 4.2.* Schematic layout of power system for a PEMFC vehicle. (From R. Ahluwalia, X. Wang, A. Rousseau, R. Kumar (2004). Fuel economy of hydrogen fuel cell vehicles. J. Power Sources **130**, 192-201. Used by permission from Elsevier.)

In assessments of energy efficiency, it is important to include all components of the system. Each energy converting device is characterised by an energy efficiency (energy output over energy input) as well as an exergy efficiency (free energy out over free energy in), with the latter reflecting the quality of energy (Sørensen, 2004a). The fuel-to-electricity efficiency of various fuel cell types is 45-70% (Chapter 3), but to this comes the upstream efficiency of producing the fuel and the downstream efficiency of using it. Hydrogen production from fossil fuels or biofuels has an efficiency of 45-80%, while production from electricity is at 60-90% efficiency (Chapter 2). To this comes the efficiency characterising the production of electricity in the first place. If the basis for electricity is fossil resources, an additional 30-45% efficiency factor must be added. For renewable energy sources, such as wind power or photovoltaic panels, the conversion efficiency is usually not included in this context because the primary source is "free". Downstream efficiencies range from 35-45% for automobile traction (Chapter 6), while for electric light and appliances, nearly the entire scale of efficiencies is met in various actual devices. It follows that the overall efficiency from primary energy to an energy service at the end-user, such as mobility, may be as low as 5%. The positive message contained in this fact is that there is plenty of room for improvement by applying human ingenuity.

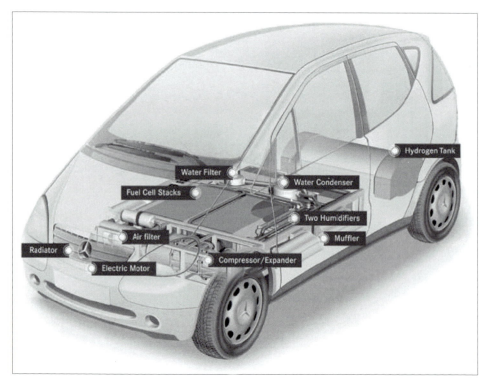

*Figure 4.3.* Placement of fuel cells, hydrogen tank and auxiliary equipment in the DaimlerChrysler prototype car *Necar 4.* (Based on G. Friedlmeier, J. Friedrich, F. Panik (2001). Test experiences with the DaimlerChrysler fuel cell electric vehicle NECAR 4. *Fuel Cells* **1**, 92-96. Used by permission from Wiley.)

## 4.1.2 PEM fuel cell cars

As most current automotive fuel cell efforts use proton exchange membrane (PEM) fuel cells, these will be described in a little more detail in this section and will be used as templates for the performance calculations presented in section 4.1.3. A typical passenger car PEM fuel cell system is depicted in Fig. 4.2. Included are heaters for bringing the equipment from ambient temperatures to the operating temperature of around 80°C and humidifiers for ensuring the level of water in the membrane and electrode areas required for operation (cf. Chapter 3, Figs. 3.39, 3.40 and 3.47). The water management equipment includes a condenser, which is integrated into a conventional radiator operating, however, at much lower temperatures than that of combustion engine cars.

The fuel cell equipment adds considerably to the weight of the vehicle, implying lower efficiency but also the possibility of increased stability, achieved by placing the heavy equipment low in the vehicle construct. Figure 4.3 shows the under-floor placement of fuel cells, storage tanks and auxiliary equipment in the *Necar 4* hydrogen-fuel prototype fuel cell car of DaimlerChrysler, making it much more stable than its combustion engine basis, the commercial Mercedes-Benz A-class car. While the *Necar 4* has a 75-kW fuel cell stack, the newer DaimlerChrysler 0-series *f-cell* car (see Chapter 6, section 6.2.4) uses a 85-kW Mark-902 fuel cell stack from Ballard Power Systems.

*Figure 4.4.* The General Motors skateboard concept with all power equipment and physical controls placed in a flat frame construction placed below the passenger cabin, and all driving control instructions transmitted electronically from the cabin to the board equipment. (From M. Herrmann and J. Meusinger (2003). Hydrogen storage systems for mobile applications. Presented at 1st European Hydrogen Energy Conf., Grenoble. Used by permission from GM.)

A step further advanced is the concept proposed by General Motors (Fig. 4.4), where not only all fuel cell equipment is placed below the passenger cabin, but this "skateboard" is isolated from the cabin and receives all instructions electronically (for steering, for braking, for accelerating). Optimum control is helped by having not just one, but four electric motors, one at each wheel. The concept will be used in the prototype car *Hy-wire*. Other car manufacturers, such as Toyota, in their prototype fuel cell vehicles place the fuel cell stacks more conventionally in what traditionally is the engine space in front of the passenger cabin (Takimoto, 2004).

The methanol-reformer systems (e.g. DaimlerChrysler's *Necar 5*) were for some time developed in parallel with the direct hydrogen fuel cell vehicles, considering that the advantages of only needing minor changes in fuelling infrastructure would outweigh the somewhat lower overall efficiency (Boettner and Moran, 2004). However, technical problems with the reformer performance have currently brought this line of development to a halt. Direct methanol fuel cell vehicles are under development, e.g. by Daimler-Chrysler (Lamm and Müller, 2003).

## 4.1.3 Performance simulation

In parallel with actual tests on the road, simulations are used by both scientists and auto manufacturers as a means of getting a first orientation at low cost, both before a new car is actually constructed and also during the testing and revision phases. Here, a brief simulation study will be made in order to illustrate the capability of simplified, but fairly realistic model assumptions.

It is possible to simulate the behaviour of various vehicle types with conventional propulsion systems such as Otto or Diesel engines, pure electric or pure fuel cell vehicles as well as hybrids, either by a detailed physical modelling or by a semi-empirical method, where different processes are simply parametrised and the parameters are adapted to measured data. The latter method is presented here, based on the software ADVISOR developed for the US government (Markel *et al.*, 2002). Parametrised models exist for fuel cell stacks, electric motor, battery energy storage, power-train control in fuel cell cars with battery, exhaust control, power behaviour of the wheel/axle system under prescribed driving conditions (slope, road surface and resistance, etc.) and also auxiliary electric energy use in the vehicle. The programme uses the required driving speed at a given time to calculate torque, rotational speed and power in each drive-train element, a procedure called a backward-facing simulation approach. However, this is combined with a forward-facing approach based on the control logic, and the simulation proceeds forward in time, but with backward consistency checks at each step.

The fuel cell modelling is either a simple one where power output and efficiency are linked by an empirical curve such as the one shown in Fig. 4.5, or a combination of relationships for each component in the fuel cell system (Fig. 4.2). For the simulations reported below, the two methods gave similar results. More detailed models, e.g. treating water management, are available (Markel *et al.*, 2002; Maxoulis *et al.*, 2004). For the battery, additional calculations were made with the battery assumed unable to be recharged by the fuel cell, as well as with batteries in various hybrid configurations. Both conventional lead-acid and advanced lithium ion batteries have been modelled, using internal resistance battery models.

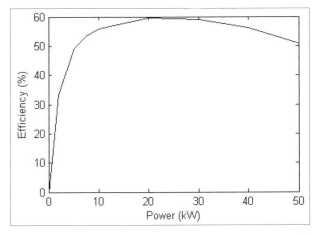

*Figure 4.5.* Power curve for a 50-kW PEM fuel cell used in simulations.

The fuel cell vehicle considered is an artefact loosely modelled over the Volkswagen Lupo TDI-3L (see Chapter 6, Table 6.6), so for simplicity, I shall call it *Little Red Ridinghood*. The 45-kW diesel engine is replaced by a fuel cell engine in a hybrid configuration with battery power, ranging from pure fuel cell operation to a pure electric vehicle. Suitable component rating is determined on the basis of efficiency for a given driving cycle rather than (poorly known) cost, sometimes leading to several systems of comparable performance.

The total mass of the each *Little Red Ridinghood* configuration has the distribution shown in Table 4.1 (compare to Chapter 6, Table 6.6 for the Diesel Lupo 3L).

| *Component mass (kg)* | *pure FC* | *hybrid* | *pure EV* |
|---|---|---|---|
| Basic vehicle (incl. start battery) | 604 | 604 | 604 |
| Fuel cell equipment (30, 20 and 0 kW) | 163 | 133 | 0 |
| Exhaust management | 8 | 5 | 0 |
| Lithium ion batteries (2-3 times more for Pb) | 0 | 113 | 681 |
| Electric motor | 44 | 58 | 70 |
| Transmission (manual 1-speed equivalent) | 50 | 50 | 50 |
| Passengers and cargo (average) | 136 | 136 | 136 |
| **Total** | **1005** | **986** | **1541** |

*Table 4.1.* Mass distribution for *Little Red Ridinghood* fuel cell-electric hybrid vehicles.

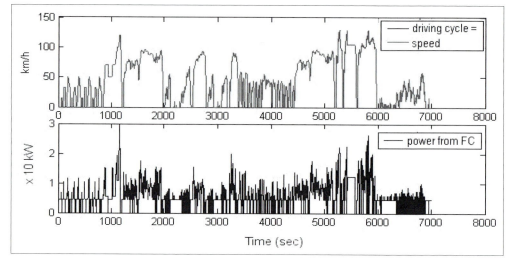

*Figure 4.6.* Simulation results for the *Little Red Ridinghood* pure 30-kW fuel cell vehicle under a mixed driving cycle (matching achieved speed). Equivalent gasoline fuel use is 2.7 litre per 100 km. See text for further discussion (Sørensen, 2004h).

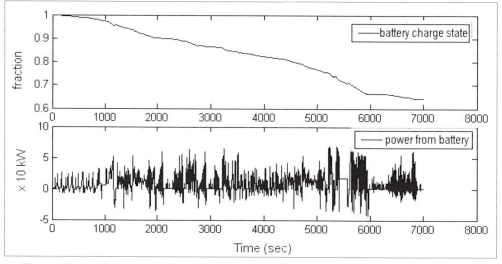

*Figure 4.7.* Simulation results for *Little Red Ridinghood* in a pure electric vehicle version with 1650 kg lead-acid batteries. The equivalent gasoline fuel use is 2.7 litre per 100 km, and 2.5 and 1.7 1/100 km for corresponding simulations using nickel metal hydride and lithium ion batteries (Sørensen, 2004h).

As the driving cycle used in the simulations, an 89 km sequence of city and highway standard cycles have been used. The driving cycle, which is shown at the top of Fig. 4.6, comprise the New European Driving Cycle (at the beginning, shown separately in Section 6, Fig. 6.8) and the US Urban Dynamometer Driving Schedule (UDDS) from 3100 to 4400 sec. The cycle does not include slopes, but has more frequent decelerations and accelerations during the highway parts, making it as demanding as a graded path.

Figure 4.6 shows the simulation results for a pure fuel cell car. The driving cycle-prescribed speeds at top are also the achieved speeds. The second graph shows the power delivery from the fuel cell to the electric motor, peaking at 30 kW at the moments of maximum acceleration. The rated fuel cell power is taken as 30 kW, which is sufficient for all periods of the driving cycle used. Actually, lowering the rated power to 20 kW makes so little difference in respect to achieved speed, that it is not visible on a similar graph. However, the fuel cell is then working much of the time at high power levels, and any smaller fuel cell ratings do not allow the driving cycle speeds to be reached at all times. The fuel consumption is 76 MJ of hydrogen for the 89-km cycle, corresponding to an equivalent gasoline fuel economy of 2.7 l/100 km (2.3 l/100 km for the 20 kW simulation). In order to have a range of 600 km, the car must then carry 4 kg of hydrogen. This agrees well with the 1.8 kg carried in two modest size compressed gas containers by the prototype Daimler-Chrysler f-cell car (cf. Chapter 6, Table 6.6) for a range of only 150 km and with a prototype Ballard fuel cell of lower than the goal efficiency expected to be reached within a few years. The 4 kg hydrogen would have a volume of 178 litres at 30 MPa.

Figure 4.7 shows the simulation results for a pure electric vehicle built upon the same concept. Three versions have been modelled, using lead-acid, nickel metal hydride or lithium ion battery technologies, and Fig. 4.7 shows the results for a 17-kWh Pb battery, adding 1650 kg of mass to the vehicle and yet achieving a range of no more than 250 km. With a 17-kWh NiMH battery, the range is increased to 350 km, with a battery mass of 1670 kg. The problem with pure battery operation is that adding battery capacity strongly influences total weight and thus creates a further need for traction power in a vicious circle. If, alternatively, the recently developed large lithium ion-type batteries are introduced, the additional mass can be kept down to 680 kg, with a range of 300 km and 12.3 kWh of stored energy, for the different characteristics of Li ion batteries (Sørensen, 2004a). The fractional discharge state and the actual power output of the batteries are nearly identical for the three simulation runs, of which the Pb battery results are used in Fig. 4.7. The maximum discharge rate of the batteries is near 70 kW. This is very different from the 30 kW for the pure fuel cell power system, because of the different discharge characteristics of batteries.

Finally, Fig. 4.8 shows the simulation results for a hybrid fuel cell-electric vehicle with 20 kW of fuel cells and a 5-kWh, 113 kg Li ion battery. The two power sources are operating in series, and particularly during the high-speed highway part of the driving cycle (at 5000-6000 sec), both contributes to propulsion, peaking at 19 and 47 kW but with simultaneous power limited to 58 kW by the motor. The fuel cells have surplus power to charge the batteries during periods of low power requirement (the low-speed segments of the driving cycle), but exhibit a net discharge during high-speed segments. The charge state of the battery at the end of the 89-km cycle is slightly higher than at the beginning (as it should be because the final speeds are low).

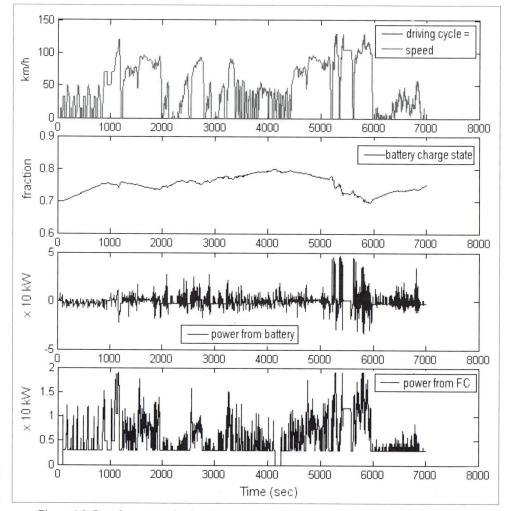

*Figure 4.8.* Simulation results for the *Little Red Ridinghood* fuel cell-electric hybrid vehicle with 20 kW of fuel cells and 113 kg of Li ion batteries. Equivalent gasoline fuel use is 2.4 litre per 100 km. See text for further discussion (Sørensen, 2004h).

Lower than 20 kW fuel cell rating leads to insufficient battery charging. Increasing the fuel cells to 30 kW, with a lowering of the Li ion battery size to 3 kWh (mass 69 kg), gives an overall performance similar to that of Fig. 4.8 (but of course with larger variations in battery charge state). In fact, even with a 1 kWh battery, the 30 kW fuel cell can handle the demands of the driving cycle, so there is no clear optimum configuration, but a range of technical options depending on e.g. requirements to performance in extreme situations not covered by the selected driving cycles.

An electric motor of efficiency 0.92 and a manual 1-speed transmission have been used for all the simulations. The transmission often constitutes a significant loss factor, and losses in using 5-speed or automatic transmissions (Cuddy, 1998) are higher than for the 1-speed gearbox. Because the actual Lupo 3L has an electronically operated automatic transmission with higher efficiency than the corresponding manual transmission, the 1-speed model was selected in order to avoid a distorting impact on the simulation results.

# 4.2 Bus, lorry

At an early stage, interest in incorporating fuel cells in larger vehicles arose, due to the easier fitting of the (then) bulky equipment as compared with installation in passenger cars. While freight companies have shown little interest, bus companies (often under public control) were among the first to volunteer to test fuel cell technology. Presently, there are about 50 fuel cell buses driving in regular route patterns in various cities of the world. One positive consideration is that the fixed route driving and use of dedicated filling stations have made it easy to accommodate the limited range of current fuel cell buses and to establish dedicated hydrogen filling stations at suitable locations in the test cities. Typical fuel cell bus layouts have the stacks in the back and compressed hydrogen tanks on top, as illustrated in Fig. 4.9.

Practical experience is accumulating through the ongoing fuel cell bus programmes. Data have been obtained, e.g. for a 13-t (including passengers), 52-passenger SCANIA hybrid fuel cell bus with a fuel cell rating of 50 kW and 44 standard 12-V lead-acid batteries capable of delivering power to two 50-kW wheel hub motors (Folkesson *et al.*, 2003). Figure 4.10 shows the energy flows in the system, based on the Braunschweig driving cycle (a multiple stop-go bus cycle with maximum speeds of 35-60 km/h). The bus is equipped with a regenerative breaking system that reduces the fuel consumption with a diesel equivalent of 9.9 l/100 km to 25.8 l/100 km. The special characteristics of the driving pattern cause the modest size power supply system to be sufficient, even with an air conditioning system installed.

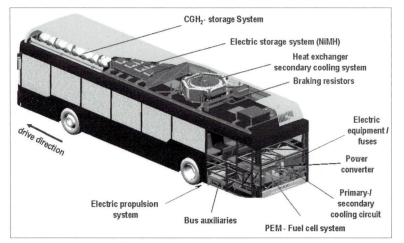

Figure 4.9. Hybrid fuel cell bus, showing placement of power equipment (68 kW PEM cell, 2×75 kW motor and NiMH batteries) (MAN, 2004; used by permission.)

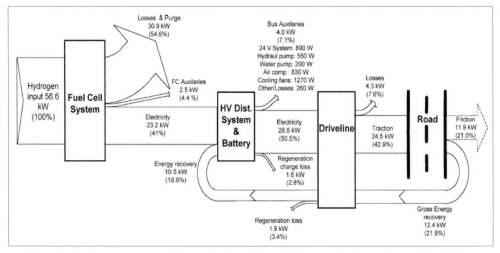

*Figure 4.10.* Average energy flows during a bus driving cycle for a SCANIA hybrid fuel cell bus (based on lower heating value of hydrogen). (From A. Folkesson, C. Andersson, P. Alvfors, M. Alaküla, L. Overgaard (2003). Real life testing of a hybrid PEM fuel cell bus. J. Power Sources **118**, 349-357. Used by permission from Elsevier.)

The largest fleet of fuel cell buses presently in operation is made up of 30 DaimlerChrysler Evobus *Citaro F* buses (200-kW PEM fuel cells, 1629 l of $H_2$ compressed to 35 MPa), in route application in European cities, including a range of hydrogen production schemes delivering to the associated filling station (Mercedes-Benz, 2004).

# 4.3 Ships, trains and airplanes

The use of fuel cells on ships, or direct hydrogen combustion, has been suggested, but so far not realised except for small pleasure boats. A European project has looked at environmental impacts from using hydrogen on ferries (see Chapter 6), and various proposals for use of PEM fuel cells have been forwarded. One such proposal considers submarines, where air-independent propulsion systems are of particular importance (Sattler, 2000). Such a submarine would need to carry both hydrogen and oxygen in containers, presumably in liquid form in both cases. Figure 4.11 shows the possible layout of such an arrangement, suitable for retrofitting of existing submarines.

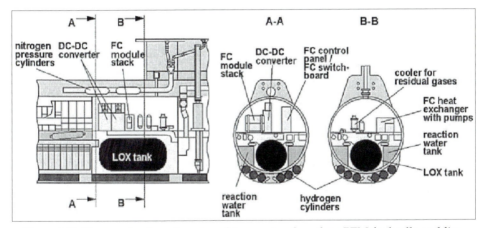

*Figure 4.11.* Proposed submarine propulsion system based on PEM fuel cells and liquid hydrogen and oxygen. (From G. Sattler (2000). Fuel cells going on-board. *J. Power Sources* **86**, 61-67. Used by permission from Elsevier.)

A project to explore the possibilities for using hydrogen for trains has been undertaken by Tokyo Gas Co. (2004). So far the arrangement for fuelling has been studied, but not the propulsion system for the trains themselves.

The use of hydrogen in aircraft gas turbines has been proposed several times over the past decades (Jensen and Sørensen, 1984). Within the European *Cryoplane* project, it has been suggested that lower altitude cruising will make propulsion by hydrogen advantageous with respect to environmental impacts (Svensson *et al.*, 2004). However, it is doubtful if this suggestion is practical in a world where air corridors are becoming overcrowded and new approaches to making better use of airspace are being sought.

A different approach is to reconsider the airship as a means of air travel. A first approach to this is considering an airship for high-altitude cruising (or as a stratospheric platform) powered by photovoltaic panels and using a reversible fuel cell system to store surplus solar power and use it when the sun is not visible. In this way, carrying possibly heavy batteries may be avoided. The envisaged relative shares of direct use of solar power, of electrolyser operation and of fuel cell power production are shown in Fig. 4.12. So far, testing of the equipment sketched in Fig. 4.12 has been performed on a 1-kW scale in the laboratory and in simulated airship conditions.

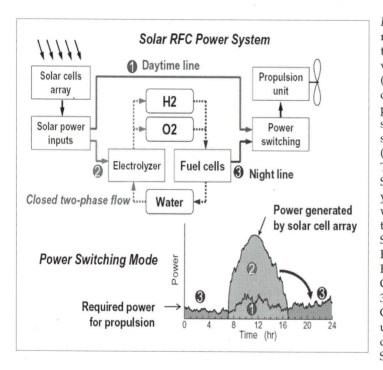

*Figure 4.12.* Power modes and their time-shares for reversible fuel cell (RFC) plus solar cell system proposed as a power source for a stratospheric airship. (From K. Eguchi, T. Fujihara, N. Shinozaki, S. Okaya (2004). Current work on solar RFC technology for SPF airship. In Proc. 15th World Hydrogen Energy Conf., Yokohama. 30A-07, CD Rom. Copyright attributed to the Hydrogen Energy Soc. Japan.)

# 4.4 Power plants including stand-alone systems

Stationary power generation on a large scale may use either low- or high-temperature fuel cell systems, and several systems rated at up to a few hundred kW have been operated (Barbir, 2003; Bischoff *et al.*, 2003; Veyo *et al.*, 2003). The systems comprise the basic units of PEMFC, MCFC or SOFC as described in Chapter 3, combined with fuel preparation and exhaust clean-

ing equipment where necessary. Placement in conventional power plant locations allows convenient accumulation of all the processes required, or access to hydrogen pipeline where this is preferred, without having to cope
with the limited space constraints of vehicles. Scenarios for the expansion of
this type of system are dealt with in Chapter 5, including the special requirements in cases where the hydrogen plants serve as stores for primary
energy sources such as wind or solar power. Also discussed in Chapter 5 is
the incorporation of large hydrogen stores in this centralised set-up. Similar
discussions of the requirements for large penetration of stationary fuel cell
technologies have been taken up for Japan by Fukushima *et al.* (2004). In the
transitional phase between the present and a fuel cell-based system, it has
been suggested to lower the cost of hydrogen for vehicles by generating it
from off-peak electricity from the present system (Oi and Wada, 2004). This
idea is similar to the use of surplus wind power for the production of hydrogen, as described in Chapter 5, section 5.5. In conjunction with intermittent
renewable energy sources, the long up-start times for high-temperature fuel
cell systems may be a problem, although short-term forecasting of e.g. wind
power production can be done rather accurately (Meibom *et al.*, 1999).

In order to give stationary fuel cell systems an optimum performance,
short-term storage in the form of ultra-capacitors has been studied (Key *et
al.*, 2003). The fast response of these storage devices allows a very precise
load matching. On the other hand, it is generally assumed that a mature fuel
cell technology will enable power systems to respond fast enough for all
normal operations. Small-scale capacitor storage is perhaps more likely to
become incorporated into electric appliances, in order to make them more
resilient. Because many new appliances with growing penetration in the
market (such as laptop computers) include on-board battery storage and stabilising circuits, the demand for high power-quality (low frequency and
voltage excursions) has generally been declining in recent years, which is
good news for many of the emerging power systems, including both primary
converters for wind and solar energy and also intermediate converters such
as fuel cells. One already emerging niche application of fuel cells is for
emergency power system (or UPS = "uninterruptible power systems").

A different kind of stationary system application is for remote power supply. Such systems are most likely based on primary energy from renewable
energy converters, because long-distance transportation of hydrogen by road
will increase the current problem of transportation cost, making remote
power much more expensive than the same type of power in densely inhab-
ited areas, due to the cost of transporting fossil fuels in trucks over land. Locations that can be served by waterways may not have this problem. The fuel
cell system will not be particularly different from what is used in other locations, except that low maintenance requirements will have a higher priority.
Use of modest lifetime MCFCs for peak-shaving has been tested (MPS, 2004).

# 4.5 Building-integrated systems

Smaller, building-integrated fuel cell systems, notably of the PEM type, have attracted much attention over recent years. This is partly related to the motion to more decentralised energy systems, where traditional energy supply has maintained a distinction between heat, which often has been decentralised (individual building oil or gas burners), and electric power supplied from central sources. The third important energy type, that of fuels for vehicles, has been delivered through supply chains ending at communal filling stations, and small-scale, portable power has been entirely supplied through the purchase of small batteries (where only the rechargeable ones have offered personal control). Fuel cells offer the possibility for individual buildings and thus their owners to become their own electricity providers and possibly also to provide individual filling stations for vehicles parked in building-attached garages (Sørensen, 2000). Finally, fuel cell technology may also replace small batteries for portable applications, allowing individuals to control all their energy supplies for heat, vehicle fuels and stationary or portable electricity uses.

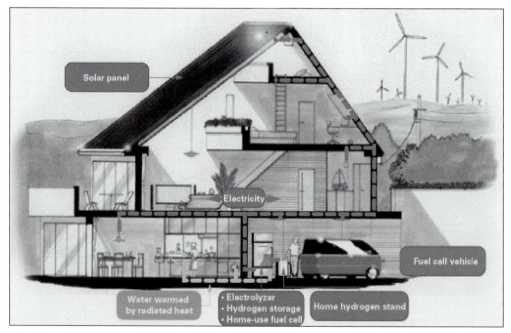

*Figure 4.13.* Vision of building-integrated fuel cell system supplying heat, power and hydrogen as vehicle fuel. (From Honda (2004), used with permission.)

These possibilities are incorporated in some of the scenarios presented in Chapter 5. Figure 4.13 gives a vision for building-integrated fuel cell systems deriving their primary energy from renewable energy supply through electric power networks. An alternative would be hydrogen supply through pipelines in analogy to natural gas distribution or district heating networks.

A first approach to the building-integrated use of hydrogen may be replacing the natural gas boiler unit used in many countries for home heating and hot water needs by a PEM fuel cell unit with reformer and thus capable of using the existing supply of natural gas. Such fuel cell plus reformer units have already been developed and may reach a competitive cost within the next one or two decades (Vaillant, 2004; Osaka Gas Co., 2004). Test results have been forthcoming for a number of prototype installations, such as a 4.0 plus 6.8 kW H-Power combined heat and power producing unit operated in Italy (Gigliucci *et al.*, 2004). In the next phase (assuming that resources of natural gas are such that alternatives should be sought and/or that greenhouse gas emission issues are taken seriously), similar units based on a hydrogen supply infrastructure would be introduced (Erdmann, 2003; Kato and Suzuoki, 2004). Development of 1- to 2-kW combined cycle SOFC modules, e.g. based on disc-shaped sub-stacks and integrated into a small unit with a fuel cleaner (e.g. for removing sulphur from natural gas), an afterburner (to reduce fuel in exhaust) and a battery storage, is ongoing (EnBW, 2004; CFCL, 2004). Detailed operating experience will be forthcoming.

Most building-integrated systems would greatly benefit from being able to use reversible fuel cells (cf. Chapter 3, section 3.5.5), rather than having to install two expensive components: fuel cell and electrolyser. The problem has been that reverse operation of a fuel cell optimised for power production used to have a low efficiency (around 50%), making it less efficient than conventional alkaline electrolysis. However, a breakthrough in technology (Ioroi *et al.*, (2004), discussed in connection with Fig. 3.52 in Chapter 3, suggests that reversible PEM fuel cell technologies may in the future live up to the expectations made in visions of widespread in-building use, such as the decentralised scenarios presented in Chapter 5, section 5.5.

An important issue for building-integrated fuel cells is the infrastructure serving the building. It might consist of an electricity grid connection and attachment to a hydrogen pipeline network. If the building is receiving hydrogen, it can produce electricity and associated heat. If more electricity is produced than can be used in the building, it may be exported to the electricity grid. In this way the fuel cell capacity is used to a fuller degree, and the need for additional power stations not part of the building-based system becomes smaller. However, another mode of operation (see Chapter 5, Fig. 5.2) is to receive electricity from the grid and to produce hydrogen to be dispensed to vehicles parked in or at the building or to be stored for later regeneration of power and associated heat. This option is most relevant for primarily renewable energy systems with intermittent power input, because then the hydrogen can be used to produce power and heat when no primary production is taking place. This points to a further possibility in case a reversible fuel cell system is used, namely that the building has a connection to an electricity grid, but not to any hydrogen network. It would then generate hydrogen from excess electricity not used immediately in the building, store it and use some of it for regenerating power in case of low supply and the rest of it as fuel for vehicles.

In such a vision, storage of hydrogen should be possible in close connection with the building. Only when there is a hydrogen network, does a central hydrogen store make sense. Building-integrated hydrogen storage may be in the form of storage tanks in vehicles that happen to be occupying the garage (if present) or parked near the house. However, this option does not guarantee availability of sufficient storage space in all cases where required. It is therefore necessary to think of other hydrogen storage options dedicated to the building. These could be compressed gas containers or, as some would prefer for safety reasons, metal hydride stores. As the scenario in Chapter 5 will show, only modest volumes of such storage (about a third m$^3$ for a family home) are required for handling the fluctuating energy production of a wind-solar primary energy production system, including consideration of losses in all components of the system. Aki *et al.* (2004) have also pointed out the systems stability advantages obtained by using a local hydrogen pipeline grid to interchange hydrogen between individual buildings.

# 4.6 Portable and other small-scale systems

Consuming patterns have over recent years seen a dramatic increase in the use of portable equipment for entertainment and work (such as music and video players, laptop computers and mobile phones with multi-functionality). This has increased the demand for batteries, but at the same time found limitations of the battery technology that seem difficult to avoid even with increasing conversion efficiencies. Fuel cells with small-scale stores are an obvious solution to these problems, because the technical performance is already far beyond that of batteries (e.g., operating a state-of-the-art laptop computer for a few days rather than a few hours). The difference between these otherwise similar technologies is the external storage of chemicals for a fuel cell versus the internal storage in batteries.

A cellular phone in use will need less than 200 mW of electric power, a video camera will need less than 6 W, and a laptop computer or a portable

CD player will need less than 20 W. More specialised portable equipment includes field environmental monitors, medical mobile life-support systems and soldier communication and signalling devices used in military operations (Palo *et al.*, 2002). Power requirements for such equipment would be in the range of 10-400 W. A standard size Li ion battery has a storage capacity of 750 mAh, and the largest one convenient for a laptop computer has a capacity of 3600 mAh. For a portable video camera, the standard Li ion battery may deliver 5 W at 7 V for about 1 h, while for the laptop at 12 V, the larger battery provides about 4 h at an average power consumption of 10 W.

The high price paid for Li ion batteries today (except for power users, the cost may be some 15-25 euro or US$ per kWh for typical average operational use of cameras and laptops over an assumed 4-year lifetime of the battery) makes portable applications an inviting niche market for new technologies such as fuel cells.

The options that may be considered for portable fuel cell applications are PEM fuel cells with compressed hydrogen canisters, metal hydrides or methanol plus reformer as the source of hydrogen, and direct methanol fuel cells. Compressed hydrogen at 30 MPa and the best metal hydrides have an energy density on a volume basis (Chapter 2, Table 2.4) of 2.7 and 15 GJ m$^{-3}$, while that of methanol is 17 GJ m$^{-3}$. These should be compared to the energy density of Li ion batteries, which is 1.4 GJ m$^{-3}$ (Sørensen, 2004a). The implication is that by adding a 50% efficient PEM fuel cell, the performance of a device using stored compressed hydrogen at up to 30 MPa plus a fuel cell is not any better than that of a Li ion battery.

While the consideration above has been to keep the size (volume) unchanged, there would be many portable applications where it is more interesting to compare energy densities by mass, as humans have to physically carry the devices around (a tautology for "portable" equipment). In fact, lower mass would be an advantage for certain laptop computers that in their present form are more "towable" than "portable". Without the mass of the container, the energy density of hydrogen in any form is 120 MJ kg$^{-1}$, but for storage in metal hydrides, the overall density goes down to under 9 MJ kg$^{-1}$ (Chapter 2, Table 2.3). For methanol the value is 21 MJ kg$^{-1}$, and for Li ion batteries it is 0.7 MJ kg$^{-1}$ (Sørensen, 2004a). The comparison on a mass basis makes most fuel cell solutions preferable to Li ion batteries (and much more to lead-acid batteries 5 times lower in energy density), whenever the volume can be tolerated.

For the best metal hydride stores, the capacity (hours of operation) could be increased 5-fold over Li ion batteries of the same volume, and this option has attracted some attention. Güther and Otto (1999) describe a Siemens-Nixdorf experimental laptop computer with such a store, obtaining a 3-fold increase over a Li ion battery. Non-metal hydrides should be able to reach better values.

The use of methanol and a PEM fuel cell in combination with a miniature reformer has been contemplated in a US army project (Palo *et al.*, 2002). For a 15-W, 1-kg portable power device, an energy density of 2.6 MJ $kg^{-1}$ was found. An auxiliary battery is still required in order to start the reformer system up.

Very good energy density improvements can obviously be achieved by using methanol as a storage medium. However, not wanting the complication of a reformer, this means using a direct methanol fuel cell. These were described in Chapter 3 section 3.6 (see Fig. 3.54), and in addition to being further from the commercial marketplace, they also are less efficient than the PEM fuel cells. This means that a larger direct methanol fuel cell (DMFC) has to be used, but not so much that it seriously affects the energy density advantage. For this reason it is believed (Meyers and Maynard, 2002) that methanol/DMFC devices are the most likely technology to win the portable niche market (which in terms of possible profits may be quite substantial as well as expanding). This view is reflected by the investments of several computer companies aimed at speeding the portable DMFC development up.

*Figure 4.14.* Laptop computer powered by a direct methanol fuel cell. The methanol cartridge is at the back. (From Y. Kubo, NEC Corp., (2004). Micro fuel cells for portable electronics. In Proc. 15th World Hydrogen Energy Conf., Yokohama. 01PL-22, Hydrogen Energy Soc. Japan. Used with permission.)

One of several recently presented prototypes of a notebook personal computer powered by a micro-DMFC is shown in Fig. 4.14. A replaceable methanol cartridge is placed behind the computer, and the 280-cm$^2$ DMFC is placed under the laptop keyboard. The fuel cell produces 14 W at 12 V, and if the methanol volume is 30 cm$^3$ (judged from photo), then the storage capacity is 142 Wh. This would allow the computer to be operated for 14 h at an average consumption of 10 W. Many computer manufacturers, notably in Japan, have recently demonstrated similar DMFC notebook prototypes, including some with a more elegantly integrated methanol container than the one on the early model shown in Fig. 4.14.

# 4.7 Problems and discussion topics

### 4.7.1

How much hydrogen must an air plane be able to carry for a London to Tokyo flight?

### 4.7.2

Discuss the optimum rating of fuel cell and battery for a hybrid car, as a function of the relative price of batteries and fuel cell equipment. Note that the energy rating of a battery is a poorly defined parameter. The reason is that the energy released from a battery depends on the discharge rate and hence on the driving cycle (Jensen and Sørensen, 1984). Manufacturers rarely mention this when they quote energy stored $E$ in kWh or charge, $C = E/V$, stored in Ah. Here $V$ is the potential across all units connected in series (some systems use a combination of parallel and series connection of individual battery module units).

### 4.7.3

Write a list of those power-consuming activities in private or work life that you believe cannot ever be expected to become powered by batteries or portable fuel cells.

Based on the fraction these activities constitute of the total power usage, estimate which percentage of total electricity use it would be possible to base on renewable energy sources such as wind or photovoltaic power, used solely to charge either batteries or fuel cells (two separate estimates).

CHAPTER

# 5

# IMPLEMENTATION SCENARIOS

# 5.1 Infrastructure requirements

In this section, the different components required for using hydrogen in various systemic environments are viewed as a network of infrastructure that has to be convenient and feasible as judged by technical and economic analysis. In particular, the additional components required to connect the system components are identified and assessed.

## 5.1.1 Storage infrastructure

Traditional energy sources may be divided into those that employ storage and those that do not. Fuels like wood, oil, natural gas or coal are conventionally stored at various stages from production (extraction) to use: central stores, intermediate stores at retail traders as well as decentralised ones at the end-user. Because of its gas form (implying higher storage volumes), natural gas has in its early years of use been distributed (by pipeline) without storage, whereas presently, containers (for intermediate storage on land or for liquefied natural gas during ship transport) or underground storage is usually employed to manage mismatch between the most economical rate of production and the patterns of use. On the other hand, electricity is largely produced, distributed and used without any storage components (except where hydro stores are available), implying that power plants have to be able to cope with the variations in demand. Increasingly, small-scale users

demand storable power, implying a widespread use of batteries (simple lead
-acid or advanced high power density versions such as lithium ion batteries),
originally only for portable applications, but increasingly also for electronic
equipment used in quasi-stationary environments (from transistor radios to
laptop computers used at distances from grids of a few tens of metres).

The situation changes dramatically if energy sources either with variable
availability or which need production methods preferring constant production are put into use. With renewable energy sources such as wind power or
photovoltaic power an autonomous system requires integration of storage,
and, if the system is not local, also transmission. Such storage is the same as
that considered with conventional electricity systems, in that it is not kinetic
energy or radiation which is stored, but produced electricity that is converted into an energy form storable, in order to be ready for later regeneration of electric power. Hydrogen is one of several possible storage options
catering to electricity storage, by storage as a fuel. It is in competition with
mechanical and electrochemical storage options, or direct storage of electricity in superconducting rings (see Sørensen, 2004a).

Alternatively, one can view hydrogen as the basic energy carrier of a
given system and then ask how intermittent sources can be used to produce
the hydrogen, and how hydrogen can be used to cover society's variable
demands of different forms of energy. The layout of such systems determines if it may be convenient to consider centralised storage facilities or if
storage is better done locally near the end-user. For automotive uses, there
clearly has to be onboard storage of hydrogen if this is the energy carrier
used by the car system.

The implications for infrastructure are that hydrogen produced centrally
will typically be divided into two parts: one going directly into pipelines to
points of use and another being stored at suitable (large) storage facilities.
These are likely to be compressed gas stores, because liquefaction entails
considerable energy loss and storage types such as batteries, hydrides or
other chemical storage are unsuited for large-scale storage. When a hydrogen transmission pipeline system exists, there is no particular reason to store
hydrogen near the point of use, and availability of cheap storage options is
the overriding consideration. Here the solution is likely to be the same used
for natural gas, i.e., that of underground storage in suitable geological formations. In many regions, it is possible to find such formations that allow inexpensive formation of the cavity for holding hydrogen, implying that the
volume of the store becomes less important and therefore the pressure can
be kept low in order to minimise leakage and avoid expensive lining to
tighten the cavity.

Figures 2.48a and c in Chapter 2 showed two such options, both currently
in use for natural gas storage. One takes advantage of the salt intrusions
found in many areas characterised by ice age clay deposits over areas once

covered by oceans. The Danish installation of the type shown in Chapter 2, Fig. 2.48a was formed by flushing the salt out of the intrusion over a period of a year or more, at very low energy expenditure. The underground cavity formed allows $4.2 \times 10^8$ m$^3$ of gas to be stored at a pressure of 23 MPa (DONG, 2003). If used for hydrogen, a pressure of 5-10 MPa would be employed. Several dozen additional underground salt domes of this kind exist in Denmark as well as in other countries with glacier deposit geology, for possible future use.

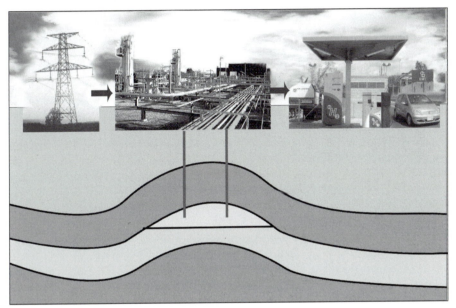

*Figure 5.1.* Hydrogen storage in aquifer, with electrolysis and hydrogen recovery plant above, receiving electric power and delivering hydrogen to filling stations, e.g. through pipelines.

The underground formation shown in Chapter 2, Fig. 2.48c and (in a system setting) in Fig. 5.1 is an aquifer, i.e., a water-carrying (sand-based) layer trapped between two impenetrable (clay) layers. Such aquifers exist in most of the world and usually bend up and down in a way that in some cases allow them to store gases in an upward bend, again with very modest capital investments. Factors determining the suitability include the motion of water along the aquifers. A Danish installation of the type shown in Fig. 2.48c can store $3.5 \times 10^8$ m$^3$ of natural gas at a pressure of 17 MPa (this being the pressure at extraction, with higher pressures at the bottom of the cavity, as in the salt dome case; DONG, 2003). Again for hydrogen, it would be prudent to employ a lower storage pressure, because of the higher permeability of the surrounding geological formations for hydrogen than for natural gas. The two Danish underground stores can hold the equivalent of six months of natural gas use in Denmark.

Decentralised storage may invoke other of the storage solutions described in Chapter 2, section 2.4. For building-integrated storage of hydrogen, compressed gas containers are already in use industrially. However, for home and high-rise city buildings, safety considerations may make metal hydride or similar stores preferable, should they become technically and economically viable. For a family dwelling, a hydride store may be buried under the house or under the garden, much in the same way as oil containers are today (Sørensen *et al.*, 2001, 2004).

For automotive applications, all the storage options mentioned in Chapter 2, section 2.4 are in consideration, although most of the presently emerging fuel cell vehicles use compressed gas stores. Some prototypes have used liquefied hydrogen (combustion engine hydrogen cars typically requiring larger amounts of fuel) or cryogenic activated carbon (acid fuel cell cars). Acceptable sizes of 30-MPa compressed hydrogen stores in small passenger cars are considered to be at most 5 kg of hydrogen, currently giving driving ranges of about 450 km, which is a little lower than the average of the present stock of gasoline-driven cars. This has spurred an effort to construct containers for around 70 MPa pressure with the same level of safety as the 30-MPa units (which are in widespread use today for industrial hydrogen and have an excellent safety record). For buses and lorries, it is usually possible to integrate enough hydrogen storage to match the driving ranges of the current fleets of such vehicles. Metal hydrides are unlikely to be used in mobile applications due to high weight, but some of the chemical hydrides under study could be considered if they are found viable in the technical and economic sense.

Storage at filling stations poses problems of a nature similar to the other stationary stores mentioned above. Current hydrogen filling stations (most of which are part of demonstration programmes, e.g. for fuel cell city buses) mostly use compressed gas storage.

## 5.1.2 Transmission infrastructure

Hydrogen transmission techniques were discussed in Chapter 2, section 2.5. Leaving alone intercontinental trade of hydrogen (the ship transport option), long-range transport from the area of production (e.g., a park of wind turbines) to the area of large users or hubs of distribution to end-users (such as filling stations) is most likely done by pipeline. This could be new, dedicated hydrogen pipelines, or it could be natural gas pipelines upgraded to use for hydrogen at a time when natural gas is no longer available or in demand

(Sørensen *et al.*, 2001). The Danish scenario discussed in section 5.5 takes advantage of the existing natural gas pipeline transmission system consisting of the major grid shown in Fig. 5.2 and minor pipelines to the filling stations also indicated in Fig. 5.2. The two storage facilities mentioned above in section 5.1.1 are also shown. Clearly, if hydrogen and natural gas systems are in use at the same time, two parallel transmission systems are required. Small amounts of hydrogen (around 10%) can easily be added to the natural gas transmitted in present lines, but this is not of interest for use of hydrogen in the transportation sector.

Discussion of transmission infrastructure in connection with surplus wind power or other off-peak power generation has also been made for Ireland (González *et al.*, 2003), for Japan (Oi and Wada, 2004) and in connection with coal-based hydrogen production for China (Feng *et al.*, 2004).

## 5.1.3 Local distribution

As in the case of general transmission lines, the local natural gas distribution lines reaching individual buildings in urban areas can with little change be used for hydrogen. This allows a decentralised hydrogen scenario such as the one described in section 5.5. Key cost items will be the in-building installations, which include interface with hydrogen stores, with fuel cell equipment, and possibly with the storage tanks of vehicles parked in a garage at the building. There is little doubt that the cost of placing such installations decentralised in buildings will be higher than for more centralised facilities covering the same needs, but traditionally, consumers have been willing to pay a premium price for what is perceived as individual control over the technology (here energy supply).

## 5.1.4 Filling stations

Hydrogen filling stations for buses, passenger cars and special vehicles currently exist world-wide in connection with demonstration projects (some 20-30 units), but of course have to be present of approximately the same numbers as gasoline and diesel fuel filling stations today in a future with extensive use of hydrogen in the transportation sector. It is sometimes argued that the current number of filling stations is larger than required, and the number is on its way down in several countries. Whether this means that the more expensive hydrogen filling stations can be reduced in number is unclear, as the vehicle range is presently lower, indicating a larger need for filling stations. The issue is thus linked to questions of storage pressure and mass of hydrogen storage in future commercial fuel cell vehicles. Figure 5.2 shows an example of filling station density, based on the assumption that the fuel cell vehicle driving range can be kept at the same level as for the current average (but not aiming for the larger range of over 1000 km characterising the fuel tank size of the most fuel-efficient current passenger cars, e.g. Volkswagen Lupo 3L).

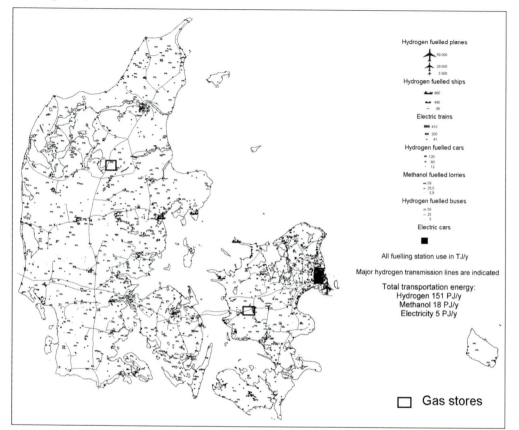

*Figure 5.2.* Infrastructure envisaged in a 2050 scenario for hydrogen use in Denmark, with emphasis on the transportation sector, including two underground hydrogen stores, pipeline transmission and distribution of hydrogen to motorcar filing stations, seaports and airports, as well as to centralised fuel cell power plants. The scenario, which is further discussed in section 5.5, uses electricity for railroad operation and electric cars for driving within the only large city of Copenhagen (Sørensen, 2003a).

## 5.1.5 Building-integrated concepts

As mentioned in section 5.1.1, building-integrated hydrogen systems may possess storage in one or both of the following forms. (1) stores in vehicles parked in or at the building and (2) stationary storage in the basement or under the building or its surroundings, such as metal hydride storage or pressurised containers. Decentralising storage means that equipment for filling and extracting hydrogen from the stores must be available in each building possessing such stores. Although use of the tanks in the vehicles parked at the building may seem attractive, the additional cost of providing a filling station for pressure upgrade of pipeline hydrogen (presumably at considerably lower pressure than the automobile storage containers) would presumably be too high. The problem would be much smaller for metal hydride stores, that need only a modest pressure above ambient for the charging and discharging operations (in the latter case also a modest temperature increase).

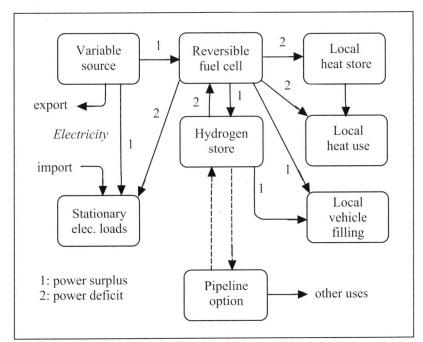

*Figure 5.3.* Layout of a decentralised, building-integrated hydrogen and fuel cell system based on intermittent primary power sources (such as wind or solar energy), reversible fuel cells and local stores, including stationary and maybe vehicle-based stores, and possibly capable of interchanging hydrogen with users in other buildings through pipelines (Sørensen, 2002a).

The building integration concept used in the decentralised scenario of section 5.5 is shown in Fig. 5.3. The concept is based on availability of reversible fuel cell technology, which as mentioned in Chapter 3, section 3.5.5 seems a realistic assumption, at least for a future scenario. At an increased cost, the reversible fuel cell could of course be replaced by a pair of a power-producing PEM fuel cell and an electrolyser cell. Both storage options (underground or in vehicle) are left open, and the heat from the fuel cell operation is envisaged as contributing to or covering the heat loads of the building. Full coverage of space and hot water demands would be possible even with highly efficiency fuel cells (less excess heat) if the building is insulated and otherwise energy-optimised to the best current standards (Sørensen, 2004a).

Finally, portable fuel cell applications may entail infrastructure requirements in terms of networks of locations (e.g., shops) selling hydrogen storage units for exchange like batteries or offering to recharge hydrogen or methanol canisters. As for vehicle hydrogen containers, high-pressure filling is unlikely to be decentralised to home level (in contrast to battery recharging).

# 5.2. Safety and norm issues

## 5.2.1 Safety concerns

The history of technology reveals that safety and risk issues have often been handled with a mixture of attitudes and approaches, including consideration of the following (Sørensen, 1982).

- Direct risk, defined as the probability of an accident (or other risk-related event) times the magnitude of the consequences.
- Social risk, defined as the damage inflicted upon society by an accident or risk-related event.
- Perceived risk, defined as the average public perception of the severity of a given risk.

The difference between direct and social risks is particularly clear in cases where the magnitude of the consequences is such that a given society has difficulties in handling them (for instance, a disruptive accident with many casualties affecting the functioning of the administrative headquarters of a given society). Perceived risks are of course as real as the technical ones when it comes to influencing political decisions regarding technology choices. This is independent of whether the perceived risks are based on real or false information.

A well-known case is the detrimental influence on the development and use of hydrogen in the transportation industry caused by the Hindenburg accident in 1936. The Zeppelin company chose to blame hydrogen for the accident, although they seemed to have known that the cause was the use of a highly flammable compound in waterproofing the blimp cloth, which was further electrically isolated from the airship frame, so that frictional spark ignition during weather patterns with thunderstorms could start an explosive fire. It has recently been concluded that the accident would have evolved identically had the airship been using helium rather than hydrogen (Bain and van Vorst, 1999).

Public perception has also in recent years ground the expansion of nuclear power to a halt. Also in this case, the defensive attitude of the industry involved probably played as decisive a role as the public criticism. The turning point here was the Chernobyl accident in the former Soviet Union, which nullified all previous industry statements of "the worst statistically significant accident being only of local consequences", by causing a global fallout of nuclear radiation-active substances and bans on particular isotope-accumulating food many thousands of kilometres from the accident site (Sørensen, 1979b, 1987). In this case, the public perception turned out closer to reality than the theoretical accident models used to calculate direct risk for similar installations (Rasmussen, 1975).

It is clear that public attitude to "small" and frequent accidents is different from the one towards catastrophic accidents. Killing thousands of people in a large number of individual traffic accidents every year has not lead to a ban on motorcars or to a halt on technology development, as did the 35 people killed in the Hindenburg accident. The logic may be difficult to appreciate, when one looks at airplane accidents which over the past 50 years have diminished from near $10^{-4}$ to $10^{-6}$ chance of perishing in an airplane accident per trip taken. Airplane accidents are also in many cases catastrophic, claiming several hundreds of victims at a time. Yet there has been no reaction in the direction of closing down air travel in general. It is sometimes claimed that this has been due to the already established acceptance of automobile transportation (and that this might not have been forthcoming, had the true accident tolls been known around 1900, when the automobile made its entry), but this does not explain the negative effect of the Hindenburg accident on hydrogen development. By 1936 the death tolls of automobile driving were well established.

Although some of these questions have to be left unanswered, they clearly have significant implications for a second attempt to introduce hydrogen in the energy system. Are there risk-related events that could halt the introduction of hydrogen and fuel cells in the transportation and the domestic sectors, even if the problems related to economy and technical performance have found a solution? This has considerable bearing on the issues, discussed briefly below, of common norms and standards to be applied worldwide, but it also has non-technical dimensions related to openness on the part of the industries and controlling authorities involved and on improving the understanding of the factors influencing not only direct and social but also perceived risk.

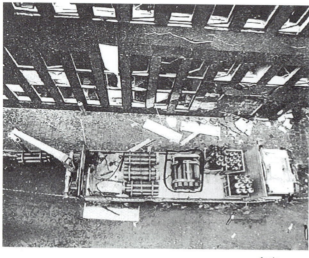

a
Photo from accident report No. 2 (1984) by Svenska Miljöverket; the photo is in public domain as confirmed by Stockholm County Police. Hydrogen canisters are encircled in black.

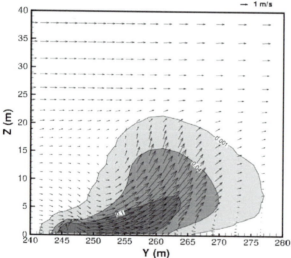

b
*Figure 5.3.* Hydrogen accident in Stockholm street 1983 (a, above), and analysis (b) of flow pattern and $H_2$ concentration (shaded areas) after 10 s in a vertical plane halfway between the canisters on the truck and the building wall. (From A. Venetsanos, T. Huld, P. Adams, J. Bartzis (2003). *J. Hazardous Mat.* **A105**, 1-25. Used by permission from Elsevier.)

Industrial use of hydrogen has experienced a few accidents, which subsequently have been useful in defining norms and procedures. One occurred in a narrow Stockholm street in 1983, where 13.5 kg of $H_2$ escaped from a set of 20-MPa pressure tanks with defect connections and exploded (Fig. 5.4a): 16 people were injured, and 10 cars and the adjacent building were heavily damaged. This accident has recently been modelled by computational fluid dynamics methods, giving the distribution of $H_2$ velocities and concentrations as shown in Fig. 5.4b (lower flammability limit is at a volume concentration ratio of 0.04 in air, delimiting the mid-grey area in the figure).

A hydrogen road accident occurred in Germany in 2001, with a truck crashing into a trailer loaded with hydrogen canisters. Escaping hydrogen ignited and caused a fire in the truck, killing the driver. Firefighters arrived quickly and were able to prevent further hydrogen release by spraying water onto the remaining canisters for several hours (Wurster, 2004).

## 5.2.2 Safety requirements

The approach to handling safety issues for hydrogen applications in the energy sector, affecting society broadly in a number of different ways, requires consistent standards in several areas from production to use and also requires a way of handling the unanticipated safety-related events bound to occur in a technology that is significantly different from that currently used.

Starting with production of hydrogen, this is a conventional industrial activity already in place, including safety prescriptions related to production by steam reforming or by alkaline electrolysis. The implications of raising the pressure of electrolyser to a proposed 12-20 MPa have spurred new safety investigations, focussed on hydrogen-oxygen explosive reactivity (Janssen *et al.*, 2004). For biological production of hydrogen, little safety-related work has been performed, as the would-be industrial processes are largely unknown.

The majority of hydrogen safety studies focus on the properties of hydrogen as such, in relation to burning, to explosions and to diffusion into unwanted locations. The basic properties of hydrogen were compared to those of other fuels in Table 2.3 in Chapter 2, section 2.3.3. Some of these numbers are still being refined by new measurements, e.g., the flammability safety limits (Chan *et al.*, 2004).

Confined and semi-confined (vented) hydrogen gas explosions are being studied in experimental settings (Carcassi *et al.*, 2004), and by simulation. Of particular concern is hydrogen transport trough tunnels, because of the large number of vehicles and people potentially involved (Breitung *et al.*, 2000). A recent study looks at leakage from a hydrogen tank in a vehicle parked in a closed garage, as well as on flame characterisation by punctuation of a hydrogen container (Hayashi and Watanabe, 2004). Figure 5.5 shows two examples of simulation results for the tunnel release and garage leakage. As expected, the hydrogen accumulates at the ceiling of the tunnel or room. Similar computational fluid dynamics calculations are performed for hydrogen installations in buildings (Tchouvelev *et al.*, 2004). Alternatively, scale experiments may be performed in wind tunnels, a method used to assess the hazards of spills at hydrogen fuelling stations (Chitose *et al.*, 2004).

As a consequence of several tunnel accidents in Europe (not related to hydrogen) and simulation results such as those mentioned above, the release of hydrogen and possible ignition in the confined space of road tunnels are seen as a key safety concern. Queues of cars may be trapped in tunnels unable to escape from the walls of fire travelling through the narrow confinements of typical tunnel structures. The simulations of hydrogen dispersion in a simulated hydrogen car accident within a tunnel shown in Fig. 5.5a takes departure in the liquefied hydrogen containers used by, e.g., BMW in their combustion engine hydrogen prototype cars (FZK, 1999). Hydrogen concentrations initially around 15 vol% are found to disperse by several car lengths in a few hundred seconds, before igniting.

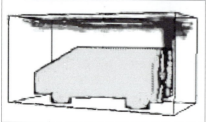

a: top, b: left

*Figure 5.5.* Simulated hydrogen diffusion in two accident situations. Top: tunnel accident releasing 7 kg of $H_2$ in 15 min; the distribution of hydrogen is shown at 900 s, just before ignition. The volume $H_2$ concentrations drop from 9 to 4% from the first to the third car to the right of the release point. Left: Closed space release in $6 \times 3 \times 2.8$ m$^3$ garage at a flow rate of 5.6 litre min$^{-1}$. (a) from Breitung *et al.* (2000). Numerical simulation and safety evaluation of tunnel accidents with a hydrogen powered vehicle. In Proc. 13$^{th}$ World Hydrogen Energy Conference, Beijing, pp. 1175-81, used by permission from the Int. Assoc. Hydrogen Energy; (b) based on data from Hayashi and Watanabe (2004). Hydrogen safety for fuel cell vehicles. In 15$^{th}$ World Hydrogen Energy Conf., Yokohama.

All components of hydrogen systems for vehicles (fuel container, handling equipment, fuel cell) must be assessed for safety under the conditions of both normal operation and accidents involving collisions between the car and other stationary or moving objects. It is well known, that road traffic gives rise to a very large number of accidents, including many that cost lives. Table 5.1 gives an idea of the fatality rates in present traffic systems.

The numbers in Table 5.1 are from Germany and imply a person fatality rate of $10^{-4}$ y$^{-1}$ or $4.8 \times 10^{-5}$ y$^{-1}$ per car registered in Germany (EEA, 2002). A similar distribution is found in the United States, except that the per capita fatality rate is 1.6 times higher than in Germany. Since most accidents are caused by driver errors, auto-pilot technology could presumably reduce the rate of accidents, but it is clear that introduction of new automobile technology such as hydrogen and fuel cells should not aggravate the outcome of a given accident, relative to the present car technology. It seems reasonable to aim for reducing the hydrogen-specific accident types involving leakage, fires or explosions to negligible levels, rather than to just make them rare compared to the "conventional" accidents, as these are already unacceptable to many observers.

| *German traffic fatalities 1999:* | *7842* |
|---|---|
| Solo accidents: | 40% |
| Collisions with pedestrians: | 13% |
| Collisions with other vehicles: | 47% |

*Table 5.1.* German fatalities in car accidents 1999 (out of a total number of road accidents with person injury amounting to 397 689) (with use of data seen in Daimler-Chrysler, 2001).

Specific safety analyses are made for hydrogen vehicles used in public transportation, such as city buses, as it is generally accepted that per passenger accident and fatality rates should be lower in public transport than for individual driving (Perrette *et al.*, 2003).

Also, the non-hydrogen components of the hydrogen fuel cycle should be assessed for risks, such as, for instance, the reformer and its fuels. There has been a debate on the health impacts of methanol, in addition to questions of $CO_2$ emissions, primarily related to the possibility of inhaling fumes of methanol during filling of fuel into a vehicle. Similar risks exist with gasoline or diesel fuels, which have led to the introduction of cuffs around the fuelling nozzles. The toxicology of methanol has recently been reassessed (Center for the Evaluation of Risks to Human Reproduction, 2004), which should offer a basis for a new assessment of the risks from using methanol as an intermediate fuel.

## 5.2.3 National and international standards

The introduction of a new technology is very much eased if the same norms and standards are applicable in all the markets of interest. In the past, differences in norms and standards have cost the consumers large sums of money (for example, in connection with video recorders and their tape media), and particularly in the automotive market, there are still no world-covering rules allowing an unmodified model to be sold everywhere.

In order to minimise the problem of introducing hydrogen vehicles, international norms and standards should be established in the early phase of finalising the main technical issues, with eventual refinement as the technology matures. The same is of course true for other uses of hydrogen and fuel cells, such as home units replacing a natural gas boiler.

The international scene for establishing common codes and standards is comprised of a number of *United Nation* work programmes (with regional sub-organisation, such as the *Economic Commission for Europe*) and the *International Standards Organisation* technical committees (Dey, 2004). Some individual countries have long traditions for such work, and many national standards working groups exist (e.g., as the basis for the *Deutsche Industrie Norm*). Often, smaller countries in a region lean towards these lead countries when establishing norms and standards. Finally, many industry sector organisations and professional organisations conduct work on norms and standards. In Europe, this type of work is performed for hydrogen by the *European Integrated Hydrogen Project*, globally by the US proposed *International Partnership for Hydrogen Economy* (Ohi and Rossmeissl, 2004), and by the *Society of Automotive Engineers*. In Japan, the *Japan Automobile Research Institute* performs norm work based on a number of specific legislative initiatives in the areas of collisions, container and fire safety, underground parking lots and road tunnels, etc. (Hayashi and Watanabe, 2004; Kikuzawa *et al.*, 2004).

As in other industrial areas, the establishment of a safety conscious culture is a very important goal. Efforts to make designs with high degrees of fail-safe operation are important, as are frequent reviews and formal hazard modelling and evaluations, in areas such as hydrogen leak detection, spontaneous ignition, and flame and shock wave progression, for a variety of different types of confinement and accident definitions.

# 5.3 Scenarios based on fossil energy

## 5.3.1 Scenario techniques and demand modelling

Scenarios are snapshot pictures of a future situation, with sufficient detail to allow them to be checked for consistency. Scenarios are not forecasts, but rather an exploration of various different visions considered attractive or for some other reason being discussed by a group of people in society. Well thought-out scenarios are important tools for decision-makers by offering the possibility of orderly change from one situation to a new one, an option that conventional economic theory is unable to offer, as it is based on trend forecasts departing from analysis of the past and application of economic rules also derived from past behaviour.

Visions of hydrogen as a fundamental energy carrier have been held for a long time (Bockris, 1972; Sørensen, 1975). Hydrogen implementation road-maps specifying plans for research, development, demonstration and com-mercialisation are available in abundance (e.g., USDoE, 2002a, 2004; Industry Canada, 2003; UKDTI, 2003), but where these explore the early steps to-wards establishing a hydrogen energy system, scenarios aim at a fairly com-prehensive description of the future system once implemented, in order to be able to test its viability and identify problem areas (that may have repercus-sions on the introductory phase actions).

Several scenarios for global energy demand and supply by the year 2050 are discussed below. For ease of comparison, they are based on the same as-sumptions regarding energy demand: concern for human welfare and envi-ronmental sustainability are assumed to lead to highly efficient energy con-sumption patterns based on increased concern for materials conservation and emphasis on non-material ("information society") types of activity.

The scenario year 2050 is selected sufficiently far into the future to allow the necessary changes in infrastructure and technology to be introduced in a smooth manner, without premature discarding of existing equipment before the end of its economic life. The use of a geographical information system (GIS) makes it possible to express all quantities on an "energy flow per unit of land area" basis, a methodology which offers new perspectives on energy supply and use, as compared with the conventional country-based statistics. This method is used for demand and for renewable energy supply, where the decentralised production makes an area-based assessment interesting. For fossil and nuclear energy with centralised production facilities, the area-based method would lead to point-like production maps, and here I use the conventional country averages.

Scenarios for a particular section of society, such as the energy sector fo-cussed upon here, require a general vision of the main characteristics of the global 2050 society, but one that need only be a broad-brush description of some general conditions. A detailed specification on the general shape of all aspects of society is clearly a formidable task, which is here approximated by the two-resolution approach. By necessity, the energy system is based on as-sumed developments for presently known technologies, but only as a means of proving the feasibility of the scenario: if better options emerge with time, as they almost certainly will, then they will replace some of the scenario technologies. However, the choice made at least constitutes a default and as such proves that there exists a possible system of the kind envisaged.

The actual development may comprise a combination of some reference scenarios selected for analysis, with each reference scenario being a clear and perhaps extreme example of pursuing a concrete line of political and tech-nological preference. It is important that the scenarios selected for political consideration be based on values and preferences judged as important in the society in question. The value basis should be made explicit in the scenario construction. Although all analysis of long-term policy alternatives is effectively scenario analysis, particular studies differ in the comprehensive-ness of the treatment of future society. For example, most of the current analysis makes normative scenario assumptions only for the aspects of society directly having a direct influence on the energy sector.

The economic viability of technology choices made for the scenarios can be assured only within considerable uncertainties, because the final cost of emerging technologies is based on necessarily uncertain and sometimes too optimistic estimates. However, for the end-use energy demand scenario, the conservative technology assumption made below is that the average 2050 technology is identical to the best technology available today, in terms of energy efficiency. It may in connection with assuming that such technologies are also economically viable be necessary to refer to inclusion of externality costs (i.e., indirect costs not currently included in market prices, as discussed in connection with life-cycle analysis in Chapter 6).

A global population scenario is important for determining energy demand. I use a recent United Nations population study estimating a year 2050 world population of $9.4 \times 10^9$ in its middle variant. Details of the population model are discussed in Sørensen (2004a), and the assumed population distribution is shown in Table 5.2, also offering migration patterns (mostly movements into cities) and a country list used as a basis for the figures given in the next subsections.

| COUNTRY | CONTINENT | REGION | POPULATION 1996 | POPULATION 2050 | Urban % 1992 | Urban % 2050 |
|---|---|---|---|---|---|---|
| Afghanistan | Asia | III | 20883 | 61373 | 18.92 | 55.97 |
| Albania | Europe | II | 3401 | 4747 | 36.34 | 72.65 |
| Algeria | Africa | VI | 28784 | 58991 | 53.34 | 89.65 |
| Andorra | Europe | II | | | | |
| Angola | Africa | VI | 11185 | 38897 | 29.86 | 75.80 |
| Anguilla | North America | IV | 8 | 13 | 0.00 | 0.00 |
| Antarctica | Antarctica | - | | | | |
| Antigua & Barbuda | North America | IV | 66 | 99 | 35.56 | 60.94 |
| Argentina | South America | IV | 35219 | 54522 | 87.14 | 90.00 |
| Armenia | Europe | II | 3638 | 4376 | 67.98 | 89.11 |
| Aruba (Netherlands) | North America | IV | 71 | 109 | 0.00 | 0.00 |
| Australia | Australia | II | 18057 | 25286 | 84.94 | 90.00 |
| Austria | Europe | II | 8106 | 7430 | 55.44 | 77.52 |
| Azerbaijan | Europe | II | 7594 | 10881 | 54.96 | 83.15 |
| Azores (Portuguese) | Europe | II | | | | |
| Bahamas | North America | I | 284 | 435 | 84.76 | 85.18 |
| Bahrain | Asia | III | 570 | 940 | 88.62 | 90.00 |
| Bangladesh | Asia | III | 120073 | 218188 | 16.74 | 57.62 |
| Barbados | North America | IV | 261 | 306 | 45.84 | 53.15 |
| Belarus | Europe | III | 10348 | 8726 | 68.56 | 90.00 |
| Belgium | Europe | II | 10159 | 9763 | 96.70 | 90.00 |

| COUNTRY | CONTINENT | REGION | POP. 1996 | POP. 2050 | Urban 92 | Urban 50 |
|---|---|---|---|---|---|---|
| Belize | North America | IV | 219 | 480 | 47.28 | 69.64 |
| Benin | Africa | VI | 5563 | 18095 | 29.92 | 68.73 |
| Bermuda | North America | IV | | | | |
| Bhutan | Asia | III | 1812 | 5184 | 6.00 | 28.85 |
| Bolivia | South America | IV | 7593 | 16966 | 57.80 | 90.00 |
| Bosnia & Herzegovina | Europe | II | 3628 | 3789 | 49.00 | 84.15 |
| Botswana | Africa | VI | 1484 | 3320 | 25.10 | 77.65 |
| Brazil | South America | IV | 161087 | 243259 | 72.04 | 90.00 |
| British Virgin Islands | North America | I | 19 | 37 | 0.00 | 0.00 |
| Brunei Darussalam | Asia | III | 300 | 512 | 57.74 | 79.29 |
| Bulgaria | Europe | II | 8468 | 6690 | 68.90 | 90.00 |
| Burkina Faso | Africa | VI | 10780 | 35419 | 21.62 | 90.00 |
| Burundi | Africa | VI | 6221 | 16937 | 6.78 | 31.77 |
| Cambodia | Asia | V | 10273 | 21394 | 18.84 | 63.06 |
| Cameroon | Africa | VI | 13560 | 41951 | 42.14 | 85.83 |
| Canada | North America | I | 29680 | 36352 | 76.64 | 89.58 |
| Cape Verde | Africa | VI | 396 | 864 | 48.24 | 68.91 |
| Cayman Islands | North America | IV | 32 | 67 | 100.00 | 100.00 |
| Central African Republic | Africa | VI | 3344 | 8215 | 39.00 | 74.15 |
| Chad | Africa | VI | 6515 | 18004 | 20.86 | 52.74 |
| Chile | South America | IV | 14421 | 22215 | 83.54 | 90.00 |
| China | Asia | V | 1232083 | 1516664 | 27.84 | 75.58 |
| Colombia | South America | IV | 36444 | 62284 | 71.08 | 90.00 |
| Comoros | Africa | VI | 632 | 1876 | 28.96 | 48.36 |
| Congo | Africa | VI | 2668 | 8729 | 55.62 | 90.00 |
| Cook Islands | Oceania | IV | 19 | 29 | 0.00 | 0.00 |
| Costa Rica | North America | IV | 3500 | 6902 | 48.14 | 84.80 |
| Croatia | Europe | III | 4501 | 3991 | 61.64 | 90.00 |
| Cuba | North America | IV | 11018 | 11284 | 74.56 | 90.00 |
| Cyprus | Europe | III | 756 | 1029 | 52.48 | 65.70 |
| Czech Republic | Europe | III | 10251 | 8572 | 65.10 | 84.26 |
| Denmark | Europe | II | 5237 | 5234 | 84.96 | 90.00 |
| Djibouti | Africa | VI | 617 | 1506 | 81.54 | 90.00 |
| Dominica | North America | IV | 71 | 97 | 0.00 | 0.00 |
| Dominican Republic | North America | IV | 7961 | 13141 | 62.08 | 90.00 |
| Ecuador | South America | IV | 11699 | 21190 | 56.24 | 90.00 |
| Egypt | Africa | VI | 63271 | 115480 | 44.26 | 75.44 |
| El Salvador | North America | IV | 5796 | 11364 | 44.38 | 75.35 |
| Equatorial Guinea | Africa | VI | 410 | 1144 | 38.30 | 90.00 |
| Eritrea | Africa | VI | 3280 | 8808 | 17.00 | 50.39 |
| Estonia | Europe | III | 1471 | 1084 | 72.32 | 90.00 |
| Ethiopia | Africa | VI | 58243 | 212732 | 12.74 | 43.08 |
| Falkland Islands | South America | IV | | | | |
| Fiji | Oceania | IV | 797 | 1393 | 39.86 | 75.26 |
| Finland | Europe | II | 5126 | 5172 | 62.12 | 86.52 |
| Macedonia (former Yugosl.) | Europe | III | 2174 | 2646 | 58.64 | 85.64 |
| France | Europe | II | 58333 | 58370 | 72.74 | 89.02 |
| French Guiana | South America | IV | 153 | 353 | 75.36 | 83.52 |
| French Polynesia | Oceania | IV | 223 | 403 | 56.40 | 80.30 |
| Gabon | Africa | VI | 1106 | 2952 | 47.42 | 87.11 |
| Gambia | Africa | VI | 1141 | 2604 | 23.76 | 68.12 |

| COUNTRY | CONTINENT | REGION | POP. 1996 | POP. 2050 | Urban 92 | Urban 50 |
|---|---|---|---|---|---|---|
| Georgia | Asia | III | 5442 | 6028 | 57.00 | 86.88 |
| Germany | Europe | II | 81922 | 69542 | 85.78 | 90.00 |
| Ghana | Africa | VI | 17832 | 51205 | 34.92 | 75.48 |
| Gibraltar | Europe | II | 28 | 28 | 100.00 | 100.00 |
| Greece | Europe | II | 10490 | 9013 | 63.64 | 90.00 |
| Greenland | Europe | II | 58 | 72 | 78.90 | 86.11 |
| Grenada | North America | IV | 92 | 134 | 0.00 | 0.00 |
| Guadeloupe | North America | IV | 431 | 634 | 98.86 | 90.00 |
| Guam | Oceania | IV | 153 | 250 | 38.08 | 59.03 |
| Guatemala | North America | IV | 10928 | 29353 | 40.24 | 78.48 |
| Guinea | Africa | VI | 7518 | 22914 | 27.32 | 72.45 |
| Guinea Bissau | Africa | VI | 1091 | 2674 | 20.82 | 63.32 |
| Guyana | South America | IV | 838 | 1239 | 34.64 | 77.45 |
| Haiti | North America | IV | 7259 | 17524 | 29.80 | 72.33 |
| Honduras | North America | IV | 5816 | 13920 | 41.98 | 80.68 |
| Hungary | Europe | III | 10049 | 7715 | 63.14 | 90.00 |
| Iceland | Europe | II | 271 | 363 | 91.00 | 90.00 |
| India | Asia | V | 944580 | 1532674 | 26.02 | 59.38 |
| Indonesia | Asia | IV | 200453 | 318264 | 32.52 | 82.58 |
| Iran | Asia | V | 69975 | 170269 | 57.38 | 88.35 |
| Iraq | Asia | V | 20607 | 56129 | 72.92 | 90.00 |
| Ireland | Europe | II | 3554 | 3809 | 57.14 | 81.50 |
| Israel | Asia | III | 5664 | 9144 | 90.42 | 90.00 |
| Italy | Europe | II | 57226 | 42092 | 66.66 | 83.08 |
| Ivory Coast | Africa | VI | 14015 | 31706 | 41.68 | 80.91 |
| Jamaica | North America | IV | 2491 | 3886 | 52.38 | 83.35 |
| Japan | Asia | II | 125351 | 109546 | 77.36 | 90.00 |
| Jordan | Asia | III | 5581 | 16671 | 69.40 | 90.00 |
| Kazakhstan | Asia | III | 16820 | 22260 | 58.44 | 87.55 |
| Kenya | Africa | VI | 27799 | 66054 | 25.24 | 70.52 |
| Kiribati | Oceania | IV | 80 | 165 | 35.04 | 61.33 |
| Korea Dem. People's Rep. | Asia | V | 22466 | 32873 | 60.40 | 86.06 |
| Korea | 8.6 | Asia | 45314 | 52146 | 76.80 | 90.00 |
| Kuwait | Asia | III | 1687 | 3406 | 96.34 | 90.00 |
| Kyrgyzstan | Asia | III | 4469 | 7182 | 38.48 | 71.03 |
| Laos | Asia | V | 5035 | 13889 | 22.00 | 62.42 |
| Latvia | Europe | III | 2504 | 1891 | 71.84 | 90.00 |
| Lebanon | Asia | III | 3084 | 5189 | 85.16 | 90.00 |
| Lesotho | Africa | VI | 2078 | 5643 | 20.88 | 66.79 |
| Liberia | Africa | VI | 2245 | 9955 | 43.26 | 81.47 |
| Libya Arab Jamahiriy | Africa | VI | 5593 | 19109 | 83.84 | 90.00 |
| Liechtenstein | Europe | II | | | | |
| Lithuania | Europe | III | 3728 | 3297 | 70.12 | 90.00 |
| Luxembourg | Europe | II | 412 | 461 | 87.42 | 90.00 |
| Madagascar | Africa | VI | 15353 | 50807 | 25.12 | 68.85 |
| Malawi | Africa | VI | 9845 | 29825 | 12.48 | 46.79 |
| Malaysia | Asia | IV | 20581 | 38089 | 51.36 | 89.39 |
| Maldives | Asia | V | | | | |
| Mali | Africa | VI | 11134 | 36817 | 25.08 | 68.88 |
| Malta | Europe | II | 369 | 442 | 88.28 | 90.00 |
| Marshall Islands | Oceania | IV | | | | |

| COUNTRY | CONTINENT | REGION | POP. 1996 | POP. 2050 | Urban 92 | Urban 50 |
|---|---|---|---|---|---|---|
| Martinique | North America | IV | 384 | 518 | 91.62 | 90.00 |
| Mauritania | Africa | VI | 2333 | 6077 | 49.60 | 90.00 |
| Mauritius | Africa | VI | 1129 | 1654 | 40.54 | 71.23 |
| Mexico | North America | IV | 92718 | 154120 | 73.68 | 90.00 |
| Micronesia | Oceania | IV | 126 | 342 | 27.04 | 49.82 |
| Moldova | Europe | III | 4444 | 5138 | 49.36 | 87.39 |
| Monaco | Europe | II | | | | |
| Mongolia | Asia | V | 2515 | 4986 | 59.16 | 88.76 |
| Morocco | Africa | VI | 27021 | 47276 | 47.02 | 80.38 |
| Mozambique | Africa | VI | 17796 | 51774 | 29.76 | 84.67 |
| Myanmar | Asia | V | 45922 | 80896 | 25.36 | 63.39 |
| Namibia | Africa | VI | 1575 | 4167 | 34.10 | 86.65 |
| Nauru | Oceania | IV | 11 | 25 | 0.00 | 0.00 |
| Nepal | Asia | V | 22021 | 53621 | 12.02 | 50.65 |
| Netherlands | Europe | II | 15575 | 14956 | 88.82 | 90.00 |
| New Caledonia | Oceania | IV | 184 | 295 | 60.78 | 76.98 |
| New Zealand | Australia | II | 3602 | 5271 | 85.32 | 90.00 |
| Nicaragua | North America | IV | 4238 | 9922 | 61.04 | 90.00 |
| Niger | Africa | VI | 9465 | 34576 | 15.92 | 51.21 |
| Nigeria | Africa | VI | 115020 | 338510 | 36.84 | 81.06 |
| Niue | Oceania | IV | 2 | 2 | 0.00 | 0.00 |
| Northern Mariana Islands | Oceania | IV | 49 | 92 | 0.00 | 0.00 |
| Norway | Europe | II | 4348 | 4694 | 72.66 | 89.08 |
| Oman | Asia | III | 2302 | 10930 | 11.88 | 49.00 |
| Pakistan | Asia | V | 139973 | 357353 | 33.08 | 75.12 |
| Palau Islands | Oceania | IV | 17 | 35 | 0.00 | 0.00 |
| Panama | North America | IV | 2677 | 4365 | 52.34 | 83.38 |
| Papua New Guinea | Asia | V | 4400 | 9637 | 15.40 | 44.58 |
| Paraguay | South America | IV | 4957 | 12565 | 50.42 | 88.35 |
| Peru | South America | IV | 23944 | 42292 | 70.76 | 90.00 |
| Philippines | Asia | IV | 69282 | 130511 | 50.96 | 90.00 |
| Poland | Europe | III | 38601 | 39725 | 63.38 | 89.08 |
| Portugal | Europe | II | 9808 | 8701 | 34.34 | 70.65 |
| Puerto Rico | North America | IV | 3736 | 5119 | 72.14 | 77.17 |
| Qatar | Asia | III | 558 | 861 | 90.50 | 90.00 |
| Reunion | Africa | VI | 664 | 1033 | 65.46 | 82.23 |
| Romania | Europe | III | 22655 | 19009 | 54.14 | 83.77 |
| Russian Federation | Asia | III | 148126 | 114318 | 74.80 | 90.00 |
| Rwanda | Africa | VI | 5397 | 16937 | 5.80 | 21.97 |
| Saint Lucia | North America | IV | 144 | 235 | 46.84 | 52.39 |
| San Marino | Europe | II | | | | |
| Sao Tome & Principe | Africa | VI | 135 | 294 | 0.00 | 0.00 |
| Saudi Arabia | Asia | III | 18836 | 59812 | 78.46 | 90.00 |
| Senegal | Africa | VI | 8532 | 23442 | 40.80 | 78.06 |
| Seychelles | Africa | VI | 74 | 106 | 51.68 | 75.09 |
| Sierra Leone | Africa | VI | 4297 | 11368 | 33.80 | 78.09 |
| Singapore | Asia | IV | 3384 | 4190 | 100.00 | 100.00 |
| Slovakia | Europe | II | 5347 | 5260 | 57.42 | 86.56 |
| Slovenia | Europe | III | 1924 | 1471 | 60.80 | 90.00 |
| Solomon Islands | Oceania | IV | 391 | 1192 | 15.60 | 40.91 |
| Somalia | Africa | VI | 9822 | 36408 | 24.80 | 62.06 |

| COUNTRY | CONTINENT | REGION | POP. 1996 | POP. 2050 | Urban 92 | Urban 50 |
|---|---|---|---|---|---|---|
| South Africa | Africa | VI | 42393 | 91466 | 49.84 | 83.52 |
| Spain | Europe | II | 39674 | 31755 | 75.84 | 90.00 |
| Sri Lanka | Asia | V | 18100 | 26995 | 21.80 | 59.06 |
| St.Kitts & Nevis | North America | IV | 41 | 56 | 40.72 | 57.03 |
| St.Vincent & Grenadine | North America | IV | 113 | 174 | 0.00 | 0.00 |
| Sudan | Africa | VI | 27291 | 59947 | 23.34 | 63.17 |
| Suriname | South America | IV | 432 | 711 | 48.66 | 86.17 |
| Swaziland | Africa | VI | 881 | 2228 | 28.32 | 78.73 |
| Sweden | Europe | II | 8819 | 9574 | 83.10 | 90.00 |
| Switzerland | Europe | II | 7224 | 6935 | 60.02 | 84.59 |
| Syrian Arab Rep. | Asia | III | 14574 | 34463 | 51.08 | 84.33 |
| Taiwan | Asia | IV | 2087 | 2583.706 | 0.00 | 0.00 |
| Tajikistan | Asia | III | 5935 | 12366 | 32.20 | 63.48 |
| Tanzania | Africa | VI | 30799 | 88963 | 22.24 | 67.52 |
| Thailand | Asia | IV | 58703 | 72969 | 19.22 | 53.98 |
| Togo | Africa | VI | 4201 | 12655 | 29.42 | 69.11 |
| Tonga | Oceania | IV | 98 | 128 | 37.50 | 59.47 |
| Trinidad & Tobago | South America | IV | 1297 | 1899 | 70.18 | 90.00 |
| Tunisia | Africa | VI | 9156 | 15907 | 55.82 | 87.77 |
| Turkey | Europe | III | 61797 | 97911 | 64.06 | 90.00 |
| Turkmenistan | Asia | III | 4155 | 7916 | 44.90 | 73.20 |
| Turks And Caicos Islands | North America | IV | 15 | 32 | 0.00 | 0.00 |
| Tuvalu | Oceania | IV | | | | |
| US Virgin Islands | North America | I | 106 | 158 | 0.00 | 0.00 |
| Uganda | Africa | VI | 20256 | 66305 | 11.72 | 42.09 |
| Ukraine | Europe | III | 51608 | 40802 | 68.62 | 90.00 |
| United Arab Emirates | Asia | III | 2260 | 3668 | 82.20 | 90.00 |
| United Kingdom | Europe | II | 58144 | 58733 | 89.26 | 90.00 |
| United States | North America | I | 269444 | 347543 | 75.60 | 90.00 |
| Uruguay | South America | IV | 3204 | 4027 | 89.46 | 90.00 |
| Uzbekistan | Asia | III | 23209 | 45094 | 40.88 | 72.73 |
| Vanuatu | Oceania | IV | 174 | 456 | 18.82 | 38.47 |
| Vatican City (Holy See) | Europe | II | | | | |
| Venezuela | South America | IV | 22311 | 42152 | 91.36 | 90.00 |
| Vietnam | Asia | V | 75181 | 129763 | 20.26 | 53.20 |
| Western Sahara | Africa | VI | 256 | 558 | 41.00 | 56.82 |
| Western Samoa | Oceania | IV | 166 | 319 | 57.86 | 79.20 |
| Yemen | Asia | III | 15678 | 61129 | 30.78 | 78.62 |
| Serbia & Montenegro | Europe | III | 10294 | 10979 | 54.46 | 88.80 |
| Zaire | Africa | VI | 46812 | 164635 | 28.50 | 66.29 |
| Zambia | Africa | VI | 8275 | 21965 | 42.04 | 73.61 |
| Zimbabwe | Africa | VI | 11439 | 24904 | 32.00 | 72.42 |

*Table 5.2.* Regional assignments of countries, with population distributions. The selection of countries/territories and their names is based on the definitions of borders used in the GIS software employed (MAPINFO, 1997); population data are from UN (1996) and urbanisation percentages from UN (1997).

With respect to economic activity and energy demand, the scenario expects the development until the mid-21st century to be dominated on one hand by efforts to "catch up" by many currently poor countries, depending on several factors including education policy, issues of global equity in trade conditions, and also issues of local conflicts and government corruption. On the other hand, a determinant will be the nature of the "new activities" being added to or replacing the current inventory of energy demanding undertakings. New, information-related activities often have smaller energy requirements than the activities they displace, leading to a de-coupling of economic growth and energy demand: industrialised country energy demands have in recent years risen much less than measures of economic activity such as gross national products. The expectation is that this trend will continue, and due to technical requirements, the energy intensity of, e.g., computer-related activities will continue to decrease, while the number of installations will increase, as will the activity level in general.

As an example, the overall activity growth in Western European countries was a factor of 5.6 during the 60-year period 1930-1990, from 2200 to 12 370 euro/cap. average GNP. The growth was uneven (depression, World War II, reconstruction period, unprecedented growth 1956-1971, stabilising from 1973), but over the entire 60 years perhaps symptomatic of the technology progress achieved during this quite exceptional period in world history. The likely European growth over the next 60 years will be lower, and high growth rates will be seen predominantly in certain Asian regions. The IPCC Second Assessment (IPCC, 1996) estimates in its high-growth scenario that the growth in Western Europe will reach a GNP of 45 300 1990-US$/cap. by 2050, as contrasted to 69 500 1990-US$/cap., if the growth factor equal that seen in the period 1930-1990. It would perhaps be a more realistic approach to estimate growth during the next 60 years as being at most the same in absolute terms as between 1930 and 1990.

A recent European study assumes the emergence of an information society with two thirds of the growth de-coupled from energy and use of materials (Nielsen and Sørensen, 1998), implying in simplistic terms that the growth factor for demand of energy services should be one third of the GNP growth. The relationship between energy and GNP is complex and depends both on attitudes and on technology developments. The ratio between energy and GNP growth 1930-1990 first declined from 1.5 to 1.0, then rose to 2.0 during the exceptional period and became negative after 1973. This is partly an effect of energy (and particularly oil) prices, but also technology requirements have played a role by improving energy efficiency after 1975 in ways often exceeding a purely cost-driven transition. The change in delivered energy service may have been even lower (e.g., the improvement in service between bicycle transport and automobile transport is not always as big as the increase in energy use would suggest).

Based on factors such as the ones outlined above, the energy demand by year 2050 has been estimated for different regions in the world, using the following methodology (Sørensen, 1996b, 2004a; Kuemmel *et al.*, 1997): current energy supplied to the end-user (taken from statistics, such as OECD, 1996) is analysed with regard to the efficiency of the final conversion step, translating energy into a useful product or service. The net energy demand is taken as that used by the most efficient equipment available at present (when the scenarios published in 1996 were constructed, they included some technology available only at the prototype stage, but by 2004 that technology has reached the commercial markets). The assumption that, as briefly mentioned in the beginning of this section, applies to all scenarios discussed, is thus that by 2050 the average of any equipment used will have an energy efficiency equal to the best found in the current marketplace. This would seem a fairly cautious assumption, as it neglects the certain invention of new methods and equipment working on principles different from those currently employed and possibly improving efficiency by much more than marginal factors.

The overall consequence of these assumptions will in the demand scenario employed be a global increase to year 2050 in end-use energy amounting to an average factor of 4.8, with much larger increases in the currently poor countries. The implied average 1994-2050 per capita growth factor for end-use energy is 2.7, and the consequences in terms of primary energy input will be different for the different supply scenarios, due to different conversion efficiencies of the intermediate conversions. The de-coupling implies that the GNP per capita growth to year 2050 will be substantially larger than the factor 2.7.

The basic methodology for estimating energy demand at the end-user is taken from Sørensen (2004a, Ch. 6). Energy demand futures are sometimes discussed in terms of changes relative to historic and current patterns (such as those shown in Fig. 5.6 together with the distribution on sources of energy). This is of course a suitable basis for assessing marginal changes, while for changes over a time horizon of 50 years, it is not likely to capture all important issues. The alternative approach used is to look at human needs, desires and goals and from this build up first the material demands required for satisfying these and then the energy required under certain technology assumptions. This is a bottom-up approach based on the view that certain human needs are basic needs, i.e., non-negotiable, while others are secondary needs that depend on cultural factors and stages of development and knowledge and could turn out differently for different societies, subgroups or individuals within a society. The basic needs include those of adequate food, shelter, security and human relations, and there is a continuous transition to more negotiable needs that incorporate material possessions, art, culture and human interactions and leisure. Energy demand is associated with satisfying several of these needs, with manufacturing/constructing the equipment and products entering into the fulfilment of the needs and with procuring the materials needed along the chain of activities and products.

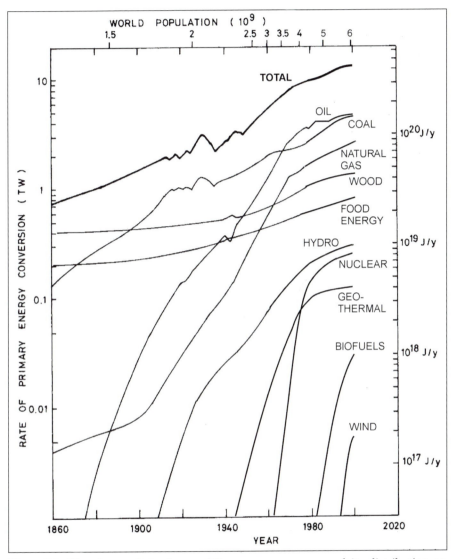

*Figure 5.6.* Historic trends in the world's gross energy use and its distribution on sources. During the early years, food energy includes an estimate of the draft animal power utilised. Wood includes straw and biofuels ethanol and biogas (updated from Jensen and Sørensen, 1984, with use of USDoE, 2003, and Sørensen, 2004a).

In normative models aimed at environmental sustainability, the natural approach to energy demand is to translate needs and goal satisfaction to energy requirements consistent already at demand level with environmental sustainability. For market-driven scenarios, basic needs and human goals play an equally important role, but secondary goals are more likely to be influenced by commercial interest rather than by personal motives. It is interesting that the basic needs approach is always taken in discussions of the development of societies with low economic activity, but rarely in discussions of highly industrialised countries.

The methodology thus first identifies needs and demands, commonly denoted human goals, and then discusses the energy required to satisfy them in a chain of steps backwards from the goal-satisfying activity or product to any required manufacture and then further back to materials and primary energy (this method is sometimes denoted "backcasting"). The evaluation is done on a per capita basis (involving averaging over differences within a population), but taking into account different geographical and social settings.

The assumption that it is meaningful to specify the energy expenditure at the end-use level without caring about the system responsible for delivering the energy is only approximately valid. In reality there may be couplings between the supply system and the final energy use, and the end-use energy demand therefore in some cases becomes dependent on the overall system choice. For example, a society rich in resources may take it upon itself to produce large quantities of resource-intensive products for export, while a society with less resources may instead focus on knowledge-based production, with both doing this in the interest of balancing an economy to provide satisfaction of the goals of their populations, but possibly with quite different implications for energy demand. The end-use energy demands will be distributed on the following energy qualities:

1. Cooling and refrigeration 0-50°C below ambient temperature.
2. Space heating and hot water 0-50°C above ambient temperature.
3. Process heat below 100°C.
4. Process heat in the range 100-500°C.
5. Process heat above 500°C.
6. Stationary mechanical energy.
7. Electrical energy (no simple substitution possible).
8. Energy for transportation (mobile mechanical energy).
9. Food energy.

The goal categories used to describe the basic and derived needs are:

A: Biologically acceptable surroundings
B: Food and water
C: Security
D: Health
E: Relations and leisure
F: Activities
       f1: Agriculture
       f2: Construction
       f3: Manufacturing industry
       f4: Raw materials and energy industry
       f5: Trade, service and distribution
       f6: Education
       f7: Commuting

Here categories A-E refer to direct goal satisfaction, f1-f4 refer to primary derived requirements for fulfilling the needs, and finally, f5-f7 refer to indirect requirements for carrying out the various manipulations stipulated. The individual energy requirement estimates are discussed in detail in Sørensen (2004a). A summary of the scenario for demand at the end-user is given in Fig. 5.7.

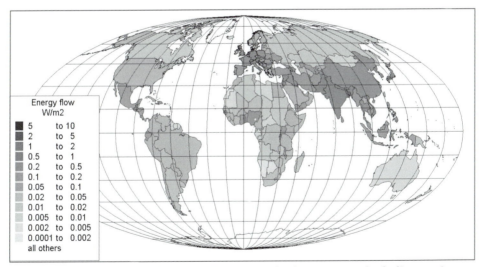

*Figure 5.7.* Total required energy delivery to final consumers (including environmental heat and food energy), for the 2050 scenarios and averaged over each country (unit: $W/m^2$; note that scale is logarithmic). (From Sørensen, 1999; the same data shown without country averaging appears in Sørensen, 2004a, Ch. 6).

## 5.3.2 Global clean fossil scenario

The interest in a future scenario based on fossil energy resources is to be able to maintain as much as possible of the current energy infrastructure, which is almost entirely based on the fossil option (Fig. 5.6). The reason for having to depart from the present practice is partly air pollution, which is becoming unacceptable, particularly for motorcar traffic in cities, and increasingly greenhouse gas emissions, the adverse effect of which on climate is becoming increasingly well established. Air pollution from power plants has been diminished by a number of technical devices, but for the small engines of motorcars with breathing height emissions, no effective means have been found to reduce the health impacts sufficiently (the reduction achieved by catalytic converters placed in the exhaust system is not nearly enough to avoid negative impacts). The greenhouse issue is seen as even more severe, placing demands of at least 60% reduction in carbon dioxide emissions for just stabilising the atmospheric content (IPCC, 1996).

There is also concern over the finite magnitude of fossil resources, which some decades ago, when energy demand increased exponentially, was predicted to cause supply problems within the next 50 years. The problem is aggravated by the move away from the most abundant fossil resource, coal, due to its higher $CO_2$ emissions per unit of energy, and because it seemed to have more air pollution impacts than, e.g., natural gas (although this depends very much on the technology used). However, the current scenarios for the mid-21st century assume that energy efficiency measures are pursued, notably on the demand side, which will stretch the availability of fossil resources considerably. How much will be discussed below. One may also remark that, particularly for natural gas, the current resource estimates may be on the low side. All together, it will be shown that with the scenario demand assumptions, the fossil resources could cover supply for at least a few hundred years, provided that techniques are introduced that allow coal to be used outside the power sector.

A number of technical options for allowing fossil fuels to be used without emission of $CO_2$ to the atmosphere are currently being explored. Research topics include primary conversion of fossil fuels before conversion or use, notably to hydrogen, for which combustion or fuel cell uses do not emit carbon dioxide, although some of the hydrogen conversion steps may. Another idea is to recover carbon dioxide from the flue gases of fossil fuel combustion and to sequester already emitted $CO_2$, e.g., by biological processes. The fossil scenario explores these technologies and makes a choice of the mix of options to use. It is clear that both carbon dioxide recovery and hydrogen use constitute a significant change in the way human societies will use fossil fuels. The shear mass of carbon dioxide to be handled is orders of magnitude above that of the pollutants $SO_2$ and $NO_x$ currently being processed. This also means that the carbon dioxide recovered before or after use of the fossil fuels will not be easy to dispose of. Options for safely depositing $CO_2$ are therefore an integral part of the technology discussion. All of the technologies considered will add to the cost of using fossil fuels. However, the extra cost should be compared with the estimated externality cost of the negative impacts of the present emission of pollutants and $CO_2$, and preliminary estimates suggest that the expenditure can in this way be defended.

This clean fossil energy system coming out of these considerations is not so similar to the present as one may have hoped. Road-based vehicles will have to use zero-emission technologies, pointing to hydrogen and/or electric vehicles. Only for air and ship transportation may oil products retain a role. Also for the power sector, conventional power plants may not emerge as the best solution, due to the limitations on removing $CO_2$ from flue gases (decreased power plant efficiency, incomplete removal), and therefore fuel cell conversion technologies are seen as providing a better solution also for stationary purposes. Some of the hydrogen produced by conversion of natural gas and coal can be used directly for process heat in industry. This all points to hydrogen as playing a major role in the clean fossil scenario. It also plays a significant role in the nuclear and the renewable energy scenarios constructed in the following sections, but not in the dominating fashion found to optimise the fossil scenario. One could therefore with good reason use the term *"hydrogen scenario"* as an alternative name for the clean fossil scenario.

### Clean fossil technologies

Removal of $CO_2$ after conventional combustion may be achieved by absorbing $CO_2$ from the flue gas stream (using reversible absorption into, e.g., ethanol amines), by membrane techniques or by cryogenic processes leading to the formation of solid $CO_2$. These techniques have the disadvantage of requiring substantial energy inputs, and the most accessible techniques (absorption) further only lead to a partial capture of $CO_2$ (Meisen and Shuai, 1997). However, with heat recycling there is hope to achieve about 90% recovery, which would be quite acceptable for greenhouse gas mitigation, and the energy requirements may be reduced to around 10% (of the power generated) for natural gas-fired units and 17% for coal-fired ones (Mimura *et al.*, 1997). In the scenario, it is further assumed than an average power plant conversion efficiency of 40% can be achieved for a modern combined power and heat-producing plant with removal of $CO_2$ from the flue gas.

An alternative "after combustion" type of $CO_2$ removal is to convert atmospheric $CO_2$ to methanol by a catalytic process at elevated temperature and pressure. Catalyst based on Cu and ZnO are used and laboratory demonstrations performed at a temperature of 150°C and a pressure of 5 MPa (Saito *et al.*, 1997). Additional reaction products are CO and water. Other options include carbon sequestering by enhanced biomass growth, where increasing forest areas can provide a long time interval between carbon assimilation and subsequent decay and release (Schlamadinger and Marland, 1996). Such options have not been incorporated into the present scenario.

The most promising option for avoiding $CO_2$ is to transform the fossil fuels to hydrogen and then use this fuel for subsequent conversions. Currently, hydrogen is produced from natural gas by steam reforming with water vapour, with a conversion efficiency around 70%, as discussed in Chapter 2, section 2.1. If the initial fossil fuel is coal, a gasification process with partial oxidation followed by the shift reaction would be used. Nitrogen from the air is used to blow oxygen through the gasifier, and impurities in the crude gas would be removed. With impurities removed, the hydrogen fuel is now of pipeline quality, ready to be transported to the points of use. Overall conversion efficiencies of about 60% are expected.

As the quantities of $CO_2$ to be disposed of following the processes above are huge, proposed storage in aquifers or abandoned wells may well be insufficient (available capacity is less than 100 Gt of coal equivalent; Haugen and Eide, 1996). This leaves ocean disposal of $CO_2$ as the only serious option. Storage here would be by dissolving liquefied $CO_2$ in sea water at depths of 1000-4000 m through special pipelines from land or from ships or by converting the $CO_2$ to dry ice form and simply dropping it from a ship into the ocean (Koide *et al.*, 1997; Fujioka *et al.*, 1997). The $CO_2$ is supposed to be subsequently dissolved into sea water, and if sites are suitably selected, it may stay in cavities or at the ocean floor indefinitely, due to its higher density.

The cost includes that of liquefaction or dry ice formation, plus operational costs and pipelines if used. Fujioka *et al.* (1997) estimate these costs to be about 0.03 US$ per kWh of fuel (0.08 $/kWh of electricity if that is what is produced) for the liquefied pipeline and ocean tanker disposal scheme, and 0.05 US$ per kWh of fuel for the dry ice scheme.

The $CO_2$-rich waters will stimulate biological growth and may seriously alter marine habitats (Takeuchi *et al.*, 1997; Herzog *et al.*, 1996). Stability of the deposits, and the subsequent fate of any escaped $CO_2$ will also have to be established, e.g. by empirical methods.

The clean fossil scenario selected includes both hydrogen produced from natural gas and from coal, with the efficiencies stated above. For use in fuel cells, the hydrogen to electricity conversion efficiency in 2050 is taken as 65%. Losses in hydrogen storage and transmission are taken as 10%, as compared with 5% for electricity transmission.

### Fossil resource considerations

Fossil resources are biomass that has undergone transformations over periods of millions of years. Their use as fuels is anyway limited to a fairly short interval in history and should be seen as a unique opportunity to smooth the road to a more sustainable energy system.

The discussion of fossil resources and their geographical distribution will be made on the basis of a simple version of the standard distinction between reserves and other resources. Three categories will be used:

- Proven reserves are deposits identified and considered economic to exploit with current price levels.
- Additional reserves are deposits that exist and may become economic, with a probability over 50%.
- New and unconventional resources are all other types of deposits, typically inferred from geological modelling or identified but not presently being considered technically or economically feasible to exploit.

The sum of all known and inferred (with reasonable probability) resources without consideration of economy of extraction is the *resource base*. The level of investigation is uneven among regions, and therefore additional amounts may be discovered, particularly in areas not well studied today. Also, extraction methods vary with time, and new techniques (e.g., enhanced oil recovery) may alter the amount of reserves assigned to a given physical resource.

Figure 5.8 gives the proven reserves of oil (including natural gas liquids), Fig. 5.9 gives the part of them located off-shore in the territorial waters of the countries indicated, and Fig. 5.10 estimates additional reserves. Figure 5.11 shows the total amounts in place (resource base). Figure 5.12 gives the proven resources of oil shale and natural bitumen, Fig. 5.13 gives the estimated additional reserves, and Fig. 5.14 gives the known amounts in place, for these possibly exploitable resources. The rate of new oil finds are currently lower than depletion by production, which has led to predictions of peak production occurring within the next few decades. It is by no means certain that usage will follow the underlying Hubbard bell-shaped curve, and production levels are influenced by many factors other than those related to resource depletion. Political choices could lead to high production levels continuing until a final rapid decline. The scenario is based on the consideration that new hydrogen schemes should not be based on the politically unstable outlooks for oil.

Figures 5.15 and 5.16 show the country distribution of proven reserves of bituminous coal ("hard coal") and other coal (sub-bituminous coal or lignite), and Figs. 5.17 and 5.18 give the additional reserves for the same two categories. Finally, Fig. 5.19 gives the total amount of coal estimated in place (i.e., the resource base).

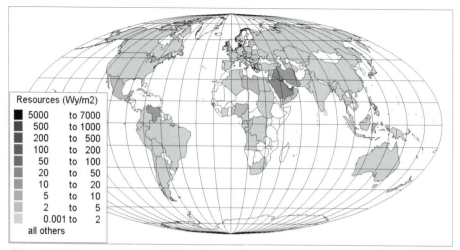

*Figure 5.8.* Proven reserves of oil and natural gas liquids (unit Wy/m², i.e., for each country, the average number of years for which an energy flow of 1 W per m² of land surface could be derived at 100% energy extraction efficiency). The resources are distributed over the country land areas, although many of the reserves and resources actually occur off-shore, cf. Fig. 5.9 (based upon data from World Energy Council, 1995; with area-based layout from Sørensen, 1999).

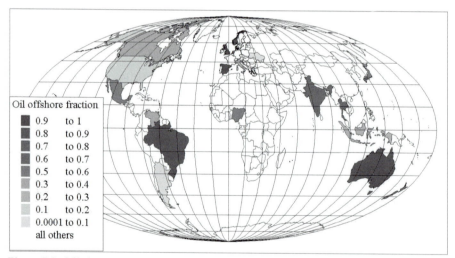

*Figure 5.9.* Off-shore fraction of the reserves given in Fig. 5.8 (based upon data from World Energy Council, 1995; with area-based layout from Sørensen, 1999).

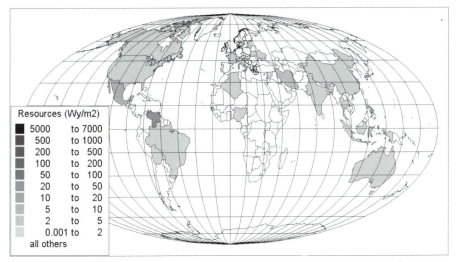

*Figure 5.10.* Additional reserves of oil and natural gas liquids (unit Wy/m², i.e., for each country, the average number of years for which an energy flow of 1 W per m² of land surface could be derived at 100% energy extraction efficiency) (based upon data from World Energy Council, 1995; with area-based layout from Sørensen, 1999).

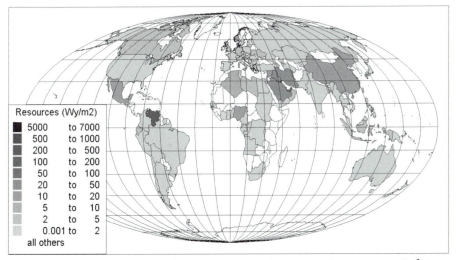

*Figure 5.11.* Total resources of oil and natural gas liquids in place (unit Wy/m², i.e., for each country, the average number of years for which an energy flow of 1 W per m² of land surface could be derived at 100% energy extraction efficiency) (based upon data from World Energy Council, 1995; with area-based layout from Sørensen, 1999).

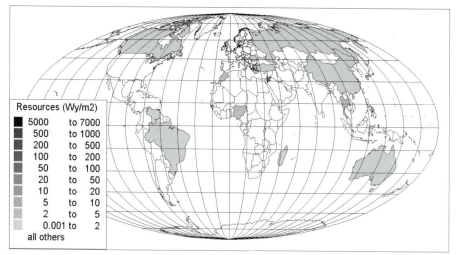

*Figure 5.12.* Proven reserves of oil shale and natural bitumen (unit Wy/m², i.e., for each country, the average number of years for which an energy flow of 1 W per m² of land surface could be derived at 100% energy extraction efficiency) (based upon data from World Energy Council, 1995; with area-based layout from Sørensen, 1999).

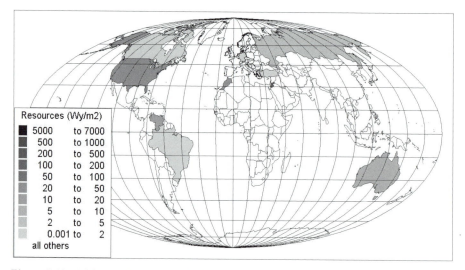

*Figure 5.13.* Additional reserves of oil shale and natural bitumen (unit Wy/m², i.e., for each country, the average number of years for which an energy flow of 1 W per m² of land surface could be derived at 100% energy extraction efficiency) (based upon data from World Energy Council, 1995; with area-based layout from Sørensen, 1999).

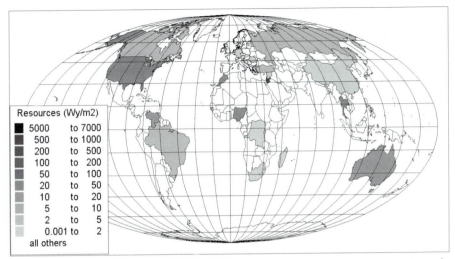

Resources (Wy/m2)

| | |
|---|---|
| 5000 | to 7000 |
| 500 | to 1000 |
| 200 | to 500 |
| 100 | to 200 |
| 50 | to 100 |
| 20 | to 50 |
| 10 | to 20 |
| 5 | to 10 |
| 2 | to 5 |
| 0.001 to | 2 |
| all others | |

*Figure 5.14.* Total resources of oil shale and natural bitumen in place (unit Wy/m², i.e., for each country, the average number of years for which an energy flow of 1 W per m² of land surface could be derived at 100% energy extraction efficiency) (based upon data from World Energy Council, 1995; with area-based layout from Sørensen, 1999).

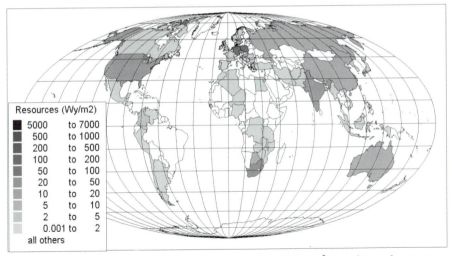

Resources (Wy/m2)

| | |
|---|---|
| 5000 | to 7000 |
| 500 | to 1000 |
| 200 | to 500 |
| 100 | to 200 |
| 50 | to 100 |
| 20 | to 50 |
| 10 | to 20 |
| 5 | to 10 |
| 2 | to 5 |
| 0.001 to | 2 |
| all others | |

*Figure 5.15.* Proven reserves of bituminous coal (unit Wy/m², i.e., for each country, the average number of years for which an energy flow of 1 W per m² of land surface could be derived at 100% energy extraction efficiency) (based upon data from World Energy Council, 1995; with area-based layout from Sørensen, 1999).

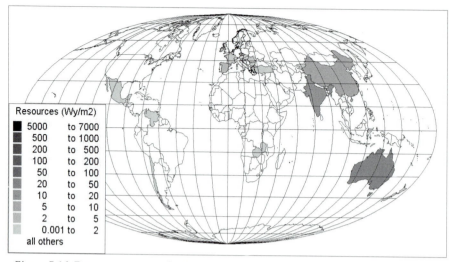

*Figure 5.16.* Proven reserves of sub-bituminous coal and lignite (unit Wy/m², cf. caption to Fig. 5.15) (based upon data from World Energy Council, 1995; with area-based layout from Sørensen, 1999).

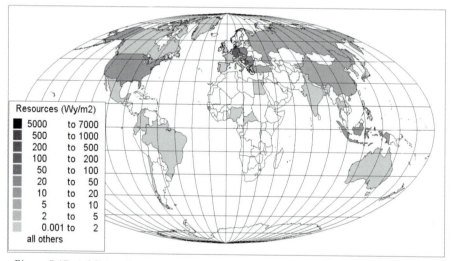

*Figure 5.17.* Additional reserves of bituminous coal (unit Wy/m², cf. caption to Fig. 5.15) (based upon data from World Energy Council, 1995; with area-based layout from Sørensen, 1999).

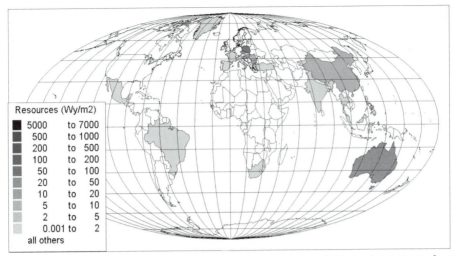

*Figure 5.18.* Additional reserves of sub-bituminous coal and lignite (unit Wy/m², cf. caption to Fig. 5.15) (based upon data from World Energy Council, 1995; with area-based layout from Sørensen, 1999).

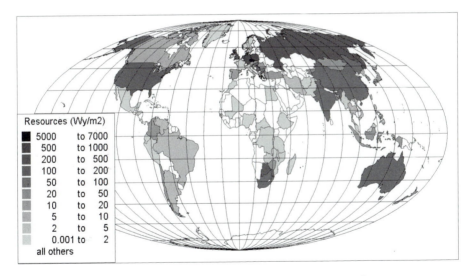

*Figure 5.19.* Total coal and lignite resources in place (unit Wy/m², cf. caption to Fig. 5.15) (based upon data from World Energy Council, 1995; with area-based layout from Sørensen, 1999).

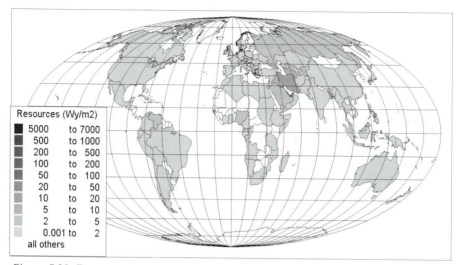

*Figure 5.20.* Proven reserves of natural gas (unit Wy/m², cf. caption to Fig. 5.15) (based upon data from World Energy Council, 1995; with area-based layout from Sørensen, 1999).

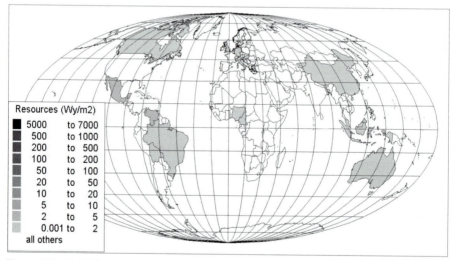

*Figure 5.21.* Additional reserves of natural gas (unit Wy/m², cf. caption to Fig. 5.15) (based upon data from World Energy Council, 1995; with area-based layout from Sørensen, 1999).

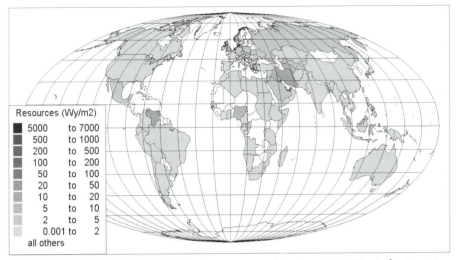

*Figure 5.22.* Total resources of natural gas estimated in place (unit Wy/m², cf. caption to Fig. 5.15) (based upon data from World Energy Council, 1995; with area-based layout from Sørensen, 1999).

Figure 5.20 shows the proven reserves of natural gas, Fig. 5.21 shows the additional reserves, and Fig. 5.22 shows the total amounts in place.

It is clear that estimates of reserves and resources are uncertain, and more uncertain, the less exploration has been made. Assessments found in the literature therefore differ among themselves, and the reporting of national agencies to international compilations may be of varying quality, for reasons of incomplete knowledge or for commercial reasons. Figures 5.8-5.22 are based on estimates collected by the World Energy Council (WEC, 1995). They agree fairly well with other estimates for coal and oil, but are substantially lower for natural gas than what as a global total has been found most likely in the assessment made for IPCC by Nakicenovic *et al.* (1996). For example, the oil-producing countries in the Middle East quote very small associated gas reserves, which may just indicate that such resources are not considered economic due to current limitations on transmission and liquefaction capacity that could transport them to regions of demand. From geological considerations, it is expected that there is more natural gas than oil (according to some estimates much more), and the assessment in the WEC of only half as much is almost certainly an underestimate.

The rate of using coal during the 1990s was 2886 GW, for oil it was 4059 GW, and for natural gas it was 2188. This means that the proven reserves shown in Figs. 5.8, 5.15 and 5.20 at a constant rate of use equal to the present one would last 167 years for coal, 46 years for oil and 70 years for natural gas. Relative to the total estimated resource in place, the numbers would be 3554 years for coal, 210 years for oil plus another 184 years from oil shale and bitumen, and 96 years for natural gas. The total natural gas resources may well be underestimated. In Chapter 7, an update is made and a closer look taken at the depletion of oil.

## The fossil scenario

In constructing a clean fossil scenario for satisfying the demand scenario described in section 5.3.1, the basic consideration has been to make use of the technologies described above in such a way that the resource utilisation becomes acceptable. Because oil resources are considered the most limiting constraint, and also because oil is in widespread use outside the energy sector as lubricants and feedstock for a number of industrial products including first of all plastics, it is decided not to use oil in the energy sector, except possibly for a few sectors where substitution with other fossil technologies may be difficult, notably as an aviation fuel. Non-energy uses of oil could alternatively be substituted with, e.g., biomass-based raw materials, but the way the current set of scenarios is constructed this would rather belong in the renewable energy scenario. The past trends shown in Fig. 5.6 also suggests that oil usage is levelling out and thus lend support to those who estimate declining oil production starting about a decade from now, even without considerations of political constraints, and thereafter declining production.

In regard to the balance between natural gas and coal, it is for resource reasons considered appropriate to increase the relative use of coal. Both fossil resources can be transformed to hydrogen (as could oil), for providing a carbon-free energy carrier, but the efficiency of transforming coal to hydrogen is a little lower than that of transforming natural gas, and the cost is higher. However, the price of natural gas is already above that of coal and would develop in the direction of an increased difference, as the resource constraint makes itself more felt, with the already manifest increase in use of natural gas in the present type of energy system. The increase in cost of natural gas and particularly of oil will almost certainly happen before the selected scenario year, 2050.

The scenario choice is then in the transportation sector to cover half the demand by electric vehicles (urban vehicles, trains) and the other half by hydrogen (used in fuel cells directly or with methanol from natural gas or coal as a precursor). The fuel cell vehicles are of course also electric vehicles, but with the power production taking place on board. Because de-carbonisation favours a large fraction of the fossil fuels being transformed to hydrogen, hydrogen is in the clean fossil scenario also used directly in industry for medium- and high-temperature process heat, thereby avoiding a second conversion step to electricity and the associated conversion losses.

There is in 2050 assumed a substantial demand for electricity proper, i.e., electricity for, e.g., electric apparel that cannot be powered by other sources, plus stationary mechanical energy, which for environmental reasons and efficiency optimisation is better served by electricity as input. The production of electricity in the scenario is partly performed at conventional power plants, but also by fuel cell power plants (presumably SOFCs) that can use hydrogen. Again, this is because it is easier to transform the fossil fuels to hydrogen before conversion than to recover carbon dioxide from stack emissions. In the scenario, about two thirds of the power generation is from fuel cells. The higher efficiency is partly offset by losses in hydrogen production and storage, but the versatility is higher, because the fuel cells may become decentralised and integrated into individual buildings. This is made possible by a hydrogen distribution pipeline system, which in many countries could serve as a useful continuation of using the existing natural gas distribution network. Transmission (regionally) and distribution (locally) of hydrogen and natural gas will both be needed in this scenario.

For low temperature heat, the conversion of fuels in both power plants and fuel cells provides considerable amounts of associated heat production, which can be employed after distribution by district heating lines, and in the case of building-integrated fuel cells may serve to provide heat to the same building and its activities. However, because of the considerable direct use of hydrogen and the fairly high electric power conversion efficiencies achieved, in the scenario there is not quite enough "waste" heat to cover all low-temperature requirements, and the remaining space and process heat (under 20% of the total, varying with season) are assumed to be provided by heat pumps. Also space cooling is provided by electricity-driven coolers, i.e., the same technique as used in heat pumps and in the scenario lumped together with these, assuming the same coefficient of performance (ratio of heat output, added or removed, to electricity input), COP = 3.33. This is not the highest technically possible value, but because much of the space heating is in climates where the source of environmental heat to the heat pumps is of low temperature, it is considered reasonable to use a cautious estimate.

Although the scenario aims at only using fossil resources, the existing hydro electricity production is retained. This is a renewable energy resource, but it works well together with the ingredients of the fossil system, and only existing plants and plants currently under construction are assumed to be present in the fossil scenario. The distribution of power production from hydro is shown in Figs. 5.23 and 5.24. The total potential contribution is an average production of 440 GW. Due to current debate over large hydro installations, a few words will be interjected on this subject.

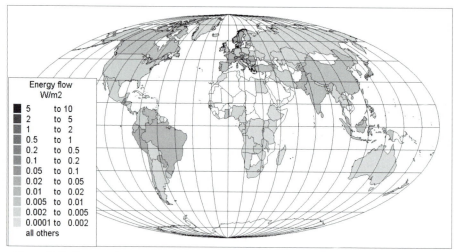

*Figure 5.23.* Potential delivery of hydropower to final consumers in the 2050 scenario (based upon data from World Energy Council, 1995; with area-based layout from Sørensen, 1999).

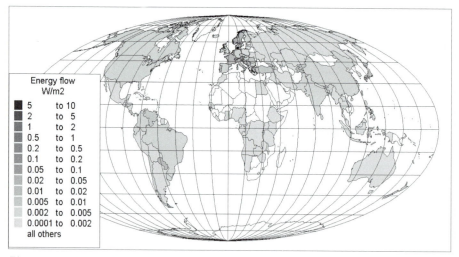

*Figure 5.24.* Annual average production of hydropower at existing hydropower plants plus estimated production from plants under construction (based upon data from World Energy Council, 1995; with area-based layout from Sørensen, 1999).

Hydropower has in the past often been used in a way not compatible with environmental considerations (flooding areas of high landscape value, dislocating people, etc.). It is here considered as a renewable energy source worthy of inclusion in any sustainable scenario, provided that it is used with proper considerations. These may include modular construction of large plants (cascading the water), allowing a highly reduced disturbed area, and planned location of reservoirs in locations where there is little conflict with preservation concerns, even if the water transport layout may become more complex and expensive.

The 2050 scenarios include all existing plants and plants under construction. It would hardly be feasible to attempt to close those existing plants that have caused human and environmental suffering. At least some studies suggest that reverting to pre-hydro landscapes will entail negative environmental effects quite similar to the original change in the opposite direction, due to the already completed adaptation of flora and fauna to the changed conditions (Tasmanian Hydro Commission, private communication, 1998).

Also included in the scenarios are small-scale hydro installations planned in several countries, considering the impacts to be acceptable for most schemes. As regards total potentials identified and actual schemes proposed but not started, a modest inclusion is made of only those for which actual proposals including environmental statements exist. As a source for such appraisals of the potential hydro generation we have used a survey made by the World Energy Council (1995). Including all their categories would roughly double the current hydropower generation. Figure 5.23 show, the distribution of this potential power generation on countries. A large fraction of such installations is or will be reservoir based, so no additional storage cycle is considered necessary (although pumped hydro would be an obvious option). Figure 5.24 shows the average generation from existing hydro plants plus those under construction, again averaged over country areas. These are the ones included in the fossil as well as in the other scenarios discussed in the following sections.

It is not attempted to show the actual reservoir areas on the maps of Fig. 5.23 or 5.24. The figures simply spread the power production evenly over each country. In any case, most hydro is a centralised resource in the sense that it is not (and largely cannot be) used by the local population. South America is the most endowed region, implying that the utilisation choices made in this region will be decisive for the global contribution of hydropower.

The clean fossil scenario will now be presented in terms of annual average energy flows between supply and demand. The handling of diurnal and seasonal variations is much the same as it is today. The only critical area is in the provision of space heating and cooling, where the co-produced heat from power plants and fuel cells on average is very close to the average demand.

For space cooling, the high season just requires more electricity input, which is provided by a sufficient installed capacity of the power plants and fuel cells. For winter heating, the co-producing power and heat plants can already with technology currently used in many places avoid a locked ratio between electricity and heat production, and by regulation of the heat to power ratio within the available range cater to any heat demand situation, without need for adding storage to the system, except possibly for diurnal storage, which is already today economic in some cases.

Figure 5.25 shows country distributions of the amount of hydrogen that would be needed to provide energy for the demand scenario medium- and high-temperature industrial process heat plus 50% of the transportation needs, making allowance for 10% losses in storage and piping of hydrogen to the end-user. Figure 5.26 gives the electricity from fossil fuels required for stationary mechanical energy, electric appliances, cooling, refrigeration and the heat pumps covering 20% of space heating needs, plus the other half of the transportation energy needs. The power provided by hydro is given first priority, leading, e.g., to no average fossil power requirement in hydro-abundant countries such as Canada, Sweden or Norway. In practice, the fast regulation of hydro plants allows them to be used in load-levelling efforts.

In Fig. 5.27, the amount of hydrogen that is needed as fuel for electricity-producing stationary fuel cells is shown, assuming a conversion efficiency of 65% (for electricity, the rest being available as low-temperature heat). Figure 5.28 similarly gives the coal input to power plants, at an electric efficiency of 40%, again with the remaining 60% being available for district heating. One third of the electricity in the scenario is assumed to come from coal-fired power plants and two thirds are from fuel cells (probably SOFCs) with hydrogen as input. In Figs. 5.29 and 5.30, the natural gas and coal inputs to produce the required amounts of hydrogen are shown. The scenario assumes 40% to be based on natural gas, with a conversion efficiency of 70%, and 60% to be from coal with a conversion efficiency of 60%.

Figure 5.31 shows the energy for space heating (20%) and cooling (100%) covered by heat pumps and coolers, while Fig. 5.32 gives the rest of the low-temperature requirements for space heating and process heat covered by district heating based on power plant and fuel cell waste heat.

Figure 5.33 gives an overview of the scenario energy system, with annual energy flows indicated. The coverage of heat with assistance from co-produced heat from power plants and fuel cells is indicated.

With the production of natural gas and coal now known, the year 2050 scenario assumes that the annual requirement is produced by each country in proportion to its total estimated resources of natural gas and coal, respectively. This creates a need for trade of these commodities, much in the same way as today.

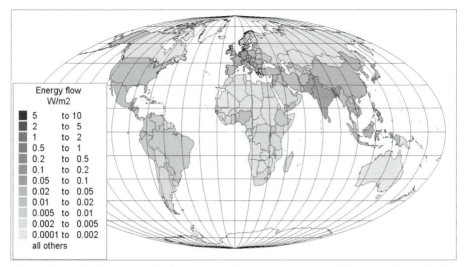

*Figure 5.25.* Hydrogen derived from fossil energy, to be used for covering half the needs in the transportation sector (using PEM fuel cells) and all requirements for medium- and high-temperature process heat (using hydrogen furnaces), after transmission and storage losses. The figure shows average flows in W per m$^2$ of land area for each country (from Sørensen, 1999).

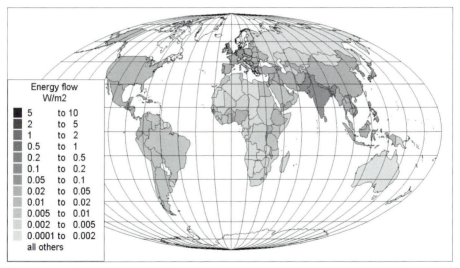

*Figure 5.26.* Electricity derived from fossil energy (by conventional power plants), to be used for covering half the needs in the transportation sector and all requirements for stationary mechanical energy, dedicated electricity and electricity for cooling, refrigeration and heat pumps, after transmission losses. The figure shows average flows in W per m$^2$ of land area for each country (Sørensen, 1999).

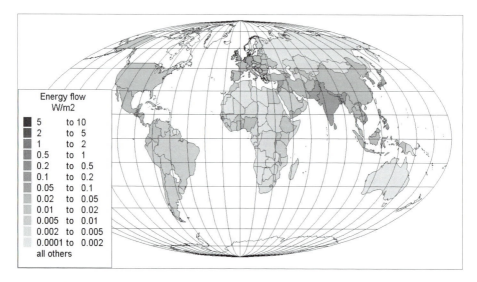

*Figure 5.27.* Hydrogen required as input to stationary fuel cells. The figure shows average flows in W per m² of land area for each country (Sørensen, 1999).

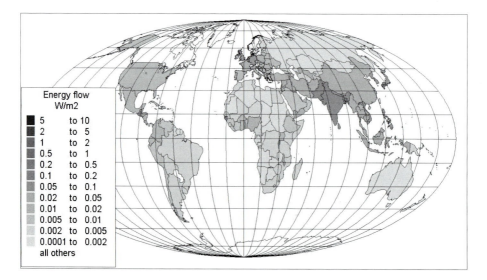

*Figure 5.28.* Coal required as fuel for conventional power plants (co-producing heat and power). The figure shows average flows in W per m² of land area for each country (Sørensen, 1999).

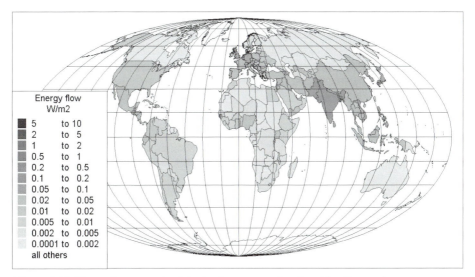

*Figure 5.29.* Natural gas required for hydrogen production. The figure shows average flows in W per m$^2$ of land area for each country (Sørensen, 1999).

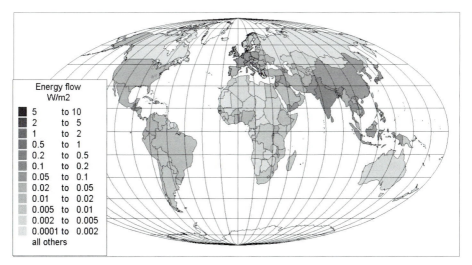

*Figure 5.30.* Coal required for hydrogen production. The figure shows average flows in W per m$^2$ of land area for each country (Sørensen, 1999).

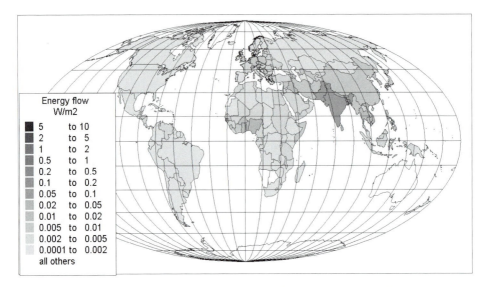

*Figure 5.31.* Space heating and cooling provided by heat pumps and electric coolers. The figure shows average flows in W per m$^2$ of land area for each country (Sørensen, 1999).

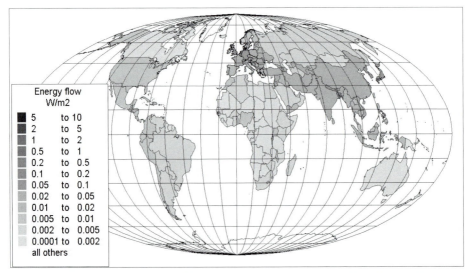

*Figure 5.32.* Space heating and low-temperature heat provided by district heating with heat input from power plants and from associated heat from fuel cell plants located near the heat load. The figure shows average flows in W per m$^2$ of land area for each country (Sørensen, 1999).

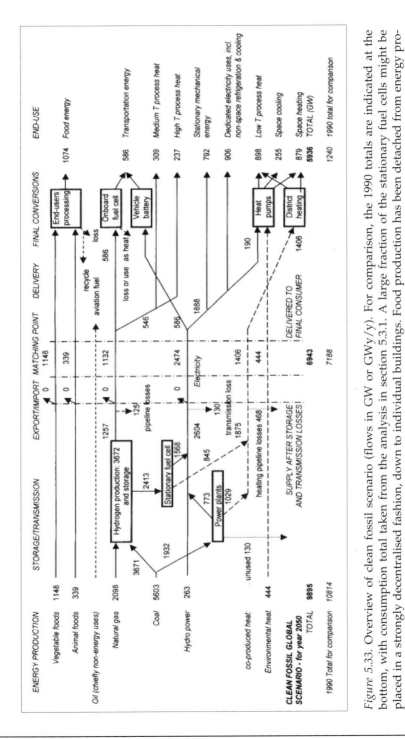

*Figure* 5.33. Overview of clean fossil scenario (flows in GW or GWy/y). For comparison, the 1990 totals are indicated at the bottom, with consumption total taken from the analysis in section 5.3.1. A large fraction of the stationary fuel cells might be placed in a strongly decentralised fashion, down to individual buildings. Food production has been detached from energy production, see discussion in Sørensen (2004a). (From Sørensen, 1999.)

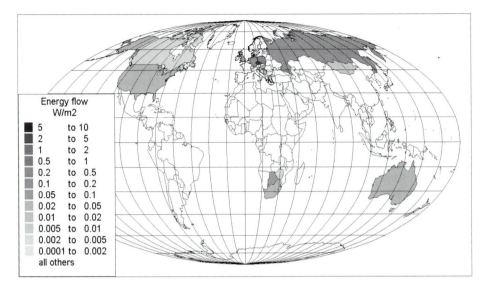

*Figure 5.34.* Coal surplus (production minus required supply for each country, whenever this quantity is positive). The figure shows average flows in W per m$^2$ of land area for each country (Sørensen, 1999).

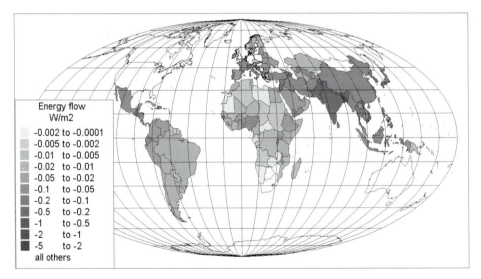

*Figure 5.35.* Coal deficit (production minus required supply for each country, if negative). The figure shows average flows in W per m$^2$ of land area for each country (Sørensen, 1999).

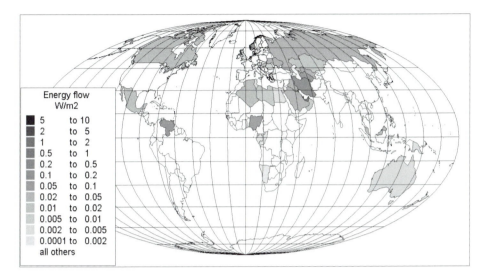

*Figure 5.36.* Natural gas surplus (production minus required supply for each country, if positive). The figure shows average flows in W per m² of land area for each country (Sørensen, 1999).

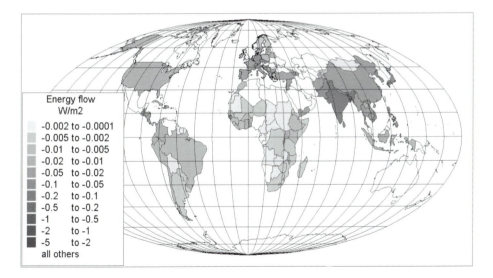

*Figure 5.37.* Natural gas deficit (production minus required supply for each country, if negative). The figure shows average flows in W per m² of land area for each country (Sørensen, 1999).

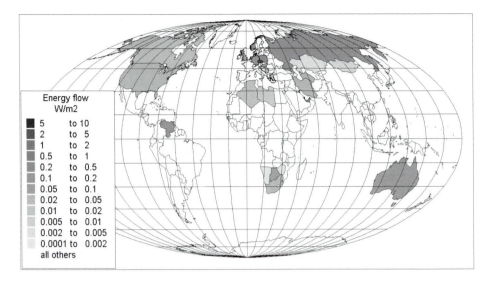

*Figure 5.38.* Total fossil fuel (coal and natural gas) surplus, i.e., production minus required supply for each country, if positive. The figure shows average flows in W per m² of land area for each country (Sørensen, 1999).

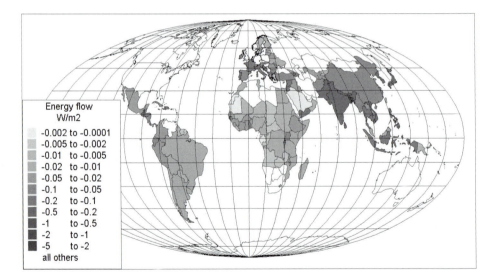

*Figure 5.39.* Total fossil fuel (coal and natural gas) deficit, i.e., required supply minus production for each country, if positive. The figure shows average flows in W per m² of land area for each country (Sørensen, 1999).

The fixed percentages of natural gas and coal inputs would in practice be modified for countries with resources primarily of one type. For example, Norway and Saudi Arabia would increase their use of natural gas in order not to have to import coal. For countries able to cover their own electricity requirements with existing hydro, e.g., Canada and Norway, the scenario construction does not initially take into account export options, and the hydro input in Fig. 5.33 is therefore less than the potential production (by 177 GW). This additional capacity will in reality be used for export of electric power to neighbouring countries and thus will slightly diminish the use of fossil fuels, on a global level.

Figures 5.34 and 5.35 give the surplus and deficit of coal production on a national basis. Figures 5.36 and 5.37 give the surplus and deficit of natural gas production. Figures 5.38 and 5.39 give the totals. It is seen that trade follows the patterns of the present situation, as one would expect. Although the fossil reserves are more evenly distributed over the globe than thought some years ago (cf. Figs. 5.8, 5.15 and 5.20), the matching with demand is most negative for countries in South America, Africa and Southeast Asia, due to the low level of identified resources and in Southeast Asia the high population density, and the assumed high standard of living in the scenario year 2050.

### Evaluation of the clean fossil scenario

Any fossil scenario has to be regarded as a parenthesis in the grand-scale history of mankind (Sørensen, 1979, 2004a). The 2050 scenario described above allows the expected future demand to be covered by a rate of fossil fuel input no greater than the current one (see Fig. 5.33). Even then, the proven reserves of natural gas and coal will sustain the scenario primary energy use only for 73 and 86 years, respectively (the similarity of these numbers was aimed at in selecting the use of gas and coal in the scenario). Taken in relation to the total resource base (whether exploitable or not), the scenario energy use can be sustained for 101 and 1830 years, for gas and coal, respectively, with the natural gas value possibly being underestimated, as discussed above.

The carbon removal and deposition technologies described in section 5.3.1 have in most cases not been demonstrated in full scale. There may be technical problems and likely new environmental problems not foreseen, leading at least to higher prices if they can at all be fixed. Particular concern is in place regarding the ocean disposal of carbon dioxide, which could produce environmental problems as large as those one is trying to avoid, if the assumed stability turns out to not be secured. The cost estimates quoted in section 5.3.1 suggest that the ocean disposal will increase the cost of energy derived from fossil fuels by a factor of 2-3. To this one must add the cost of the new technologies for producing hydrogen, whereas the current high cost of fuel cells, whether stationary or mobile, has to be assumed to decline to near the general level of conventional power plants, per unit of energy produced. All together, the cost of energy produced in the clean fossil scenario will be larger than the present by amounts having to be justified by

- The avoided greenhouse warming damage; this may be defended by referring to the higher ones among the externality estimates found in the literature, but not by the lower ones (Kuemmel *et al.*, 1997; IPCC, 1996).
- The avoided uncertainty of oil availability, for reasons of political conflicts or of declining extraction due to resource depletion.

Apart from the question mark on the environmental implications of ocean $CO_2$ disposal, the fossil scenario seems to be technically feasible. However, the original justification for continuing use of fossil fuels was to be able to use existing infrastructure. This is not really the outcome of the scenario analysis: the transportation sector will be very different from the current one (key technologies hydrogen, various types of fuel cells, batteries and electric motors), although a development which may well be needed in any case due to urban pollution considerations. District heating, which is presently common only in a few countries, will have to be expanded, and an increase must take place in the use of heat pumps for heating. The industry sector will not feel much change in going from, e.g., natural gas to hydrogen, and electricity use will generally expand.

Estimating the likelihood of the assumed technical advances of technologies for fossil gasification to hydrogen fuel and subsequent use in fuel cells, there is a general optimism, but also a realisation of the substantial further development needed in order to become economically attractive. Particularly for the fuel cell developments, the improvement goes far beyond what can be swept under the heading of "including externalities" in the cost appraisal. For the disposal technologies, and particularly the ocean disposal required for a large-scale handling of carbon dioxide, there is little doubt that the $CO_2$ can be dumped into the ocean (as ice or using suitable pipes) at an affordable cost. However, it needs to be proven by experiment that this simple procedure does in fact constitute a safe form of disposal, leaving the carbon-containing substance away from the biosphere for sufficiently long periods of time.

The known reserves of fossil fuels are too small to warrant the difficulties of a major transition from today's fossil technologies to the new "clean" ones, and only the belief that a high proportion of the general resources known to exist can indeed in the future be converted to "reserves" and exploited in practical, economic and environmentally acceptable operations will make the fossil scenario interesting.

The scenario technologies are selected from a much larger catalogue of options as being most likely to reach acceptable costs. Nobody envisages this cost to be as low as the present cost of energy, and that would also not be a proper starting point for comparison, as the current supply system has to be changed over the period considered, for reasons of the environment if not for reasons of resource depletion. Without subsidy, e.g., in the form of environmental cleanup (or accepting damage), the cost of presently delivered energy would be much higher.

The only exception from the expectation of rising energy costs is the impact derived from energy efficiency measures assumed taken in the demand scenario: they all make sense even at current energy cost. For the supply and conversion technologies, the best that can be hoped for is that the cost can be kept within reasonable limits in a frame of reference defined by adding externality costs to all energy-related activities.

One might then suggest to turn the problem around and ask that a "standard price" or "target price" be determined for each new technology, i.e., the price that would allow this technology to penetrate the market to precisely the extent assumed in the particular scenario (Nielsen and Sørensen, 1998). Again, this assumes knowledge of the future society that cannot be taken for granted. Changes in values and paradigms will alter the valuation of externalities and will likely add other externalities than those identified today. Therefore it may well turn out to be impossible to calculate such standard prices to the accuracy that would render them useful for the discussion. In other words, what can be done is only to point to the plausibility of the selected technologies as having the potential for becoming economic in the eyes of a future society and to remind of the fact that all decisions on changes so profound as the ones considered here for the energy system have in the past been made on the basis of normative convictions, i.e., by visions of determined individuals along with their persistence in pursuing specific goals. No progress was ever based upon economic evaluations, and most historic innovations had to fight claims of defying economic rationality. The rationality of the past is a concept of little use in shaping the future.

Implementation of a scenario situation to be reached by 2050 involves sketching a path of moving from the current system primarily based on fossil fuels to the very different system and identifying the conditions that have to be fulfilled to make the transition happen. These would involve economic milestones for the new technologies involved, as well as political decisions needing to be taken. It is here assumed that the social climate, in which the transitions happen, is governed by a mixture of free market competition and regulation by society, as it is the case in most regions of the world today. The regulatory part would impose requirements, such as building codes and minimum technical standards for reasons of safety and consumer protection, and would set maximum energy use for certain appliances. Another public handle is to use environmental taxes that incorporate indirect costs which otherwise would distort the competition between different solutions in the marketplace. A consistent method of estimating the environmental taxes that will make the market a fair one is life-cycle analysis of the entire energy supply chains and the technologies involved. The methodology for doing this, with examples for many of the renewable energy systems considered here, is described in detail in Sørensen (2004a) and Kuemmel *et al.* (1997) and in condensed form in Chapter 6.

# 5.4 Scenarios based on nuclear energy

## 5.4.1 History and present concerns

Electricity generated at nuclear power stations presently accounts for some 8.4 EJ $y^{-1}$ or 2% of global energy use (USDoE, 2003). The technology used is primarily light water reactors, a commercial spin-off from the submarine nuclear-powered propulsion systems introduced in the 1950s. The situation after World War II was characterised by two factors of some importance for the development of nuclear energy and the specific reactor choice:

- There was the existence of a highly skilled group of nuclear physicists with a desire to be able to employ their knowledge for peaceful purposes, as opposed to the wartime bomb making. In other words they wanted to demonstrate that nuclear techniques could be used for beneficial purposes as well as for destruction. No group of researchers and developers held in similar esteem were available for the development of competing energy technologies such as those based on renewable energy sources.
- There was general shortage of economic means, relative to the magnitude of tasks needing to be performed in rebuilding the assets destroyed during the war. This led decision-makers to prefer the cheapest solution, without regard to supply security and environmental impacts, that later became part of energy policy decisions. It meant that although alternative nuclear reactor designs were indeed suggested and developed (e.g., CANDU and high-temperature types), the economic advantage of copying an already existing military technology was very tempting, even if some critics found the design less than suited for civilian deployment.

The subsequent fate of nuclear power in the energy sector may be seen as a consequence of the special circumstances suggested by these observations. A major public dispute erupted in a number of countries with tradition for public participation, basically centred on three issues:

♦ The nuclear reactor technology selected poses risks of large accidents with major releases of radioactivity and unpredictable consequences.

♦ Operation of nuclear power plants creates volumes of radioactive waste, to be kept separate from the biosphere during intervals of time that cannot be guaranteed as they are much longer than the lifetime of economic entities or even national states.

♦ The nuclear fuel chain generates plutonium and other material useful for nuclear weapons production and thus increases the risk of such material being diverted to belligerent states or to terrorist organisations for use in actions of nuclear blackmail or war.

The early proponents of nuclear power claimed that any risk associated with the technology was zero or negligible and that the engineering and operational safety was far superior to that of any other technology. Lengthy reports were procured, which seemed to prove these statements as being virtually mathematical tautologies.

This provoked a number of independent scientists to look closer at the claims, and they concluded that there were basic flaws in both the methodologies used and the actual calculations (see overview by Sørensen, 1979b). It is therefore not surprising that when major accidents started to happen the nuclear proponents lost credibility in a way that could perhaps have been avoided with a more honest approach to the information issue.

The important accidents involving commercial plants were the Three Mile Island Reactor partial meltdown accident, which did not breach the outer containment but totally ruined the plant, and the Chernobyl meltdown accident, which caused major releases of radioactivity to the atmosphere and global fallout.

Gradually, national energy policies in most democratic countries were changed to exclude additional construction of nuclear power plants, either directly or indirectly by enforcing strong licensing conditions that could only be met with increased time intervals between the decision to build and the final commissioning, and consequently with considerable increases in costs. Concurrent efforts to make the electricity industry more competitive caused power plant operators to forego the "difficult" nuclear option, thereby avoiding to deal with its uncertainties.

A clear indication of this development is found in the current market projections of future world uranium requirements, as, e.g., stated by the US De-

partment of Energy (deMouy, 1998). The global 2015 requirements are estimated as only 85% of the current ones, as shown in Fig. 5.40, despite some
growth in countries in Southeast Asia and South America.

The purpose of the "safe nuclear scenario" is to investigate if nuclear
technologies different from those employed today may solve the problem or
strongly reduce the concerns stated above, while forming a solution that is
viable in terms of the transition time needed and resource depletion and has
reasonable economy. As it turns out, it is hardly possible to address all three
of the voiced concerns, and most improvements of nuclear technology are
aimed at solving one or at most two of the key problems, while often at the
same time improving other performance aspects.

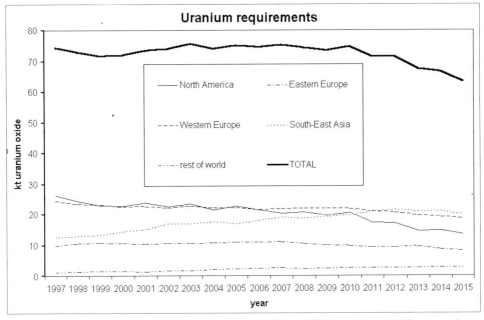

*Figure 5.40.* World uranium requirements according to US Department of Energy reference case projection (based on deMouy, 1998).

## 5.4.2  Safe nuclear technologies

The technologies selected for the safe nuclear scenario should address the
main objections to current nuclear power technologies: proliferation issues,
large nuclear accidents and long-term storage of waste. A wealth of ideas
and new technologies for avoiding or reducing these problems has been dis-

cussed for some years, but all are still fairly speculative. A few have been tested at laboratory scale, but their implementation will take further technical development and would presumably make nuclear power more expensive than today. These additional costs have to be justified by including social costs (including those external to current economic valuation) of the problems associated with current nuclear technologies.

Among the reactor types studied over the past 4-5 decades, but not reaching the stage of market introduction on a commercial basis, are high-temperature gas-cooled reactors and sodium-cooled fast breeders. Current proposals are aware of the issues raised above, but still far from deal with all of them. The reactor industry has recently concluded that a new generation of safer reactors will require substantial breakthroughs (particularly in materials science) that may push commercialisation at least 25 years into the future (USDoE, 2002b). The proposed concepts are summarised in Table 5.3.

| Technology | Coolant | Pressure | Temp. (K) | Issues |
|---|---|---|---|---|
| Conventional breeders | sodium | low | 820 | safety, cost, reprocessing |
| Supercritical water | water | very high | 800 | safety, materials, corrosion |
| Very high temperature | helium | high | 1300 | safety, materials, accidents |
| Gas-cooled breeders | helium | high | 1130 | materials, fuels, recycling |
| Lead-cooled breeders | lead-bismuth | low | 800–1100 | materials, fuels, recycling |
| Molten salt | fluoride salts | low | 1000 | materials, salts, reprocessing |

Table 5.3. Industry proposals for new generations of nuclear reactors (USDoE, 2002b; Butler, 2004).

The four proposed reactor types operating at temperatures over 1000 K can be used to produce hydrogen directly. All the concepts operate at temperatures higher than the ≈600 K of existing light water reactors and therefore would produce electric power at a higher efficiency. The likely cheapest of the systems is the extreme pressure water-cooled reactor, but it does not solve problems of large accidents and large amounts of nuclear waste. The sodium-cooled breeder has already been carried to the large-scale demonstration stage, but has had troubled operational experiences and as a concept does not solve problems of safety and cost. Like the three other types in Table 5.3 requiring reprocessing of spent fuel, it has severe weapons proliferation dangers. Helium-cooled reactors have also been researched for many years, with the most recent prototype just going critical in Japan. The very high-temperature proposals do pose materials problems expected to require many years of research and development efforts. While the Japanese prototype uses fuel pellets in a honeycomb graphite structure, future versions are expected to be of the pebble-bed type, consisting of millions of 5-10 cm diameter spheres of fuel coated with graphite (acting as moderator) and a very hard ceramics layer, to capture and encapsulate fission products. In this way, it is hoped that an accident will not lead to release of large amounts of radio-activity to the environment. However, this depends on temperature control in the event of an accident, a problem still needing to be resolved. No proposal for safeguarding plutonium created during reprocessing has yet been found credible.

Among the chemical cycles available for hydrogen production at temperatures above 1000 K, as an alternative to electricity production by a thermodynamical Brayton cycle (Sørensen, 2004a), the following set currently appears most attractive (Summers *et al.*, 2004),

at 830°C:  $H_2SO_4 + heat \rightarrow \frac{1}{2}O_2 + SO_2 + H_2O,$

at 120°C:  $I_2 + SO_2 + 2H_2O \rightarrow H_2SO_4 + 2HI,$  (5.1)

at 320°C:  $2HI \rightarrow I_2 + H2,$

where the net reaction is water splitting, and where all the chemicals may in principle be recycled between the reaction steps, while recuperated heat from the first reaction is used for the following lower temperature steps.

The reactor is proposed to be modular, but with module sizes of 500 MW units. This is chosen for convenience in sizing Brayton turbines, but is not small enough to constitute an inherently safe design as defined below.

### Inherently safe designs

This concept means that the risk of core meltdown in case the heat from fission processes cannot be led away must be absent. Two examples of proposed inherently safe reactor designs are:

- ♦ Either to reduce the size so much that core melt accidents almost certainly can be contained by the vessel used (this involves maximum unit sizes of 50-100 MW in a traditional design, while the pebble-bed reactor may circumvent this limitation, if the integrity of the pebbles can be guaranteed),
- ♦ Or to use a design in which the core of a conventional pressurised water reactor (PWR) is enclosed within a vessel of boronated water that will flood the core if pressure is lost: there is no barrier between the core and the pool of water, which in case pressure in the primary system is lost will shut the reactor down and continue to remove heat from the core by natural circulation. It is calculated that in an accident situation, replenishing of cooling fluid can be done at weekly intervals (in contrast to hours or less required for current light water reactor designs) (Hannerz, 1983; Klueh, 1986).

To avoid proliferation, fissile material such as plutonium should never accumulate in large amounts or should be difficult to separate from the stream of spent fuel. This can be addressed by using accelerators to produce fissile material at the same pace as it is used to produce energy. Accelerators are also among the options for "incineration" of current nuclear and military waste, in order to reduce waste storage time and again avoid storage of waste from which weapons material could be extracted.

There are two proposed technologies that use accelerators. One is called accelerator-breeding, as it aims at converting fertile material (e.g., Th-232) into fissile fuel to be used elsewhere in a (conventional) reactor (Lecocq and Furukawa, 1994). This does not in itself reduce accident probabilities, and to achieve this, the reactors should be of the above-mentioned inherently safe type.

The other accelerator concept proposed during the last decade (Rubbia, 1994; Rubbia and Rubio, 1996) integrates the accelerator and the reactor-like device into what is termed an *energy amplifier*. The central point in this design is that the energy amplifier does not have to be critical (i.e., the nuclear processes do not have to sustain themselves), as there is a continuous supply of additional neutrons from spallation processes caused by the accelerated protons. This is supposed to greatly reduce the risk of criticality accidents. Exactly how much needs to be further studied. The main components of the fuel chain are illustrated in Fig. 5.41. Although this concept is not supported by the existing reactor industry, it will be used here as a template for the scenario presented below, because of its unique promise to reduce all three objections to the current variety of nuclear reactors. For this reason follows a brief outline of the technology proposed.

### Technical details of energy amplifier

The proposed energy amplifier technology involves a proton accelerator (linear or cyclotron) for the purpose of producing fast neutrons. This is achieved by letting high-energy protons impinge on some heavy target (which could be lead, uranium or thorium). The process is called spallation and typically produces around 50 neutrons of energy 20 MeV per 1 GeV proton. The device proposed by Rubbia resembles a reactor or the blanket of a reactor. Fertile material such as thorium-232 ($^{232}_{90}$Th) needs bombardment with energetic neutrons to transform into nuclear isotopes capable of fissioning, i.e., splitting into two approximately equally heavy products under release of the energy difference (see, e.g., Sørensen, 2004a). Examples of fertile materials are $^{232}_{90}$Th and $^{238}_{92}$U, whereas examples of isotopes that may undergo fission already by absorption of slow (little energetic) neutrons are $^{233}_{92}$U, $^{235}_{92}$U and $^{239}_{94}$Pu. In current light water moderated reactors, the fissile isotope is $^{235}_{92}$U, and the additional neutrons released by fissioning are capable of sustaining a chain reaction (or in absence of careful control a run-away reaction, which is the criticality problem leading to accidents such the one that occurred in 1986 in the Chernobyl reactor, cf. Kurchatov Institute, 1997). In the dense core of fast breeder reactors, the number of $^{238}_{92}U$ isotopes hit by neutrons and transformed into $^{239}_{94}Pu$ is so large that the potential energy derived from the fissile material may be 60 times that of the $^{235}_{92}U$ input.

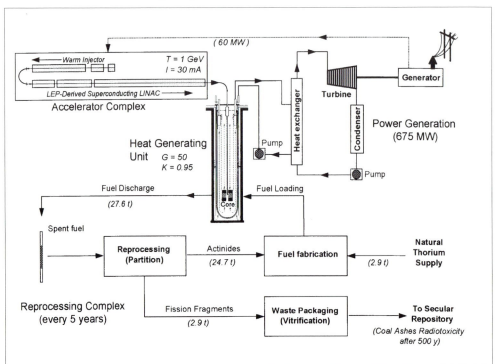

*Figure 5.41.* The energy amplifier fuel chain concept. The fraction of the generated electricity that has to be used for powering the accelerator is indicated. The reprocessing step is essential for the resource viability of the system. (From C. Rubbia and J. Rubio (1996). A tentative programme towards a full scale energy amplifier. European Organisation for Nuclear Research. Used with permission).

The accelerator-breeding efficiency depends strongly on the neutron multiplication factor $k$, which in the proposed design should be about 0.95 (as opposed to very slightly above 1 in a conventional reactor). This is equivalent to an energy gain (breeding factor) of about 50, and it is believed that the value of $k = 0.95$ is sufficient to avoid k ever exceeding the value 1, in situations of irregularity. If the accelerator is halted, the nuclear process will stop. The heat-generating unit will be placed underground, with the proposed design illustrated in Fig. 5.42.

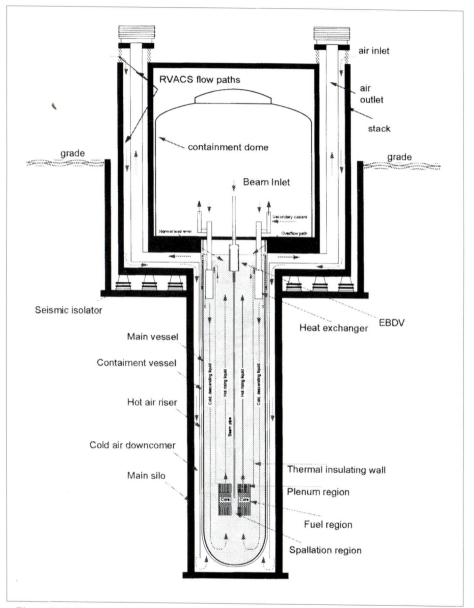

*Figure 5.42.* Layout of the central heat-generating unit of an "amplifier", a 30 m high silo to be placed underground. Note the emergency beam dump volume indicated as "EBDV", (From Rubbia *et al.* (1995). Conceptual design of a fast neutron operated high power energy amplifier. European Organisation for Nuclear Research. Used with permission.)

Sub-criticality is the first essential feature of the concept; the second is the use of a thorium cycle, based on the accelerator-induced reactions ($T_{1/2}$ is the half-life, i.e., the time after which activity has been reduced to half),

$$^{232}_{90}\text{Th} + n \rightarrow {}^{233}_{90}\text{Th} \ (T_{1/2}=23m) \rightarrow {}^{233}_{91}\text{Pa} \ (T_{1/2}=27d) + e^-$$

$$\rightarrow {}^{233}_{92}\text{U} \ (T_{1/2}=163000y) + e^-. \tag{5.2}$$

The fissionable end product $^{233}_{92}\text{U}$ is not the only outcome, as some of the $^{233}_{90}\text{Th}$ will undergo an (n,2n) reaction to $^{231}_{90}\text{Th}$ that will further decay to $^{231}_{91}\text{Pa}$ and $^{232}_{92}\text{U}$, leading eventually to $^{208}_{81}\text{Tl}$, the strong gamma-activity of which makes reprocessing difficult, although not impossible. Because of the breeding property, reprocessing of fuel elements is expected to take place only with intervals of five years or more. The thorium cycle as described above has significant advantages in terms of reducing the amounts of nuclear waste to be kept separated from the biosphere for long periods of time. Figure 5.43 gives the calculated radio-toxicity of waste from the thorium energy amplifier (one-pass through or recycling reprocessed waste) and compares it with current reactor waste toxicity (Lung, 1997; Magill et al., 1995). The accelerator-breeder will accept current reactor and military waste as input, thus offering a way of disposing obsolete nuclear weapons and high-level waste from light water reactors.

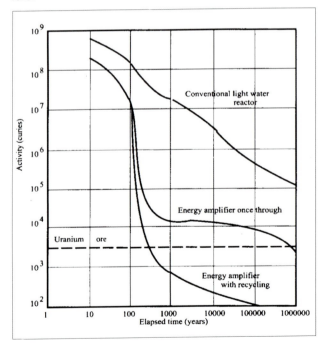

*Figure 5.43.* Activity of radioactive waste from nuclear power plants after 40 years of continuous operation, indicating the advantage of accelerator-breeder concepts. The calculations are preliminary and do not correspond to expected final designs (based on Rubbia et al., 1995; Lung, 1997).

Concern has been expressed that the interface between the accelerator and the breeder could constitute a weakness in the Rubbia design, allowing escape of radioactivity together with molten lead, in case of sudden accelerator failure (J. Maillard as quoted by Mirenowicz, 1997).

The combination of accelerator-breeding and use of the thorium cycle with possible inclusion of uranium resources is needed in the global "safe nuclear scenario" for resource reasons, as further explained in the section below: the available U-235 reserves used in conventional reactor types would only last for a short time at the rate of usage assumed in a global nuclear scenario, and although additional resources may become available, the time horizon is not going to expand significantly towards a sustainable system. Therefore, breeding is necessary in any long-term nuclear scenario, and inclusion of thorium resources would extend the resources to last some 1000-2000 years in scenarios of efficient energy usage such as the present (IPCC, 1996). Current thorium extraction is discussed by Hedrick (1998).

The ideas of accelerator-breeding and thorium-based nuclear power have been around from the start of the effort to transfer nuclear techniques from the military to the civil sector, first for resource reasons (Lawrence as quoted by RIT, 1997), later for reasons of reducing the waste disposal problem (Steinberg *et al.*, 1977; Grand, 1979; OECD, 1994) and getting rid of military nuclear waste and obsolete warheads (Toevs *et al.*, 1994; National Academy of Sciences, 1994), and finally for avoiding proliferation and reducing accident risks (Rubbia *et al.*, 1995).

The ideas involving separation of the accelerator-breeding stage and the energy production stage (Furukawa, as quoted by Lung, 1997) were used in an early version of the present scenario (Sørensen, 1996b). This is the preferred concept of the Los Alamos Group as well as the French and Japanese groups, and it would use a mixture of lithium, beryllium and thorium fluorides as input to the accelerator-breeder, allowing this molten-salt fuel to be transported to and used in graphite-moderated nuclear reactors (cf. Table 5.3), despite negative experiences with such a reactor operated for some years at Fort St-Vrain in Colorado (Bowman *et al.*, 1994; Lung, 1997). Early schemes for waste transmutation proposed the use of breeder reactors (Pigford, 1991), which do not fulfil the demand for inherent safety. The scenario presented below is based on the Rubbia idea, but it could include parallel operation of inherently safe reactors, from which spent fuel would be shipped to the reprocessing plant of the accelerator-amplifier installation. It should be stated clearly that this scenario is based on largely unproven technology and would require huge research and development investments over significant periods of time. Of course, the same can be said for the current light water technology and certainly for all the advanced concepts contemplated over the past 40 years, including those of Table 5.3.

## Nuclear resources assessment

Current estimates of the available reserves and further resources of uranium and thorium, and their global distribution, are shown in Figs. 5.44-5.50. The uranium proven reserves indicated in Fig. 5.44 can be extracted at costs below 130 US$/t, as can the probable additional reserves indicated in Fig. 5.45. Figure 5.46 shows new and unconventional resources that may later become reserves. They are inferred on the basis of geological modelling or other indirect information (OECD and IAEA, 1993; World Energy Council, 1995). The thorium resource estimates are from the US Geological Survey (Hedrick, 1998) and are similarly divided into reserves (Fig. 5.47), additional reserves (Fig. 5.48) and more speculative resources (Fig. 5.49). The thorium situation is less well explored than that of uranium: the reserves cannot be said to be "economical", as they are presently mined for other purposes (rare earth metals), and thorium is only a byproduct with currently very limited areas of use. The "speculative" Th-resources may well have a similar status to some of the additional U-reserves.

The magnitude of nuclear sources identified are similar to that of oil or natural gas, if the nuclear fuels are used in conventional non-breeder reactors (of the order of 100 years). Thus, if nuclear fission energy is to constitute a reasonably sustainable resource, some type of breeding is required. The breeding ratio for liquid metal fast breeders (which in the scenarios are excluded because the technology of liquid sodium cooling close to a very dense nuclear fuel core makes the occurrence of accidents with conventional explosions followed by fuel meltdown and possible criticality leading to nuclear explosions very high) may theoretically reach about 60, whereas for the accelerator-breeder concept outlined above, a breeding factor of 10 is assumed to be technically feasible. This brings the thorium resources up to about 1000 years (the similar magnitude of uranium resources may not be of immediate interest if thorium is selected as the principal fuel). On the other hand, once economic interest in thorium is established, the exploration of these resources will take a more serious shape, and it is possible that additional reserves will be identified. The overall resource base is quite large.

The different levels of reserves and resources are summed up in Figs. 5.50 and 5.51, giving a probable magnitude of total exploitable resources. The one for thorium will be used as a basis for the nuclear scenario below.

The highest grade of thorium resources is that found in veins of thorite ($ThSiO_4$), containing 20-60% of equivalent $ThO_2$. However, the most common mineral considered for exploitation is monazite ($MPO_4$, where M is Ce, La, Y or Th, often in combination), containing about 10% pure $ThO_2$ equivalent. A number of other minerals have been identified, which contain Th at a lower percentage (Chung, 1997).

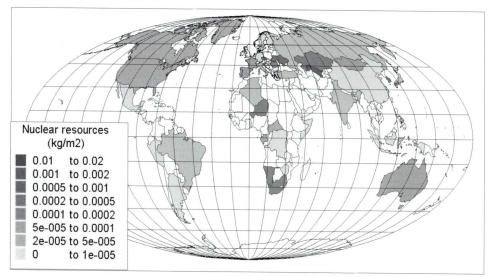

*Figure 5.44.* Proven uranium reserves given as kg of uranium oxide per m², averaged over each country  (source: OECD and IAEA, 1993; GIS layout Sørensen, 1999).

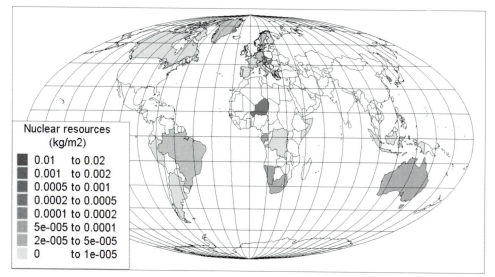

*Figure 5.45.* Estimated additional uranium reserves given as kg of uranium oxide per m², averaged over each country (source: World Energy Council, 1995; GIS layout Sørensen, 1999).

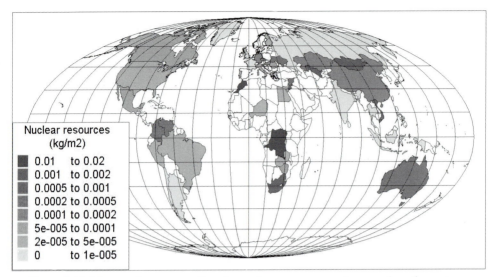

*Figure 5.46.* New and unconventional uranium resources given as kg of uranium oxide per m², averaged over each country (source: World Energy Council, 1995; GIS layout Sørensen, 1999).

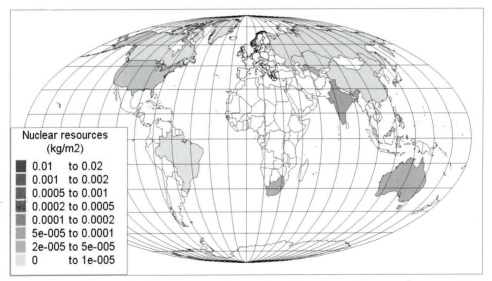

*Figure 5.47.* Proven thorium reserves given as kg of thorium oxide per m², averaged over each country (source: Hedrick, 1998; GIS layout Sørensen, 1999).

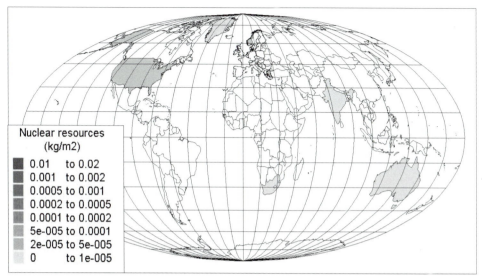

*Figure 5.48.* Additional thorium reserves given as kg of thorium oxide per m², averaged over each country (source: Hedrick, 1998; GIS layout Sørensen, 1999).

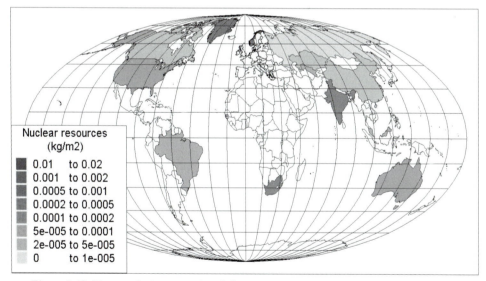

*Figure 5.49.* New and unconventional thorium resources given as kg of thorium oxide per m², averaged over each country (with use of Hedrick, 1998; GIS layout Sørensen, 1999).

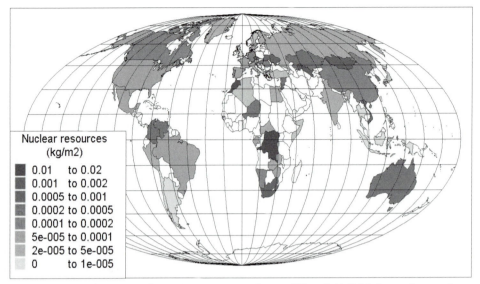

*Figure 5.50.* All estimated uranium resources (sum of Figs. 5.44-5.46), here given as kt of uranium oxide per m², averaged over each country (from Sørensen, 1999).

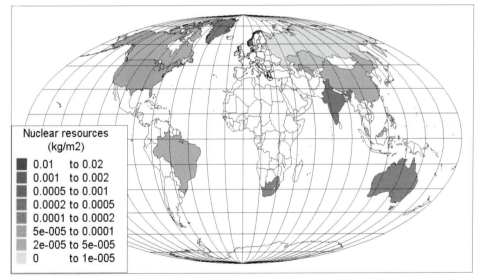

*Figure 5.51.* All estimated thorium resources (sum of Figs. 5.47-5.49), here given as kt of thorium oxide per m², averaged over each country (from Sørensen, 1999).

### Safe nuclear scenario construction

The safe nuclear scenario employs the common energy demand scenario described in section 5.4.1 and the following conversion technologies: the main source of energy is the energy amplifier described above, assumed primarily to use thorium as fuel (although some U-233 may be used during the first start-up, i.e., for the initial reactor cores). The scenario construction task is then to determine the thorium fuel annual input necessary for covering the demand. The conversion routes included are partly production of hydrogen for use in the transportation sector and possibly for industrial process heat and partly electricity for direct use or for powering heat pumps and cooling devices to provide low-temperature heating and cooling, with additional inputs of environmental heat. Whereas this is the preferred way of producing heat in renewable energy scenarios (because no sources of waste heat exist), for scenarios such as the fossil and nuclear, the associated heat production from power plants should have first priority, and the local heat production by heat pumps should only be used when distances between power plants and load points make it inconvenient or too expensive to use district heating lines to carry the heat to the end-user. In the nuclear case, "waste heat" is abundantly present, so only the transmission conditions limit their use.

The demand for food energy is covered by agriculture and the distribution of production as described in detail in Sørensen (2004a). Another renewable energy source that has been incorporated in the nuclear scenario is hydropower, where existing plants and plants under construction are retained. This is due to their long productive life and ability to operate in harmony with the nuclear power plants, both being components of a centralised system with similar requirements for transmission. The fast regulation of incorporating hydro plants in a given system also adds to the technical viability of nuclear power plants. The scenario presented below assumes that the nuclear plants can be regulated to a sufficient extent to allow following the load with minor backup from hydro and stored energy. However, should it turn out to be technically or economically inconvenient to perform this regulation, the solution is to increase the hydrogen production and use hydrogen as a storage medium, for subsequent use either directly for process heat or by regenerating electricity, presumably using fuel cell technology. This in indicated in Fig. 5.62, where lines of storage cycle supply are shown, but at the moment with zero flow.

The demand for transportation energy is assumed to be covered by equal amounts delivered to electric vehicles and to fuel cell-based vehicles. For the former, a 50% storage cycle loss associated with the entire battery cycle operation is assumed, and for the fuel cell-based vehicles, operation is assumed to be based upon either hydrogen or a more storable derivative (e.g., methanol). Fuel cells are considered to have a 50% conversion efficiency. In both cases, the minimal losses that occur in the final electric motor conversion to traction power are considered as included in the 50% overall loss. The nuclear electricity going through a hydrogen conversion process is assumed subject to a 20% conversion loss, which would be typical of state-of-the-art electrolysers. For conventional electrolysis processes in use today, the loss is more like 35% (see Chapter 2, section 2.1.3). Figure 5.52 shows the amounts of hydrogen, which in this scenario will be produced from nuclear power.

The remaining energy delivered as electricity comprises all energy for electric apparel, stationary mechanical energy and input to coolers, refrigerators and heat pumps used for providing 50% of the low-temperature heat (space heating, hot water and other process heat under 100°C). The other half of the low-temperature heat demand is covered by district heating based on nuclear waste heat. The available associated nuclear heat is much larger (cf. Fig. 5.62), but the amount included is considered the maximum amount that can be delivered to load areas at reasonable transmission distances.

The required electricity delivery for all these purposes (except that used to generate hydrogen) is shown in Fig. 5.53. This will be delivered by hydro or nuclear power, in that order of priority. Figure 5.54 shows the total electric energy input required from nuclear energy, i.e., that for hydrogen production plus the part of the ones shown in Fig. 5.53, that cannot be supplied by hydro power, and all augmented by losses in transmission. In other words, Fig. 5.54 gives the power that must leave the nuclear power plants, after in-plant uses for the accelerator-based energy amplifiers. Figure 5.55 gives the amounts of nuclear waste heat to be delivered through district heating lines (the heat leaving the power plants being some 25% higher due to transmission losses), and Fig. 5.56 shows the amounts of environmental heat drawn by the heat pumps used in locations unsuited for district heating (this may come from air, soil or waterways, and the assumed heat pump coefficient of performance COP, i.e., ratio between heat output and electricity input, is 3.33).

Having now determined to total amount of nuclear electricity required, the thorium fuel input to the energy amplifiers can be calculated from the design data of Rubbia and Rubio (1996). The thermal output from the prototype design reactor is 1500 MW, with a fuel amount of 27.6 t in the reactor (Fig. 5.42). The fuel will sit in the reactor heat-generating unit for 5 years, after which the "spent" fuel will be reprocessed to allow for manufacture of a new fuel load with only 2.9 t of fresh thorium oxide supply. This means that 2.6/5 t $y^{-1}$ of thorium fuel is required for delivery of $5 \times 1500$ MWy of thermal power over 5 years, or 675 MWy of electric power, of which the 75 MWy is used for powering the accelerator and other in-plant loads. The bottom line is that 1 kg of thorium fuel produces very close to 1 MWy of electric power, and 1 kt thorium produces close to 1 TWh$_e$.

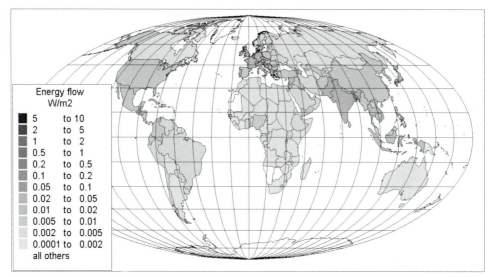

*Figure 5.52.* Hydrogen derived from nuclear energy (sufficient to cover 50% of needs in the transportation sector). The figure shows average flows in each country (Sørensen, 1999).

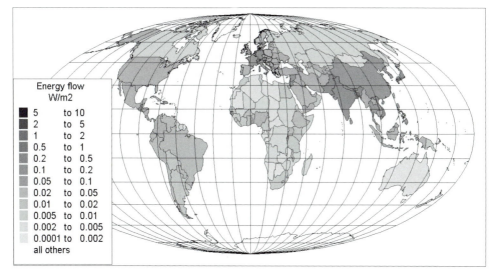

*Figure 5.53.* Electricity supply derived from nuclear or hydro energy (as delivered to the final consumer, covering dedicated electricity use including electric input to heat pumps and for refrigeration, stationary mechanical energy, medium- and high-temperature process heat). The figure shows average flows in each country (Sørensen, 1999).

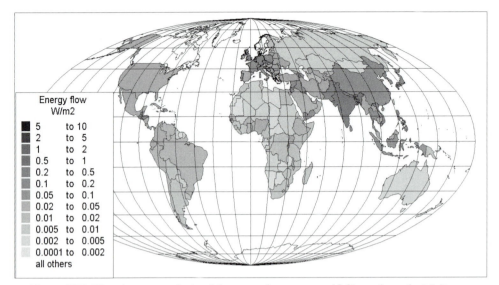

*Figure 5.54.* Electric energy derived from nuclear energy (delivered as electricity or used to produce hydrogen, and including transmission losses). The figure shows average flows in each country. Note that countries with large hydro power production need no or little nuclear energy, as hydro is given first priority (Sørensen, 1999).

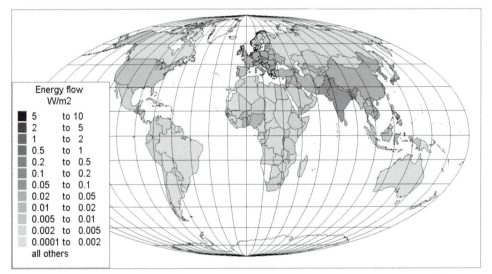

*Figure 5.55.* District heating derived from nuclear "waste heat" (50% of total low-temperature heat requirements considered suitable for piped delivery). The figure shows average flows in each country (Sørensen, 1999).

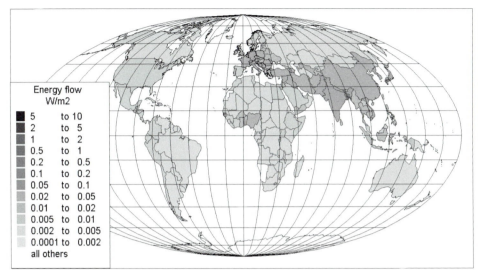

*Figure 5.56.* Environmental heat for heat pumps (that cover 50% of low-temperature demands). The figure shows average flows in each country (Sørensen, 1999).

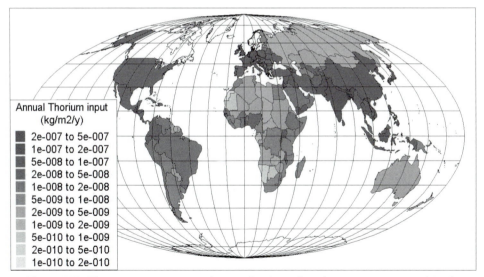

*Figure 5.57.* Thorium requirements for production of nuclear power assumed in the safe nuclear scenario, given as kt of thorium oxide per year and per m$^2$, averaged over each country and over time (Sørensen, 1999).

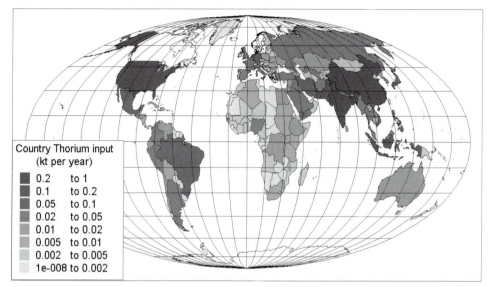

*Figure 5.58.* Thorium requirements for production of nuclear power as assumed in the safe nuclear scenario, given as kt of thorium oxide per year for each country (Sørensen, 1999).

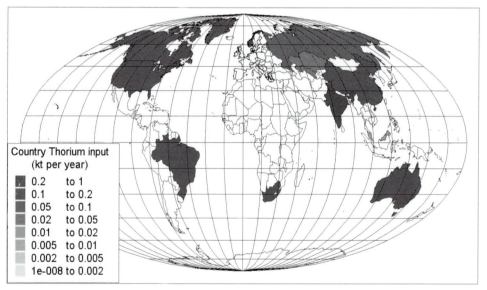

*Figure 5.59.* Thorium production assumed in the safe nuclear scenario, given as kt of thorium oxide per year for each country (Sørensen, 1999).

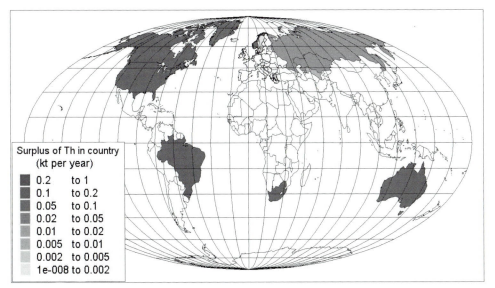

*Figure 5.60.* Thorium surplus (production minus required supply for each country, if positive) for the safe nuclear scenario, given as kt of thorium oxide per year for each country (Sørensen, 1999).

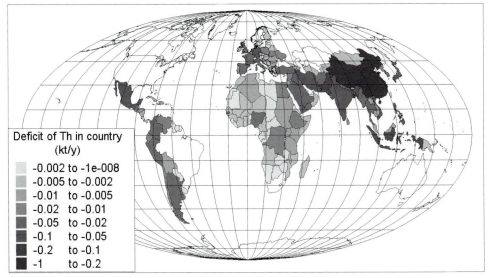

*Figure 5.61.* Thorium deficit (production minus required supply for each country, if negative) for the safe nuclear scenario, given as kt of thorium oxide per year for each country (Sørensen, 1999).

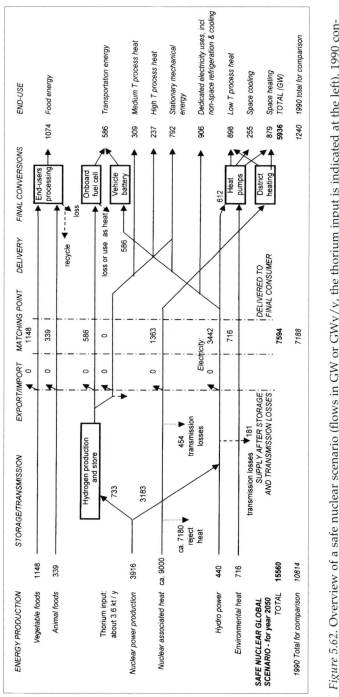

*Figure 5.62.* Overview of a safe nuclear scenario (flows in GW or GWy/y, the thorium input is indicated at the left). 1990 consumption is indicated at the bottom (Based on Sørensen. 1999).

Figure 5.57 gives the scenario requirements of thorium input per year and per m², averaged over each country, and Fig. 5.58 gives the total amounts per year for each country. It is seen that countries with abundant hydropower (such as Canada) do not need any nuclear energy in the scenario, even if they do today. The reason is the substantial increase in energy efficiency assumed, a fairly stable population in Canada, and maintaining all hydro plants currently operating or under construction. The hydro contribution is the same in all the scenarios constructed here (fossil, nuclear and renewable).

The thorium requirement per country may be compared with the assumed production in each country, shown in Fig. 5.59. It is determined by assuming that the world production of thorium matches the world requirements, and that each country produces in proportion to its total thorium resources, as estimated in Fig. 5.51.

It is now possible to determine the required trade of nuclear fuels by comparing the production and use of thorium fuels. This is done in Figs. 5.60 and 5.61, showing the surpluses and deficits in thorium balances for each country. It is seen that the world is fairly sharply divided into energy-producing countries and energy-importing countries, the latter being in South America (except Brazil), Africa, Europe (minus Sweden, Norway and Russia) and Asia (minus the former Soviet Union countries).

A summary of the scenario is shown in Fig. 5.62. The more than four times increase in energy services at the end-use level, relative to the present level, is here achieved by primary energy inputs 50% above the current ones. This is more "overhead" than in the renewable energy scenarios discussed in section 5.5, due to a large extent to the associated thermal energy from the energy amplifiers, that cannot all be made useful.

### Evaluation of the safe nuclear scenario

The nuclear scenario presented above has zero emissions of greenhouse gases. It largely avoids the accumulation of nuclear material with risk of diversion for military purposes, with the only sensitive material being the spent fuel. It is suggested to join the power production and reprocessing units at the same location (Fig. 5.41), where the reprocessing is fairly delicate due to the presence of strongly radioactive Tl-208 (cf. the technical description of the energy accelerator above). This makes it necessary to perform the reprocessing by robots in a heavily gamma-shielded environment, a technology just starting to be technically feasible and currently excessively expensive (Lung, 1997; Chung, 1997). On the other hand, this fact makes theft of the spent fuel unlikely.

As regards nuclear accidents, the sub-criticality suggests dramatic reduction in the risk (Buono and Rubbia, 1996), but it is clear that more work is required to fully assure this proposition.

With respect to radioactive waste, Fig. 5.43 suggests large advantages over current nuclear reactor technology in reducing the period in which wastes must be kept separate from the biosphere from some 10 000 years to 500 years. However, this is still a very long time to require integrity of depository sites, in fact of the same order of magnitude as predicted for nuclear fusion wastes (speculatively, as the decisive factor is the choice of materials for the fusion reactor, the design of which – should controlled commercial fusion one day become feasible – is unknown). In the first 100-year period of time, the energy amplifier waste produces a surge of radioactivity coming from the U-233 formation, which requires special precautions in storage and disposal. Figure 5.43 indicates that this "early" surge of radioactivity is still a factor of 3-10 lower than the level for light water waste. It is therefore a possible conclusion that the overall advantage of the energy accelerator in regard to waste toxicity is fairly modest compared to present reactor waste containing plutonium-uranium mixtures. A more accurate statement will involve a detailed comparison of the handling techniques for each isotope in the two different waste compositions.

When Rubbia and Rubio (1996) proposed the energy accelerator they noted that a unique opportunity to rapidly evaluate the energy amplifier concept on a realistic scale was offered by the planned year 2000 retirement of the LEP superconducting cavity electron accelerator previously in use for research at the European Organisation for Nuclear Research (CERN). This accelerator could have been modified to accelerate proton beams at a current of 20 mA to energies above 1 GeV, which would make it suitable for an energy amplifier of up to 1500 MW thermal power. The proposal for such a facility, submitted to the European Commission, was evaluated by a working group dominated by current nuclear industry representatives, assigned by the European Commission's DGXII. They recommended (Pooley, 1997), and the Scientific and Technical Committee of the European Commission adopted the view, that the project should not be realised, because it would put nuclear power "back to square one", discarding all existing technology and building a novel one (which was indeed the intention), with associated expenditures as huge as those leading to the existing nuclear technology and with little hope that the public would understand the subtle difference between the old and the new nuclear power and better accept the latter.

Unfortunately, this reasoning could equally well be directed at the other new nuclear concepts in Table 5.3, as well as against nuclear fusion, and seems to convey the fatalistic attitude that the nuclear industry has already lost so much credibility in the general populations that it is not worthwhile to spend money trying to remedy the situation. Not all funding agencies share this negative attitude, but it is probably true that the nuclear industry, most of which is privatised or in the process of being privatised, finds it difficult to imagine spending as much money on the next generation of nuclear reactors as the military institutions and governments of nuclear nations have put into the development of the light water reactor.

What the construction of a global scenario based on the energy amplifier concept, as presented here, has shown is that if the energy amplifier project could indeed be successfully developed, it would enable the use of nuclear energy on a much larger scale, globally, than ever envisaged for the present nuclear technology. The estimated resources of thorium makes it possible to sustain the thorium-cycle nuclear scenario at the level contemplated for some 1000 years, with a similar interval being possible for uranium-based concepts with a similar breeding ratio. The direct cost will certainly be higher than that of present nuclear reactor technologies, with the reprocessing and the waste management costs considerably higher and still very uncertain.

Clearly, from a resource point of view (Figs. 5.44-5.51), the present light water nuclear technology is uninteresting as a long-term solution. Furthermore, it is not useful as a short-term remedy for the current problems of unreliable oil supply, as it would only be replacing coal in power plants. The future of nuclear energy technology thus depends on developing breeder technologies that successfully address the three objections raised in the beginning of section 5.4.1.

The safe nuclear scenario is weaker than the clean fossil and renewable energy scenarios discussed in this chapter, as it depends on a technology development that has hardly even started and is still far away from the commercial markets. On the other hand, the expertise for handling both accelerator technology and a new reactor concept is today present to an extent that cannot be compared with the level characterising the early reactor development during the 1950s. Still, a deeper reason for questioning this technology development is a very real concern over whether a practical version of new concepts such as the energy amplifier will indeed get rid of the problems characterising current nuclear technologies. Would there not still be a risk that other new developments will bring the proliferation risk back, just as the centrifuge technology changed enrichment from being in the hands of a very exclusive club of huge and expensive laboratories to becoming accessible to poor countries and potentially to determined terrorist organisations? May there not turn out be novel accident routes that will become revealed only when the technology has reached a more concrete form? Because of the time required to transform a scientist's idea into an industrial product, the safe nuclear scenario is likely to have problems reaching the envisaged high share in energy supply already by the middle of the 21st century. Lessons from renewable energy tell us that the transition time required for this type of substantial technology shift may well be of the order of 25 years from the day where the new technology is technically ready for commercialisation.

# 5.5 Scenarios based on renewable energy

For the fossil and nuclear scenarios based on hydrogen as an energy carrier described above, the relevant data were available on a country basis and the geographical distributions were thus simple country maps. In the case of renewable energy, the sources are closely connected to the geography of climates, and the use of an area-based description is much more rewarding in revealing the detailed correlation between potential energy production and final energy use, with the latter being already for many forms of energy clearly area-based, as shown in Fig. 5.7. Below, the renewable energy counterpart to the fossil and nuclear scenarios is first presented on a global scale. This work is already discussed in detail in Sørensen (2004a) and will only be summarised here. However, subsequent work has expanded the insights considerably, while at the same time working on a national scale, allowing a spatial resolution of $500 \times 500$ m$^2$ and a time resolution of 1 h (in contrast to the seasonal time scale of the global scenarios). This makes the model capable of catching both realistic demand variations and also variations in fluxes of solar and wind energy at a particular geographical spot. With these levels of resolution, a realistic appraisal of energy storage requirements in connection with the fluctuations in renewable energy inputs can be made, which is important for the hydrogen aspects of the scenario, because hydrogen is used both as an energy carrier (for mobile and stationary applications) and also as an energy storage medium. The details of such a scenario are described in section 5.5.2.

## 5.5.1 Global renewable energy scenarios

Two global scenarios based on renewable energy have been constructed, in analogy to the clean fossil and safe nuclear ones. The scenarios differ by one (termed the decentralised one) allowing no centralised energy production, such as wind parks or solar megawatt arrays placed in marginal land areas (deserts, etc.), and the other including such options (termed the centralised renewable scenario). The latter turn out to be much more resilient to misjudgements of the future energy demands or errors in defining acceptable resource utilisation.

The outcome of this "centralised renewable energy scenario" is illustrated in Fig. 5.63 and 5.64, showing the surplus and deficit of demand coverage by the mix of renewable resources available to each region (and inviting trade).

A scenario summary is shown in Fig. 5.65. Evidently, city conglomerations use more energy than can be supplied by rooftop solar energy and wind energy placed within the urban setting, whereas many rural areas turn out to be able to supply far more energy than used locally, allowing an export to urban areas profitable to both.

Implementation of either of the 2050 scenarios, or a combination of them, involves sketching a path of moving from the current system primarily based

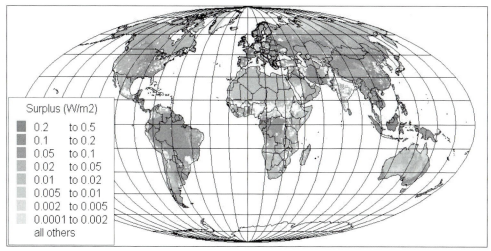

*Figure 5.63.* Surplus energy (production minus demand if positive) in the 2050 centralised renewable energy scenario, for all energy use except food energy.

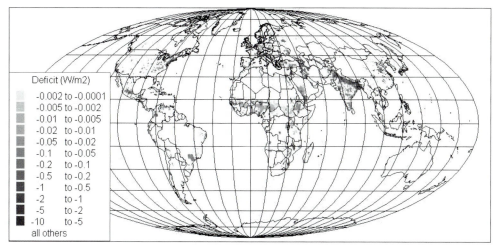

*Figure 5.64.* Deficit energy (production minus demand if negative) in the 2050 centralised renewable energy scenario. The total surplus and deficit are equal.

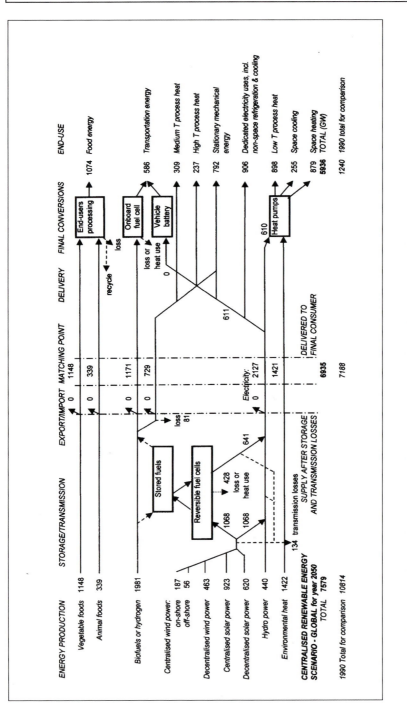

Figure 5.65. Overview of centralised renewable energy scenario (from Sørensen *et al.*, 2001).

on fossil fuels to a very different system and identifying the conditions that have to be fulfilled to make the transition happen. These could be economic milestones for the new technologies involved or they could be political decisions needing to be taken. It is here assumed that the social climate, in which the transitions happen, is governed by a mixture of free market competition and regulation by society, as is the case in most regions of the world today. The regulatory part would impose requirements, such as building codes and minimum technical standards for reasons of safety and consumer protection, and maximum energy use for appliances. Another public handle is environmental taxes that incorporate indirect costs which otherwise would distort the competition between different solutions in the marketplace. A consistent method of estimating the environmental taxes that will make the market a fair one is life-cycle analysis of the entire energy supply chains and the technologies involved. The methodology for doing this, with examples for many of the renewable energy systems considered here, may be found in Kuemmel et al. (1997). The key methodology is outlined in Chapter 6.

In a fair market, the price that new technologies have to measure up against is the price of currently used coal, oil, natural gas, hydro and nuclear technologies, all supplemented with the externalities not included in the present market prices and reflecting, as well as our knowledge allows, the negative impacts that the new scenarios propose to remedy. Renewable technologies such as wind power, the cost of which today is only slightly above current fossil fuel-based systems, will clearly be economic if externalities are included (as these are very small for wind power and at the same order of magnitude as the actual price for fossil fuels). Also, many other renewable technologies such as biofuel production, which today involves a cost about twice that of fossil fuels, would in the fair market be able to enter by standard competitive forces. For photovoltaic power and the new conversion and storage technologies such as those based upon fuel cells, present costs are higher than what can be remedied by introducing externalities in the comparison with fossil fuels. Therefore, these have to be assumed to pass through a technical development over the 50-year transition period considered that would bring the price down to below the threshold value. Subsidies may be contemplated in order to speed up this process in the initial phase, but already the political readiness to include externalities in prices would constitute a strong motivation for the development of alternative solutions. For the alternatives of coal gasification and advanced nuclear energy technologies, a fair comparison will similarly include externality costs. The extra cost of coal gasification may in this way become acceptable, whereas the ocean disposal of carbonates requires further environmental assessment. For nuclear energy such as high-temperature generation of hydrogen or the accelerator technologies, there exist preliminary cost estimates exhibiting the same kind of over-optimism that characterised the early development of nuclear power ("too cheap to meter") (Fernandez *et al.*, 1996). The real cost of the safe nuclear fuel cycle cannot be realistically estimated today, but is likely to be at least 2-3 times the current energy costs.

The assumptions that the future transition will be driven by fair market rules are somewhat at variance with the present situation. On one hand, there are hidden subsidies in many regions (e.g., to fossil and nuclear energy, where society pays for environmental and health impacts and assumes the responsibility for risk-related events), and on the other hand, monopolies and generally differences in size and power of the energy industries involved in different technologies make the actual price setting likely not to follow those prescribed by the life-cycle analysis in a fair market philosophy.

An obvious solution is to regulate the market not by general taxation (as it is currently done, e.g., in Europe) but by legislation, requiring, e.g., power providers to use specific technologies. This makes it unnecessary to accumulate tax money at the state level (avoiding this is by some nations seen as a positive feature), but makes the system rather stiff, as each technical change has to be followed up by possibly altering the legislative regime. The taxation method where tax levels are set to reflect life-cycle impacts is more flexible, and once the level of environmental tax is decided by governments (who thereby cut through scientific uncertainty), the market functions exactly as before, but should give the manufacturers of the new technologies with smaller environmental impacts a good chance to compete, even if they are initially smaller than the established market players. It is important that externalities are set politically, because there will always be a continued debate over the uncertainties of scientific assessment, which itself is not static.

A problem is the possible differences in tax levels that different nations may see as fair. International synchronisation is highly desirable, as in all policy aimed at reducing global threats. Depending on the degree of planning tradition in different societies, the energy transition might also benefit from "setting goals" and continually monitoring if they are fulfilled. If, for example, the market does not respond well enough to the price signals set by the environmental taxes, it is then possible to adjust the size of the imposed externality (within limits which would not violate the scientific basis for it) or introduce specific legislation to remove the obstacles to a free and fair market.

Remaining technical problems include the determination of the optimum share of different energy forms in the total and the requirement for energy storage and infrastructure implied. The next section gives an example of how these issues may be approached on a national level. Scenarios focussing on a single form of energy, such as fossil, nuclear or renewable, may be illustrative in disclosing the nature of problems posed by the different solutions, but in practice, there may of course be an advantage in combining the energy sources in the actual energy policy of a given nation, unless the "purity" of the solution is a demand of the public in a given democracy.

## 5.5.2 Detailed national renewable energy scenario

Two detailed scenarios for 2050 will be discussed in this section. They pertain to Denmark, a country already having a high share of renewable energy in its energy system, and a policy plan to become totally based on renewable energy by 2050, with a 50% share of renewables by the year 2025 (Danish DoE, 1998). The scenarios use hydrogen either in a centralised way or in a decentralised, building-integrated mode. The descriptions will start with a demand projection, followed by a resource assessment and finally the supply-demand matching that constitutes the two scenarios.

### Danish energy demand in 2050

The energy demand depends on population size and level of activities. The Danish population is expected to decline slightly, but will likely be compensated by immigration. As integration policy is difficult to predict, the assumption of unchanged total population has been made. This is the same outlook as the one upon which United Nations (1996) population predictions for Denmark is based. The current population data are thus used, with a small correction based on the tendency to decentralise economic activities by moving businesses and production factories away from cities. This feature of highly developed societies is in contrast to that of the developing nations, where migration is rather from rural to urban areas. As a basis for creating area-based projections, I used the Danish building database (Danish National Agency for Enterprise and Construction, 1999), which gives current buildings on a geographical basis (a 500 m × 500 m grid was used), distributed on types and use of each building (categories are detached family homes, dense-low dwellings, farm houses, public and privately owned high-rise apartments, public and privately owned office and service buildings, institutions, public and privately owned production enterprises and finally vacation and recreational buildings). This database, which includes information on present energy installations, is used for determining the distribution of energy use, but also for estimating population, for which statistics only exist on a county level. The totals for each county are distributed on the area-based grid by assuming a population distribution given by building occupancy such that the number of people is proportional to the floor area of all residential category buildings, but with recreational buildings counting only 0.33. This is the estimated use of vacation houses for permanent living. In this way, variations in dwelling area per person in different parts of the country are reflected in the model. The outcome is shown in Fig. 5.66, with the underlying dwelling floor area shown in Fig. 5.67 (total summed floor area for multi-storey buildings). The migration between urban and rural areas has been modelled by assuming a 30% growth in population for grid cells that in the year 1996 had under 50 inhabitants and a corresponding decline in those with more people in 1996, maintaining a total 2050 population of 5.2 million.

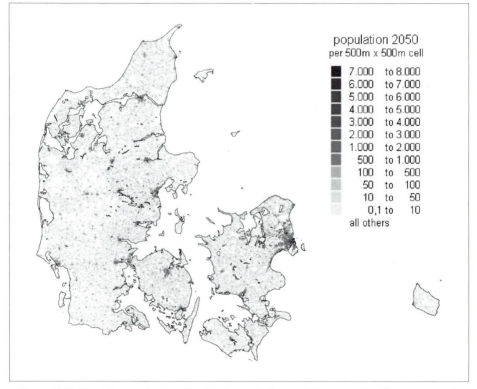

*Figure 5.66.* Danish population distribution in 2050 used in scenarios (Sørensen *et al.*, 2001).

A number of energy demand types are assumed to be directly proportional to the population, so that the energy use in a grid unit is proportional to the population density. Table 5.4 gives a summary of 2050 end-use assumptions made for the scenario construction. The space heating requirement is calculated from the floor area by use of a model taking into account hourly time series of outdoor temperatures, a wind-dependent chill-factor and solar irradiation on various building surfaces (according to their angle to the direction of direct solar radiation and with a model for the absorption of both direct and scattered solar radiation through windows). The gradual improvements of the building shell (higher insulation, control of ventilation) that have taken place over the last 30 years are assumed to continue to lower the average heating consumption per square metre of building floor area at the same rate as during the past 30 years, where the space heating requirement was reduced by 50%. The same measures make space cooling unnecessary under Danish climatic conditions. Details of the model are described in Chapter 6 of Sørensen (2004a).

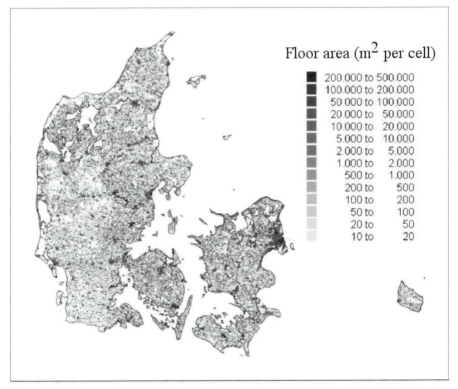

*Figure 5.67.* Geographical distribution of floor space in the Danish residential sector 1998 (note that this figure is on a logarithmic scale, in contrast to Fig. 5.66) (Sørensen *et al.*, 2001).

The calculation of space heating requirements takes into account occupancy of building space, so that office and production space is not heated outside working hours and dwelling space has zonal heating with time variations in temperature depending on occupancy (the body heat from which is also included), as controlled by computerised space management units already having a certain penetration in the building sector today. Figures 5.69 and 5.70 give the floor space geographical distribution similar to the one shown in Fig. 5.67 for living areas, but now for production (including agriculture) and commercial building areas. The geographical distribution of all these building types has been taken as unchanged from the present, although relocation of businesses and new preferences regarding residential preferences and types of services offered could lead to a number of (difficult to predict) changes in geographical distribution of floor areas.

| Energy type/quality | Annual average use (W/cap.) |
|---|---|
| Space heating (climate dependent, cf. Fig. 5.68) | 389 |
| Hot water and other low-temperature heat | 150 |
| Medium-temperature process heat (100–500°C) | 50 |
| High-temperature process heat (over 500°C) | 40 |
| Space cooling (climate dependent) | 0 |
| Other cooling and refrigeration | 35 |
| Stationary mechanical energy | 150 |
| Electric appliances and other electric equipment | 150 |
| Transportation energy | 150 |
| **TOTAL** | **1214** |

Table 5.4. End-use energy assumed in the Danish 2050 scenario (Sørensen et al., 2001).

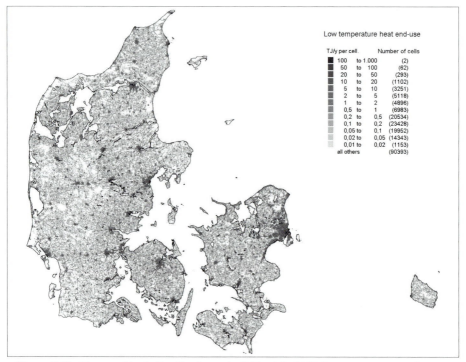

Figure 5.68. Geographical distribution of all Danish 2050 low-temperature heat requirements based on Table 5.4 and the heated floor areas shown in Fig. 5.67, 5.69 and 5.70 (Sørensen et al., 2001).

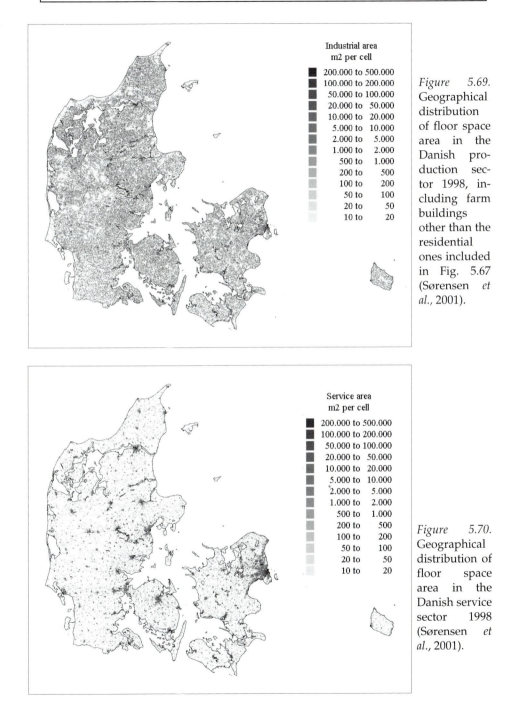

*Figure 5.69.* Geographical distribution of floor space area in the Danish production sector 1998, including farm buildings other than the residential ones included in Fig. 5.67 (Sørensen *et al.*, 2001).

*Figure 5.70.* Geographical distribution of floor space area in the Danish service sector 1998 (Sørensen *et al.*, 2001).

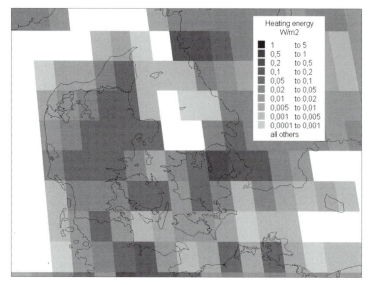

*Figure 5.71.* Geographical distribution of January average space heating requirements in the year 2050, based on satellite measurements of temperature (on a gross scale of an approximately 50-km grid) in combination with the scenario's assumed building standards. The variations thus reflect both climate differences across Denmark and differences in heated space per unit cell. The map uses the Mollweide area-preserving projection, in contrast to the straight longitude-latitude co-ordinate system of, e.g., Fig. 5.70 (Sørensen *et al.*, 2001).

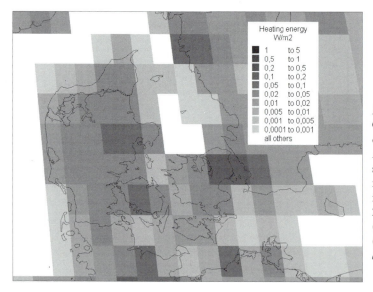

*Figure 5.72.* Geographical distribution of April average space heating requirements in the year 2050 scenario (cf. Fig. 5.71) (Sørensen *et al.*, 2001).

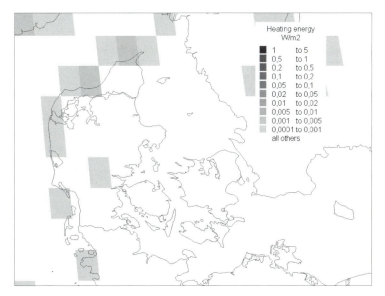

*Figure 5.73.* Geographical distribution of July average space heating requirements in the year 2050 scenario (cf. Fig. 5.71). The small summer heating requirements are in regions of colder temperatures, but with small population densities (Sørensen *et al.*, 2001).

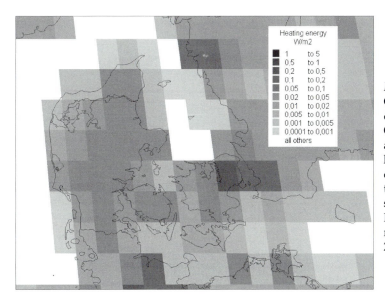

*Figure 5.74.* Geographical distribution of October average space heating requirements in the year 2050 scenario (cf. Fig. 5.71) (Sørensen *et al.*, 2001).

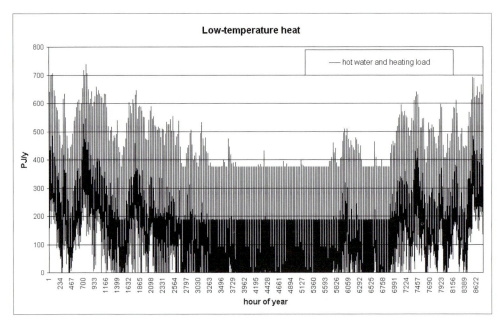

*Figure 5.75.* Hourly time variation of low-temperature heat (space heating and hot water) calculated for the 2050 scenario with use of actual temperature and wind chill data for Denmark during the year 1999 (Sørensen *et al.*, 2001).

Figure 5.68 gives the total annual requirement of end-use low-temperature heat. Figures 5.71-5.74 give the geographical variations on a seasonal basis, using a coarse grid of $56 \times 56$ km$^2$, because this is the resolution of the satellite temperature data used because temperature variations across Denmark are rather small. The actual heat use map of the model has the full 1 hour and $0.5 \times 0.5$ km$^2$ resolution, and only the calculations of heat loss use the same outdoor temperatures for the larger grid together with the building descriptions of the smaller grid data. Finally, Fig. 5.75 gives the hourly time dependence of heat requirements for the whole country.

The end-use energy given in Table 5.4 is the energy flow measured after any final conversion by the end-user. For example, that means the light (radiation) energy, not the electric power supplied to a light-producing device (fluorescent tube, etc.). For electric appliances the 2050 value is twice that for the year 2000, and for transportation it is three times the 2000 value, while as mentioned the space heating energy is lower than the present one. In Fig. 5.76, the time variation of the necessary electricity requirement for Denmark is shown. It is assumed to be the same as the measured one for the year 2000, except that the absolute magnitude will be increased.

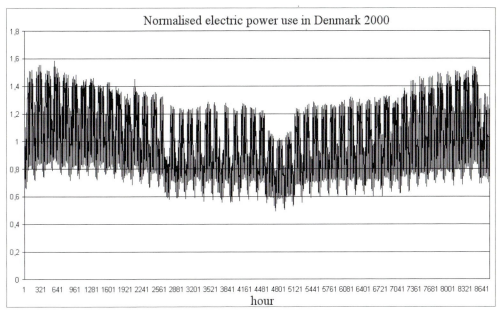

*Figure 5.76.* Normalised hourly variation of electricity use in Denmark for the year 2000, used for the 2050 scenario to model time variation of necessary electricity use (i.e., excluding use of electricity for industrial processes, etc., which may have a different time variation) (Sørensen *et al.*, 2001).

### Available renewable resources

Denmark is well endowed with renewable energy sources, with an identified wind potential far exceeding power demands, forestry and particularly agriculture producing primary food for some five times the Danish population, plus plenty of residues that could be used for energy purposes, and finally solar energy which on average is not small, but due to the high latitude (around 56°N) has a pronounced seasonal anti-correlation to loads (in addition to the diurnal mismatch).

Figure 5.77 shows the wind potential for Denmark, on land and inland seas. The off-shore potential has been estimated for a number of reserved locations, shown in Fig. 5.78. These have been selected as suitable for wind turbine parks without disturbing fishery activities, boat routes for passengers and freight, military exercise areas, etc. The total exploitable wind power is given in Fig. 5.78 for each reserved area. It by far exceeds the envisaged 2050 electricity use. In several of the areas, wind power production has already been initiated, albeit at a total level substantially lower than that of the 2050 scenario. The year 2003 installed capacity is about 3.3 GW.

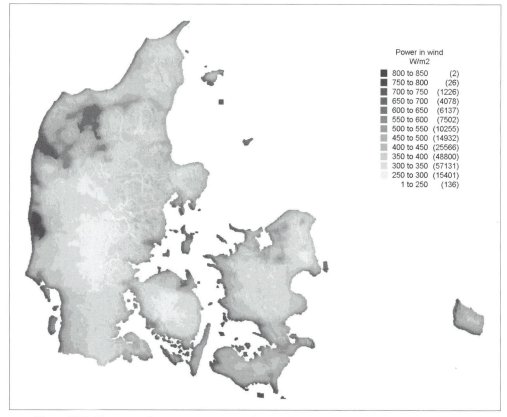

Power in wind
W/m2

| | | |
|---|---|---|
| ■ | 800 to 850 | (2) |
| ■ | 750 to 800 | (26) |
| ■ | 700 to 750 | (1226) |
| ■ | 650 to 700 | (4078) |
| ■ | 600 to 650 | (6137) |
| ■ | 550 to 600 | (7502) |
| ■ | 500 to 550 | (10255) |
| ■ | 450 to 500 | (14932) |
| ■ | 400 to 450 | (25566) |
| ■ | 350 to 400 | (48800) |
| ■ | 300 to 350 | (57131) |
| ■ | 250 to 300 | (15401) |
| | 1 to 250 | (136) |

*Figure 5.77.* The annual average power in the wind some 70 m over Danish land area, estimated on the basis of geostrophic winds (winds aloft) and the surface roughness (caused by vegetation, buildings, etc.) traced in 24 different directions at each grid point. In the legend, the number of grid cells with power in a given interval is given. A wind turbine is typically able to convert 30-40% of the power in the wind to electricity. For off-shore wind potential, see Fig. 5. 78 (based on Energi og Miljødata, 1999; Sørensen *et al.*, 2001).

Time variations in wind power are determined by passage of weather front systems. Typical passage of fronts over Denmark has a pattern exhibiting similar conditions over periods of a few days. This has implications for the amount of energy storage that would remove the mismatch between variable wind power production and demand. Lulls are followed by periods of high winds with delays of typically 3-6 days and very rarely over 12 days.

Time variability of wind is discussed in more detail in Sørensen (2004a). An impression of the situation can be obtained from looking at Fig. 5.84,

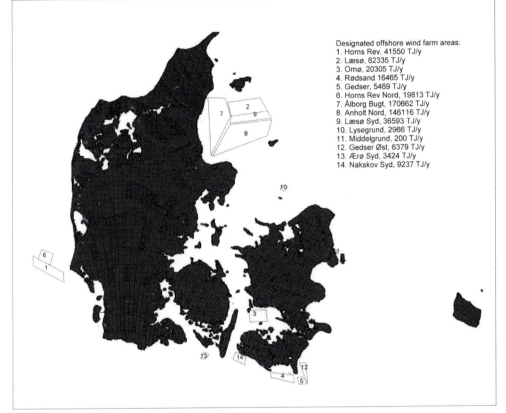

*Figure 5.78.* Off-shore areas set aside for wind power production by Danish planning legislation. For each area, the legend gives an estimate of the total annual production possible from wind parks in that area. Considerations of minimum spacing between turbines and optimal configuration enter these estimates (based on Danish Power Utilities, 1997; Sørensen *et al.* 2001).

which gives the hourly variation for the total Danish production from the wind turbine capacity needed for the 2050 scenario described in the following subsection.

For bio-energy resources, residues from households and food industry are available year round, and fresh biomass, whether harvested or collected, can usually be stored, so that conversion may take place at a time convenient for the conversion capacity or desirable from the point of view of users. The scenario use of biomass is primarily for production of biogas, liquid biofuels such as methanol or eventually hydrogen.

For solar energy, the availability in Denmark, including spatial and time variations, was used in Sørensen (2004a, Ch. 6) to illustrate system simulation techniques. In the present scenario, only modest contributions from solar energy (thermal panels and photovoltaics) are counted on, so the details will not be repeated here.

## Construction of 2050 scenarios for Denmark

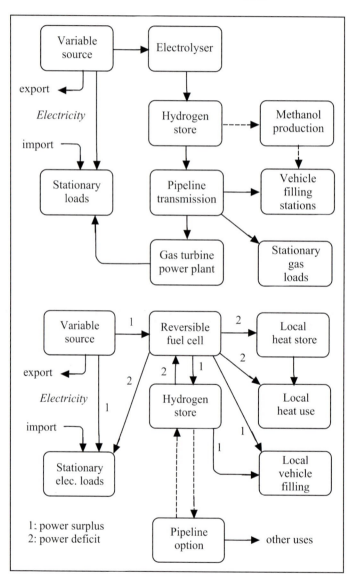

←A

*Figure 5.79.* Structure of hydrogen and renewable energy scenarios for 2050. The main source of hydrogen is surplus renewable energy (wind and solar power available at times where production exceeds demand). In the centralised scenario (A, on top), hydrogen production and storage is at central locations and pipeline transmission carries hydrogen to load areas such as filling stations. For the decentralised scenario (B, below, cf. Fig. 5.3), hydrogen production from excess power takes place in buildings, as does storage and vehicle filling. Reversible fuel cells may regenerate electricity at times of insufficient production (Sørensen *et al.*, 2004).

←B

The idea behind the two hydrogen and renewable energy scenarios considered for Denmark in the year 2050, but generally applicable for any part of the world with proper substitution of the renewable energy sources most relevant for the particular region considered, is depicted in Figs. 5.79. The centralised scenario (A) assumes hydrogen to be produced and stored in a few locations and distributed by pipeline. The decentralised scenario (B) assumes building-integrated production and storage.

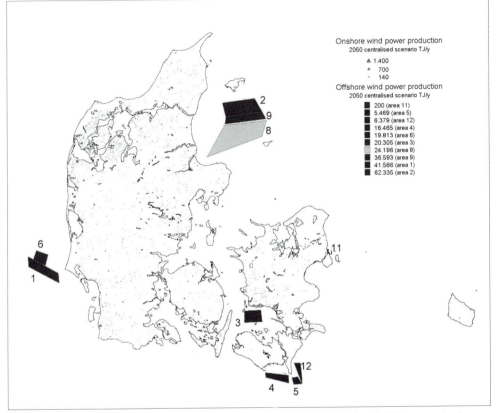

*Figure 5.80.* Total annual wind power production in the centralised 2050 scenario for Denmark. The caption indicates how much off-shore production is derived from each numbered area (Sørensen *et al.*, 2001)

*Centralised scenario*

The centralised scenario corresponding to the outline in Fig. 5.79A is based on primary energy production from a mix of renewable energy sources available in Denmark. As seen from the overview scenario energy flows in Fig. 5.83, the most significant Danish energy source is wind power, followed by bio-energy production from agricultural and forestry residues (i.e., not interfering with food or wood production by introducing dedicated energy crops).

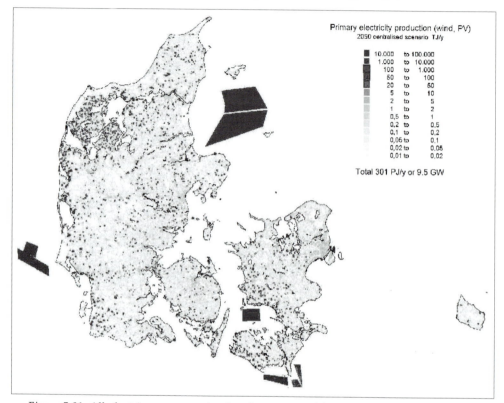

*Figure 5.81.* All electric power production from renewable sources in the Danish 2050 centralised scenario, including the wind production on-shore and off-shore (as indicated in Fig. 5.80), plus solar power from rooftop photovoltaic installations placed on south-facing building surfaces and roofs (of which about 25% of all suitable surfaces are assumed to be in use). Grid cells with less than 10 TJ/y production represent solar energy, the ones with higher cell production wind.

The renewable energy mix available in Denmark is dominated by energy conversion into electricity, making available electric power far exceed demand of this energy form. This means that electricity can be used for all medium- and high-temperature process heat and can supplement solar heat (the total production of which is given in Fig. 5.83) in case of insufficient heat production (heat stores empty). This conversion is assumed to use heat pumps for maximum efficiency. The time distribution of power production is such that large surpluses are produced during high-wind periods, as indicated in Fig.

5.84 (a fuller discussion of time variations in wind production may be found in Sørensen, 2004a). The surplus power production is assumed to be used for production of hydrogen in centralised facilities, placed approximately where there are current electric power plants (of which Denmark currently has about 30 000, including larger plants of 200-1000 MW as well as smaller combined heat and power plants, typically rated at 10-200 MW). The distribution of such hydrogen production sites is shown in Fig. 5.82.

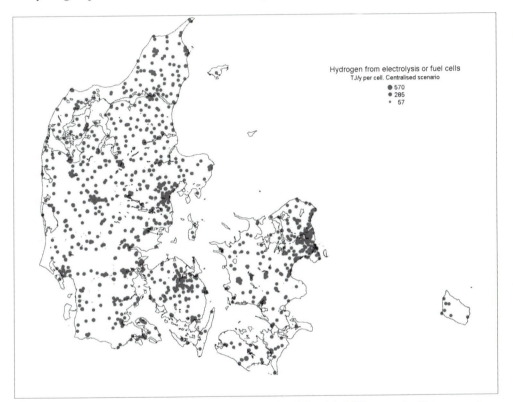

*Figure 5.82.* Hydrogen production in the centralised Danish 2050 scenario (Sørensen *et al.*, 2001).

In Fig. 5.81, the geographical distribution of the total electricity production in the 2050 scenario is shown, adding photovoltaic production (21 PJ $y^{-1}$) to that of wind (280 PJ $y^{-1}$). In climatic conditions other than the Danish one, the optimum share of photovoltaic relative to wind power may be much higher. The hourly time distributions of wind and photovoltaic power are given in Figs. 5.84 and 5.85 and are combined in Figs. 5.86 (year) and 5.87 (week), with the demand for electric power indicated (dedicated power plus other demands covered by electricity in the scenario, cf. Fig. 5.83).

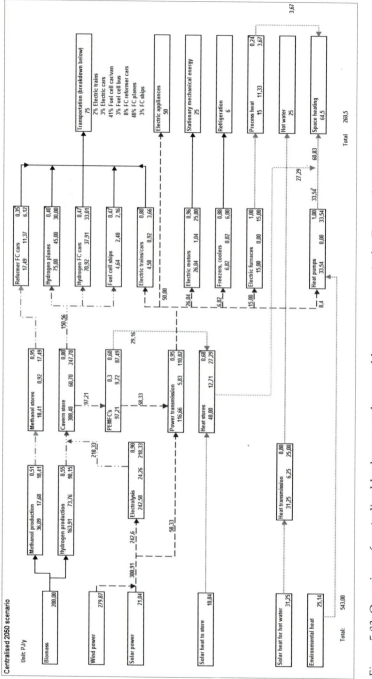

*Figure 5.83.* Overview of centralised hydrogen and renewable energy scenario for Denmark 2050. Each box represents a conversion step. The numbers in the lower left of the box is the input energy, that in the lower right the output energy. The middle number is the difference between input and output, i.e. the energy loss (all numbers in TJ/y). The number in the upper part of the box is the assumed conversion efficiency. In case of combined power and heat production there are two efficiencies. The different signatures of lines between boxes indicate energy forms. In some cases, there is recycling, e.g., of waste heat. In the upper right is a breakdown of transportation end-use energies (based on Sørensen *et al.*, 2001).

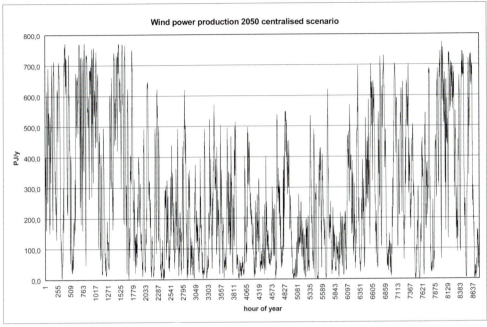

*Figure 5.84.* Hourly wind power production in the centralised 2050 scenario for Denmark.

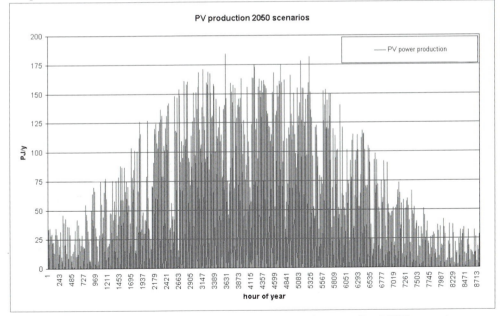

*Figure 5.85.* Hourly photovoltaic power production in the centralised 2050 scenario.

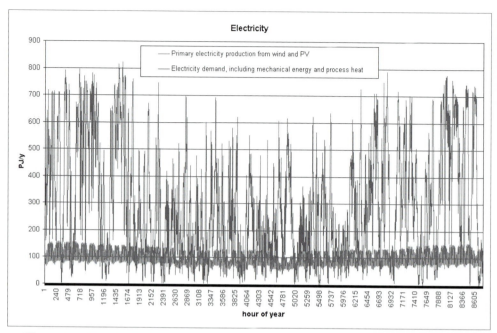

*Figure 5.86.* Hourly total power production and use in the centralised 2050 scenario (Sørensen *et al.*, 2001).

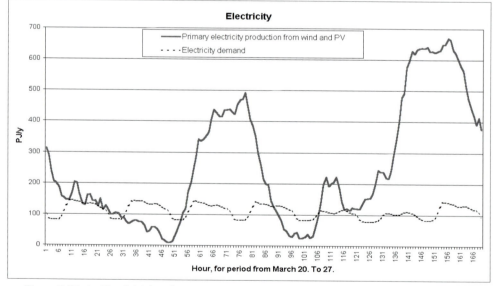

*Figure 5.87.* As Fig. 5.86, but for a single spring week.

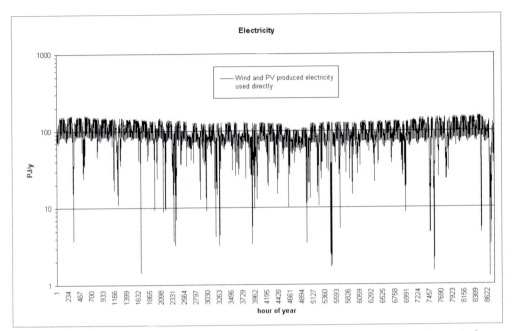

*Figure 5.88.* The amount of electricity demand that can be covered directly from wind and photovoltaic power as it is produced (Sørensen *et al.*, 2001).

The 2050 scenario assumes that as much electricity as possible produced by variable sources is used immediately in order to avoid losses associated with going through storage cycles. Dedicated electricity uses have top priority, followed by mechanical energy needs (motors) and high-temperature industrial process heat. Then follows heat pumps and lower temperature heat, in case available heat stores are at a low level of filling. The surplus still remaining is used to produce hydrogen. This still amounts to 242 PJ y$^{-1}$, implying that most of the electricity produced goes through the hydrogen cycle, either to be used as hydrogen in industry or in transportation or to be reconverted to electricity at times when production is lower than demand. Figure 5.88 shows the direct electricity delivery, showing the effect of the low wind periods. They are seen to be timewise shallow and fairly evenly distributed over the year. One reason for this is the positive seasonal correlation between demand and power production, and the other is the structure of weather system passages. As a result, the storage requirements for hydrogen are favourable, requiring no more than days and occasionally a few weeks of energy usage to be held in the hydrogen stores. The scenario assumption regarding inclusion of hydrogen stores is discussed below. The location of stores is shown in Fig. 5.2.

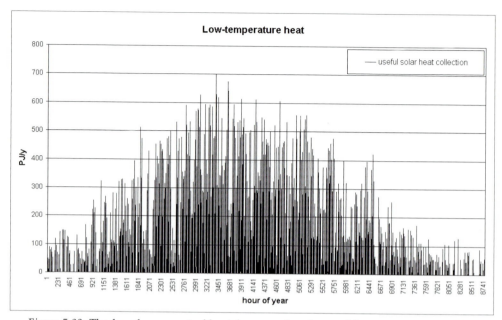

*Figure 5.89.* The hourly amount of heat for space heating and hot water produced by solar thermal collectors in the 2050 scenario.

The amount of solar heat produced by thermal collectors in central configurations or mounted on rooftops or building surfaces is shown in Fig. 5.89. Its variation over the year is similar to the power production by photovoltaic collectors mounted on similar surfaces, except for a spring-autumn asymmetry caused by the higher thermal storage and hence collector inlet temperature causing lower efficiency in autumn. Some of the installations may be combined photovoltaic-thermal collectors, where the waste heat from a photovoltaic panel is made useful (cf. Sørensen, 2004a). As power production by the photovoltaic surfaces is only around 15% of incident solar radiation, the thermal collection may still be near 50% of the energy in the solar radiation, as in the case of pure thermal collectors.

Figures 5.90 and 5.91 show the hourly collected solar heat along with the low-temperature heat demand. Building heat losses and performance of the solar heat system are modelled as described in Chapter 6 of Sørensen (2004a). It is seen as expected that demand cannot be met in winter, whereas summer heat collection with suitable heat storage (day-night or a few overcast days) is consistent with the demand, which during summer is primarily for hot water.

Figure 5.92 shows the annual time variation in use of electricity for hydrogen production, by use of reversely operated fuel cells or conventional electrolysers, along with the much smaller use for heat pumps. A detailed view for a

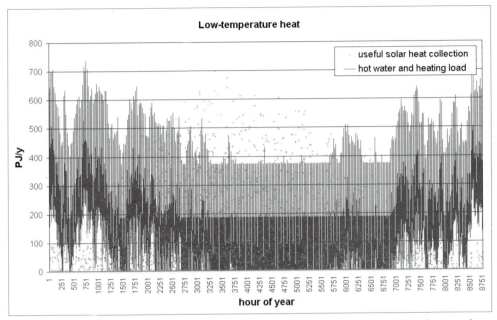

*Figure 5.90.* Comparison of low-temperature heat collection by thermal solar panels with demand, on an hourly basis, for the 2050 scenario.

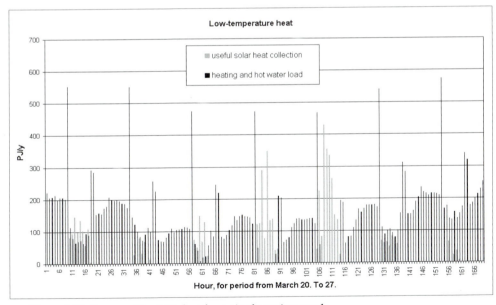

*Figure 5.91.* Same as Fig. 5.90, but for a single spring week.

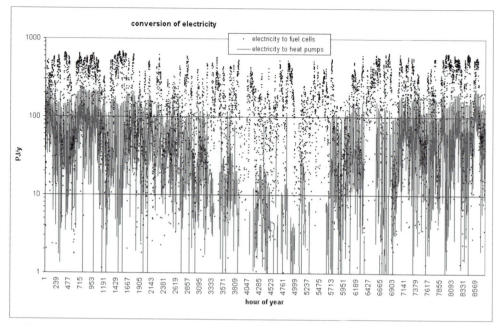

*Figure 5.92.* Hourly amounts of electricity used for hydrogen production by fuel cell electrolysers and for heat pumps in the Danish 2050 scenario.

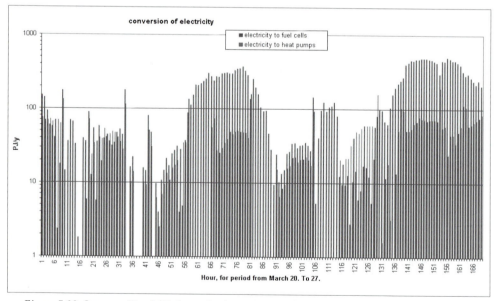

*Figure 5.93.* Same as Fig. 5.92, for a single spring week.

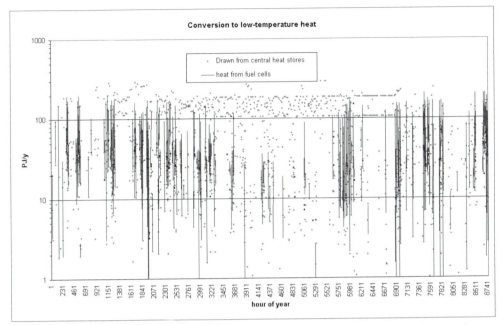

*Figure 5.94.* Sources of low-temperature heat in the 2050 Danish scenario. Time variations are shown for heat drawn from central stores (filled using heat pumps at times of excess electricity production) and waste heat from central fuel cell operation to generate power. This heat is distributed through district heating pipelines.

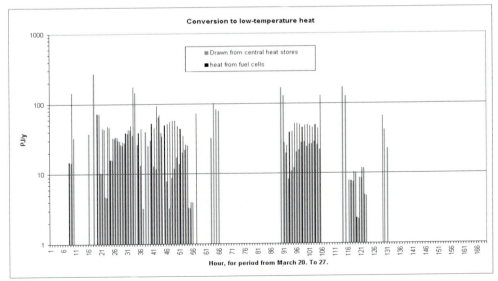

*Figure 5.95.* Same as Fig. 5.94, but for a single spring week.

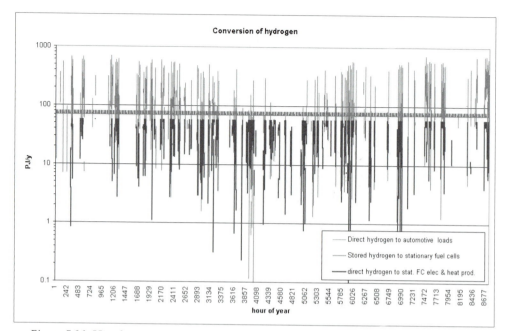

*Figure 5.96.* Hourly conversion of hydrogen in the 2050 scenario. The curve varying regularly between 70 and 80 PJ/y is the (schematic) consumption by vehicles, and the strongly varying curves are the fuel cell electricity production in cases of insufficient direct power production. The lower part is direct use of hydrogen when production is ongoing, and the top part is based on hydrogen drawn from stores.

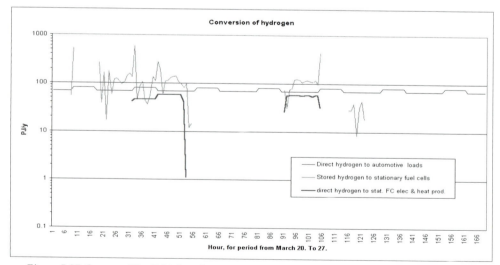

*Figure 5.97.* Same as Fig. 5.96, but for a single spring week.

single spring week of the same is shown in Fig. 5.93. Figures 5.94 and 5.95 show how waste heat from hydrogen fuel cells may provide additional low-temperature heat, although smaller amounts than those drawn from central heat stores, except for the winter period.

In Figs. 5.96 and 5.97, the time variations in the disposition of hydrogen in the energy system are depicted. The automotive sector is assumed to use hydrogen through filling stations distributed geographically as shown in Fig. 5.2. No seasonal variation has been assumed, but only a variation over the hours of the day. Regeneration of electric power by use of fuel cells appears in Fig. 5.96 (and for a single week in Fig. 5.97) as fairly narrow features. Those in the lower part of the figures are based on direct hydrogen supply during hours of excess power production. The larger amounts appearing in the upper part of the figures are based on hydrogen drawn from one of the two central hydrogen storage facilities shown in Fig. 5.2. According to the overview in Fig. 5.83, 97 PJ $y^{-1}$ of hydrogen is used to regenerate 50 PJ $y^{-1}$ of electricity, while the hydrogen used in the transportation sector is 100 PJ $y^{-1}$. The scenario assumes hydrogen used not only in passenger cars, but also in airplanes and ships. The demand scenario is based on a continued expansion of air traffic, making the energy requirement for planes and cars about equal in the year 2050.

The large Danish agricultural sector, along with managed silviculture, produces biomass residues allowing a fairly large energy production, which in the scenario is assumed in the form of methanol (18.4 PJ $y^{-1}$). This liquid fuel is taken to be used in particular subsectors of transportation, such as lorries and possibly buses. Additionally (in view of the abundance of electricity characterising a renewable energy-based future Danish situation), electric vehicles are assumed to be present, including all trains and a special fleet of small city cars available in the only large city of Denmark, Copenhagen. These details are shown in Fig. 5.2. It is also seen how the two central hydrogen stores in the scenario are conveniently located in different regions of the country. This is not surprising, since they are assumed to be modifications of the already existing natural gas stores serving the current energy system as strategic security stores created as a safeguard against rupture of the main natural gas pipeline from the North Sea.

In order to quantify the hydrogen storage requirement, a time simulation of the entire system has been made, using a specific and typical year of hourly data shown above for wind, solar radiation and various components of energy demand (scaled to 2050 usage assumptions). The outcome is a sequence of hourly data quantifying the degree of filling of the two hydrogen stores (one aquifer and one salt cavern, cf. Fig. 5.2 in section 5.1), shown in Fig. 5.98, along with variations in total hydrogen production (the upper ceiling being the rated capacity of the production facilities). The conclusion is that 60 000 PJh $y^{-1}$ of storage capacity allows the system to run smoothly at all times.

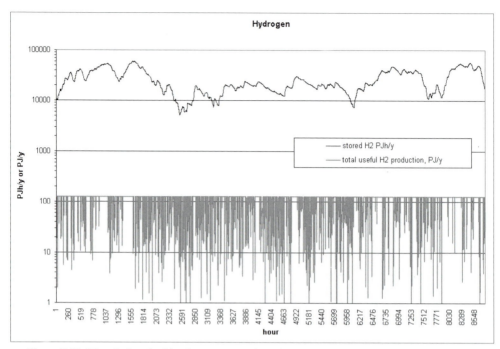

*Figure 5.98.* Hourly level of hydrogen stored in the two central underground facilities (top) and hydrogen production (lower part of figure) in the centralised 2050 renewable energy scenario for Denmark (Sørensen *et al.*, 2004).

The amount of 60 000 PJh y$^{-1}$ or around 7 PJ is actually only a fraction of the capacity of the existing natural gas stores (cf. section 2.4.1 in Chapter 2 and Figs. 5.1-5.2), implying that hydrogen may be stored at low pressure (say 5 MPa), which means low cost and little need to seal the storage caverns that are imbedded in clay layers of low hydrogen diffusivity. Figure 5.2 indicates that the centralised renewable energy hydrogen system has a fairly high infrastructure similarity to the present system.

*Decentralised scenario*
One may notice already today a tendency for decentralisation in society, replacing the centralised systems characterising early industrialisation by increasingly decentralised facilities. Much business activity (offices, sales outlets) has moved away from city centres. It has become attractive to offer employees work in rural environments for reasons of avoiding urban traffic and obtaining a relaxed environment. This has been made possible by current electronic facilities allowing communication from any location, facilitating not only industrial activities, but also residential recreation. Transport by road has

offered strong competition to earlier long-distance freight transport by rail-roads, increasing numbers of cars have substantially augmented mobility, and air transport for business and leisure has become commonplace in industrial-ised countries, where working schedules allow individual vacation trips to any destination in the world, often several times each year. Citizens of many countries have acquired personal command over energy for residential heat-ing as well as automobile transportation, and it is likely that the prospect of becoming one's own electricity producer by building-integrated fuel cells is appealing to many people. This would imply an acceptance of a possibly higher price for such installations, relative to the centralised solutions. Of course, this is true only within certain limits, but precisely fuel cell technology, at least of the PEM type, appears to offer a modular concept with rather mod-est mass production advantage. Furthermore, if the residential installation can not only provide electric power and associated heat, but also provide hydro-gen for the vehicle(s) parked in a garage, the concept does look very appealing to the paradigm of being in personal control.

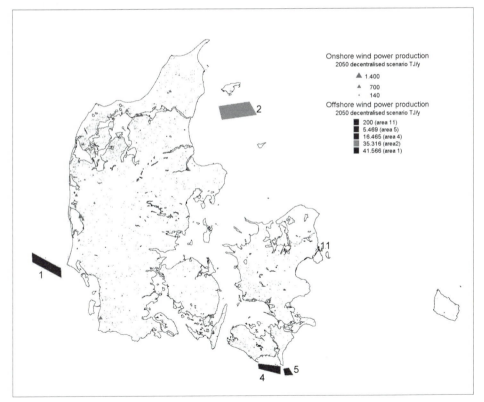

*Figure 5.99.* Wind power production in the 2050 decentralised Danish scenario.

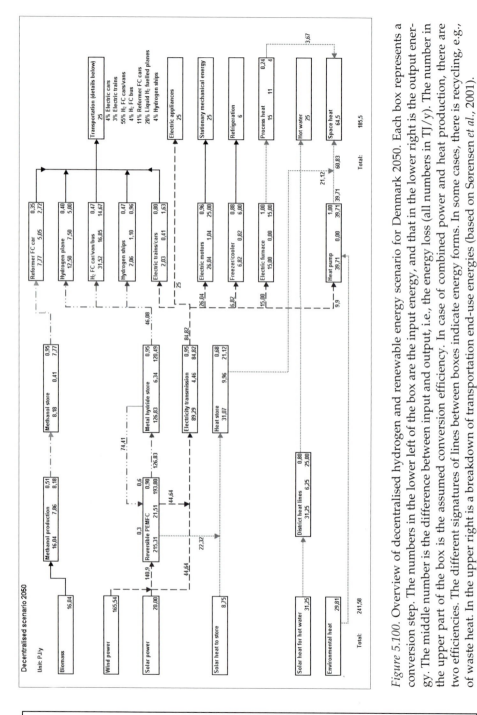

*Figure 5.100.* Overview of decentralised hydrogen and renewable energy scenario for Denmark 2050. Each box represents a conversion step. The numbers in the lower left of the box are the input energy, and that in the lower right is the output energy. The middle number is the difference between input and output, i.e., the energy loss (all numbers in TJ/y). The number in the upper part of the box is the assumed conversion efficiency. In case of combined power and heat production, there are two efficiencies. The different signatures of lines between boxes indicate energy forms. In some cases, there is recycling, e.g., of waste heat. In the upper right is a breakdown of transportation end-use energies (based on Sørensen *et al.*, 2001).

The outline shown in Fig. 5.79B indicates how this may be achieved. The figure assumes that reversible fuel cells are available in 2050, without the current problem of lower reversed operation efficiency than conventional electrolysers. The higher efficiency of fuel cells relative to current power plants implies that if waste heat should cover space and hot water heat requirements, these should be provided by more efficient means than the present ones.

Hydrogen produced by fuel cells
2050 decentralised scenario, TJ/y

| | | | |
|---|---|---|---|
| 100 | to | 1.000 | (10) |
| 50 | to | 100 | (107) |
| 20 | to | 50 | (614) |
| 10 | to | 20 | (2192) |
| 5 | to | 10 | (3841) |
| 2 | to | 5 | (5800) |
| 1 | to | 2 | (5859) |
| 0,5 | to | 1 | (8363) |
| 0,2 | to | 0,5 | (20311) |
| 0,1 | to | 0,2 | (20603) |
| 0,05 | to | 0,1 | (18406) |
| 0,02 | to | 0,05 | (14064) |
| 0,01 | to | 0,02 | (689) |
| 0,001 | to | 0,01 | (78) |

*Figure 5.101.* Geographical distribution of hydrogen production in the 2050 decentralised scenario for Denmark.

The decentralised 2050 scenario for Denmark is constructed in a somewhat different way from the centralised one, with significantly lower energy demand, particularly in the transportation sector. The centralised scenario assumes a standard activity development typical of current thinking by economists, where the transportation end-use demand by 2050 is three times higher than today. The decentralised scenario, on the other hand, considers this development as unattractive, in view of the already annoying congestion on roads and in the air, and opts for a more modest growth, as seen in the scenario overview shown in Fig. 5.100. The transportation activity level remains at the present and so does the energy use for electric appliances. The latter assumption does of course not mean that the activity level is stationary, but expresses the view that the efficiency of new electricity-using devices improves faster than the number of such devices.

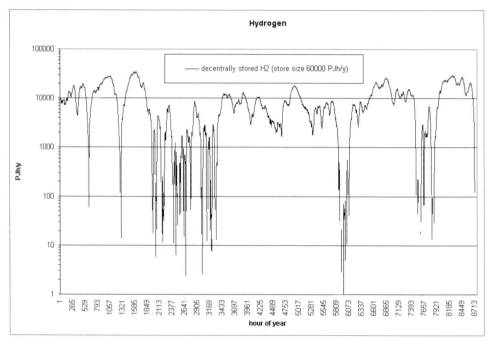

*Figure 5.102.* Hourly level of hydrogen stored in all building-integrated hydrogen storage facilities (such as metal hydride stores) in the decentralised 2050 renewable energy scenario for Denmark (Sørensen *et al.*, 2001).

The fairly extreme emphasis on energy efficiency underlying this assumption is not based on a conservationist "enough is enough" philosophy, but rather expresses the well-documented fact that there are numerous energy efficiency improvements not undertaken today, despite the fact that they are by far cheaper (per unit of energy use avoided compared with an energy unit to be purchased) than any of the new energy system components considered in the scenarios (renewable energy and hydrogen converters). At present, many of the economically viable energy and technically feasible efficiency measures are not undertaken, because they have a lower status among decision-makers and consumers than new energy provision technologies (Sørensen, 1991).

As a result of these assumptions, the required energy production is less in the decentralised scenario as compared with the centralised one, allowing wind power production to be reduced to 106 PJ y[-1] and therefore occupying fewer of the designated off-shore sites around Denmark, as shown in Fig. 5.99. The hydrogen production sites are now all buildings, as shown in Fig. 5.101, in contrast to the limited number of sites used in the centralised scenario (Fig. 5.82).

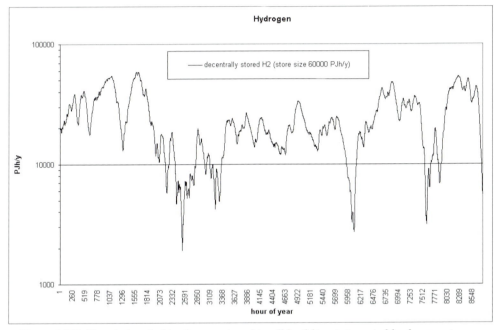

*Figure 5.103.* Hourly level of hydrogen stored in all building-integrated hydrogen storage facilities in the decentralised 2050 scenario, with off-shore wind production increased from 99 to 180 PJ/y (Sørensen *et al.*, 2001).

Figure 5.102 shows the hourly combined filling of all the hydrogen stores placed decentrally in buildings. The combined storage capacity is taken as the same as for the two central stores in the centralised scenario. It is seen that the system is more prone to having low storage filling, but still fulfils its purpose except for a few hours of the year. This is due to the higher demand on storage caused by not having hydrogen production based on biomass (omitted due to perceived concerns about removing carbon from the soil, carbon which is released to the atmosphere as $CO_2$ in compensation for carbon assimilated by the plants at an earlier time). Furthermore, the use of hydrogen in the building system is different from that of the centralised facilities. Increasing the stores does not alter this picture, and the emptying of the decentralised stores can only be avoided by increasing the production, e.g., by adding more electricity production from wind (the surplus being then converted into hydrogen). Figure 5.103 shows that completely emptying stores can in this way be avoided, although variations in degree of filling are still substantial. The extra cost of increasing production just for avoiding a few hours of hydrogen supply-demand mismatch is not considered justifiable, as the tiny amounts of extra hydrogen can be produced from imported electricity or from small reserve fossil plants being operated for only a few hours a year (e.g., gas turbines).

### Assessment of renewable energy scenarios

The renewable energy scenarios are more complex to model than scenarios based on fossil and nuclear energy. This is primarily because the efficiency of solar heat panels depends on the temperature of the water (or other medium) circulated through the solar collectors, causing a complicated coupling between the time development of heat loads, heat stores and solar collectors, where the performance of the system depends on its previous history of behaviour. The other complication is the substantial amounts of energy storage required to cope with the intermittence of renewable sources such as wind and solar energy. In fuel-based systems the fuel itself acts like an energy store. This behaviour is mimicked in the renewable energy scenarios by using intermediate fuels such as liquid biofuels and hydrogen, which can fairly easily be stored by use of the different kinds of devices discussed in Chapter 2. The scenario method has allowed a demonstration of the ability of the renewable energy systems to function in a dependable and resilient manner, under conditions of varying demand and natural variations in renewable energy inputs.

An important condition for this behaviour is the maintenance of a strong electricity grid of both regional transmission lines and local distribution lines. This grid transports renewable power to loads or to fuel cells (or other electrolysers) producing hydrogen and again transports power generated in fuel cells from hydrogen during periods of insufficient renewable energy inputs to the system. The additional grid systems available today are natural gas pipelines and district heating lines (both of which are abundant in the Danish energy system although not available to every building). These systems are supposed to be retained: the district heating lines for carrying heat to seasonal stores (which are not building-integrated) and for adding heat supply security, and the gas lines (converted to transport hydrogen) for additional flexibility in providing hydrogen for either transportation or stationary uses.

Heat demands are covered by a prioritised succession of options: first solar collectors, then heat stores and finally electric heating by heat pumps (which in areas with district heating lines may be placed centrally, for maximum efficiency). High-quality energy demands are first covered by direct renewable electricity and then by hydrogen (or methanol) or electricity regenerated from hydrogen. Surplus electricity after direct usage and heat pump demands is converted to hydrogen, and the power production capability (total rating) of wind and photovoltaic converters is chosen such that all stationary demands can be met by the above succession of options and all transportation energy needs can be met by available hydrogen, methanol or electric power. This determines the rating of the primary renewable energy collectors required, with values as given in Figs. 5.83 and 5.100 for the centralised and decentralised renewable energy scenarios. For wind power, the current 4000 turbines placed on land are supposed to be replaced by new ones before the year 2050, with a unit turbine rating of at least 2 MW. For off-shore wind parks, the decentralised scenario requires less than 2000 turbines, and the centralised scenario about 4000, if the average unit rating is 4 MW. The off-shore installations have a rated power of 6.8 GW in the decentralised scenario and 14.7 GW in the centralised one. The density of turbines on the water has been assumed to be 8 MW km$^{-2}$, in order to avoid significant "shadowing" effects (reduced power of narrowly spaced turbines). Methanol production from biomass uses technology described in Sørensen (2004a).

In describing a plan for implementation over the period from now until the year 2050, a gradual lowering of price as a function of accumulated manufacture of critical cost items such as fuel cells and solar cells must be assumed, as further discussed in Chapter 6. An important technical issue in determining cost is the lifetime of fuel cell components. Other technical issues relate to infrastructure changes implied in the transition from the present system to the scenario pictures of hydrogen filling stations, hydrogen stores and transmission. There are further areas presently very far from viability, such as hydrogen-fuelled airplanes. It would be natural during the transition period to reserve the declining fossil fuel usage to such applications. Filing station replacement suggests an implementation strategy beginning with dedicated fleets of vehicles to operate on hydrogen, and regarding hydrogen pipeline systems it may be necessary to establish new parallel ones during a transition period when natural gas pipelines are still used for natural gas. The alternative of driving hydrogen containers around on the road system is not attractive in a system aiming at using resources efficiently. While central hydrogen stores can be established fairly easily (also without using existing natural gas facilities, as many suitable geological formations are available in most parts of the world), the safer alternatives to compressed hydrogen storage in containers, such as metal hydrides, will need a few decades to reach viability, even if the technical problem solution goes smoothly. Fuel cell electricity production may be by low-temperature PEM cells (or further developments using intermediate temperatures) and later SOFCs for stationary uses. Electrolysis may be by reversely operated fuel cells, optimised for either hydrogen production or two-way (reversible) cells. Conventional electrolysis is a fallback option.

Countries such as Denmark accomplishing an early penetration of fluctuating renewable energy sources will be more interested in an early entry of hydrogen technologies. The main driving forces of increasing fossil fuel prices and delivery problems in connection with the bulk of such resources located in politically unstable regions make it less likely that early introduction of hydrogen based, during a transitional period, on fossil fuels will be successful, except in particular areas where a surplus of fuels such as natural gas may be available locally.

# 5.6 Problems and discussion topics

### 5.6.1
Try to figure the succession of events that would be needed for a smooth transition from the present infrastructure to the one required for hydrogen use in stationary and mobile sectors.

### 5.6.2
Construct a scenario for your country or region in analogy to the ones described in this chapter. First, enumerate the energy sources that would be most appropriate in your setting, and then use available data (or the scenarios in this chapter) to determine the amount of each primary energy form needed for a consistent system matching supply and demand. You may want to make different assumptions regarding future demand (activity expansion and efficiency improvement), and thereby see what restrictions the demand assumptions make on the other options.

### 5.6.3
Discuss how long it would be possible to maintain a hydrogen-based energy system if all hydrogen had to be produced from natural gas as the sole primary resource (e.g., use the natural gas reserve estimates given in section 5.3).

# SOCIAL IMPLICATIONS

## 6.1 Cost expectations

### 6.1.1 Hydrogen production costs

The cost[a] of hydrogen production from natural gas by steam reforming methods is well established once the fuel input cost is known (Amos, 1998; Longanbach *et al.*, 2002; Simbeck and Chang, 2002). At low (pipeline) pressure, it is about 1.0 US$ kg$^{-1}$ based on inexpensive natural gas available at the time of estimation (assumed at 1.5 US$ GJ$^{-1}$ or under half of the 2004 price). Filled into pressurised flasks the price is 30% higher, and liquefied is more than twice higher. On the other hand, the smaller volume makes transportation costs lower, and dispensed from a central hydrogen production plant to the final user, the hydrogen from pipeline natural gas has the highest cost (5 US$ kg$^{-1}$) and that based on liquefied gas has the lowest cost (3.7 US$ kg$^{-1}$), depending of course on the assumptions regarding transportation distances and technology. Simbeck and Chang (2002) estimate that the delivered $H_2$ cost from biomass waste is only about 2.5 US$ kg$^{-1}$ and from electricity (at 2001 US electricity costs) by electrolysis about 5 US$ kg$^{-1}$. Conventional small-scale electrolyser $H_2$ costs have been estimated by Fingersh (2003) and by Padró and Putsche (1999) as 8-12 US$ kg$^{-1}$, but larger units using surplus

---

[a] One must generally be careful in using cost estimates from different countries with different taxation and subsidy structures. In particular, one should be concerned over use of current energy costs for estimating break-even costs of new technologies.

wind power may attain a cost down to 2 US$ kg$^{-1}$. Based on coal gasification, the H$_2$ cost is estimated as over 12 US$ kg$^{-1}$. No externalities are included in these cost estimates.

If conversion of surplus electricity to hydrogen is made by a fuel cell in reverse operation, but paid for on the basis of its electricity production, and if the cost of a reversible fuel cell is similar to that of a one-way cell, then the hydrogen cost based upon fuel cell electrolysis may be as low as the cost of the power used (depending on how costs are allocated to the two process directions). The power used for hydrogen production could thus be off-peak power or surplus power from renewable energy systems such as wind and photovoltaics, which cannot be made useful for covering demand at the time of generation. Today, the cost of such power in auction pools is often lower than 2 (US- or euro-)cents per kWh, corresponding to about 6 US$ GJ$^{-1}$ of hydrogen at some 90% conversion efficiency.

The cost of hydrogen from (most likely lignocellulosic) biomass gasification and shift reaction (cf. Chapter 2) has been estimated as 9-14 US$ GJ$^{-1}$ by Faaij and Hamelinck (2002), with a long-term potential for reduction to 4-9 US$ GJ$^{-1}$. The higher value reflects inclusion of the production cost under European conditions (most likely of forestry), and if discarded residues of no economic value can be used, lower price outcomes are possible. Production of biomass-based hydrogen from photosynthetic algae involves even higher costs. Kondo *et al.* (2002) estimate hydrogen costs of 235-574 yen m$^{-3}$ or 188-460 US$ GJ$^{-1}$, for a 5000 m$^3$ day$^{-1}$ prototype system based on a *Rhodobacter sphaeroides* mutant.

In case methanol is used as an intermediate fuel (e.g. in mobile fuel cells), the cost of methanol production is of interest. Produced from fossil fuels, notably natural gas, at a price of 3 US$ GJ$^{-1}$, reforming or series reactor schemes lead to a methanol production cost estimated around 5.5 US$ GJ$^{-1}$ (Lange, 1997). Advanced micro-structured string-reactors for this concept are under development (Horny *et al.*, 2004).

## 6.1.2 Fuel cell costs

Concepts such as molten carbonate or solid oxide fuel cells are not expected to reach a commercial stage during the next decade. Preliminary cost estimates for a possible commercialisation are at 3200 US$ kW$^{-1}$ by 2010, declining to 1300 US$ kW$^{-1}$ by 2050, according to a recent assessment of available data (Fukushima *et al.*, 2004). Problem areas include the availability of La, a material currently used in high-temperature ceramics. Recycling should be pursued to a high rate.

Due to their early success, alkaline fuel cells (AFCs) are still less expensive than proton exchange membrane (PEM) fuel cells, although both are still way over the requirements for commercial use. However, the prospects for rapid cost decline envisaged (and necessary) for PEM cells do not seem to be paralleled for AFCs. McLean *et al.* (2002) explain this as due to lack of interest in further development of the AFC technology. They quote the cost of a small batch of AFCs as 1750 US$ kW$^{-1}$ with a potential for reduction to 155 US$ kW$^{-1}$ for mass production at a large but unknown scale. They quote the cost of a small batch of PEM cells as above 2000 US$ kW$^{-1}$, but with potential for reduction to 20 US$ kW$^{-1}$ for a mass-produced automotive 50-kW stack. This roughly agrees with the target of 30 euro kW$^{-1}$ to be obtained by 2025 for a level of production amounting to 250 000 fuel cell units per year (Danish Hydrogen Committee, 1998). Peripheral components (gas circulation systems, piping and electronics plus electrolyte recirculation for AFC, humidifier for PEM) are estimated to cost three times more for PEMs than for AFCs, and although these are small compared with current fuel cell stack costs, they would become more important if the stack costs reach the lower bound.

More recent estimates of the gain by mass introduction of PEM cells into vehicles arrive at a future cost as a function of both the accumulated volume of production and the two parameters: power density improvement (increasing from current 2 to 5 kW m$^{-2}$) and speed of maturing (taken as slope of an assumed logarithmic learning curve), that ranges from 15 to 392 US$ kW$^{-1}$ (Tsuchiya and Kobayashi, 2004). The lower cost estimate assumes an accumulated 5 million fuel cell vehicles by 2020, with average rated fuel cell power of 110 kW, while the upper estimate assumes 50 000 vehicles and a fuel cell power density having reached 3 kW m$^{-2}$, with the cost being suggested to be a possible minimum achievement for the year 2010.

Vehicle PEM fuel cells are currently designed for a lifetime of around 5000 h, in contrast to the 40 000 h set as the minimum for continuous stationary uses. The current semi-commercial PEM cells used, e.g., in small series of fuel cell vehicles, do not reach these lifetime goals. Lipman *et al.* (2004) estimate a break-even price for stationary fuel cells as 1200 US$ kW$^{-1}$ for 5-kW home systems and 700 US$ kW$^{-1}$ for larger, 250-kW systems (for given expectations of future fossil fuel prices). They also suggest that vehicles when parked in a garage could contribute to electricity production for the building system. This would seem at variance with the shorter lifetime of the fuel cells designed for vehicle use.

The current introduction of the first small-series, semi-commercial fuel cell units, for vehicles and for building usage, should provide a better basis for predicting cost reduction potentials in the long term. It may be instructive to compare with the empirical learning curves for other technologies in the energy sector, such as wind and photovoltaic power, or with high-density battery developments. Figure 6.1 gives the results of an analysis of the wind and photovoltaic learning curves. Economists often describe such findings in terms of straight-line logarithmic behaviour, writing the cost $Y$ as a function of accumulated volume of production $X$,

$$\log Y(X) = -r \log X + \text{constant}. \tag{6.1}$$

The slope $-r$ is sometimes quoted in terms of a "progress ratio" $PR = 2^{-r}$ or the "learning rate" $LR = 1 - PR$.

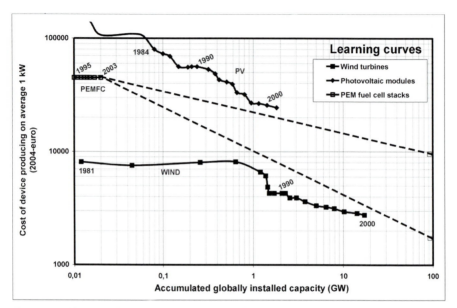

*Figure 6.1.* Observed learning curves for wind turbine and photovoltaic module costs used to suggest possible learning behaviour for PEM fuel cell stacks. The total accumulated capacity at a given time is along the abscissa, and the average cost of power production is along the ordinate. It is taken as the cost per kW of installed power divided by $c_p$, which is the ratio between average actual production and the device rated power level (i.e. the equivalent fraction of time at maximum power production). The dashed extrapolation lines for automotive PEM fuel cells correspond to learning at the rates characterising the wind and photovoltaic industry in recent years, respectively. Data for wind turbines are based on Lauritsen *et al.* (1996) and Neij *et al.* (2003), using globally installed capacities from Madsen (2002) and turbine list prices from Danish manufacturers (Energi- og Miljødata, 2002). The average production estimates are for class 1 location (best on-shore sites) and inferior to those of Danish off-shore sites or on-shore sites in locations such as the west coast of Ireland. Data for photovoltaic modules are from IEA-PVPS (2003; limited geographical coverage) and Schaeffer *et al.* (2004), based on European Photovoltaic Industry Association and Strategies Unlimited (2003) cost and shipping data. The early penetration PEM fuel cell cost is based on Danish Hydrogen Programme tenders 1998-2001 and on European Commission *Citaro F* bus project estimates 2002.

Figure 6.1 shows that quite complicated features are found in empirical learning curves (in 2004 euros by use of deflators from UK Treasury, 2004).

Looking at the wind turbine costs, the flat or rising cost curve during the 1980s reflects a period where tax evasion schemes made it possible to sell wind parks in the USA (particularly California) without worrying too much about cost. By the end of the decade, the Californian market collapsed and prices rapidly went down in an effort to penetrate new markets in other parts of the world. This was not successful, and several manufacturers went bankrupt (vertical part of the curve). After a restructuring of the industry, the diversification strategy worked better, and after a few years with a flat cost curve, the markets really took off and learning curve behaviour set in. The decade 1990-2000 exhibits a learning rate of about 10% in a steadily expanding market (Germany, India, Spain, Denmark, plus several others). The curve in Fig. 6.1 pertains to the performance at a class 1 wind condition site, where the coefficient of performance reached $c_P = 0.294$ by year 2000, compared with 0.20 by 1981. During the years 1997-2001, expansion has particularly been in off-shore wind farms, where the coefficient of performance is as high as 0.46 (Sørensen, 2004a). This is not due to any radical technical modification, but rather to a change in design philosophy away from aiming at maximum annual energy production to one where more annual operational hours have a higher priority (in practice obtained by keeping the blade design and pitch angle roughly as it were for the on-shore turbines).

For photovoltaic panels, the straight-line approximation works better, but closer inspection of the curve reveals a jumping from one plateau to a lower one every 3–5 years. This reflects the introduction of better technology each time new production equipment is introduced, but a competing influence is from the many public subsidy programs (particularly in Germany and Japan). These make it less attractive for the manufacturers to decrease prices, but they are eventually forced to do so by the tender condition set by the large public programmes. The overall learning rate is slightly above 20%, e.g., considerably higher than for the more mature wind technology. The coefficient of performance is around 0.16, meaning that on average over a year the panels produce energy corresponding to 16% of the one they would have produced if operated in full solar radiation at their maximum (rated) power all seasons and day and night.

Turning now to PEM fuel cells, the emerging technology nature does not allow even a starting price to be determined with much accuracy. In 1995, fuel cells were sold to research laboratories and prototype vehicles at prices lower than the starting price shown in Fig. 6.1. However, these cells were without any warranty and often worked for less than a year. When the technology started to take off around the year 2000, with PEM fuel cell buses and passenger cars coming into operation by 2003-2004 in small series of 30-80 units, most fuel cell manufacturers entered contractual agreements with automobile manufacturers at prices not generally known. Efforts to purchase PEM fuel cells by national programmes faced difficulties, and the few manufacturers willing to quote prices demanded much more than in 1995. The market is still for test series, and the cells do not yet have the goal of 5 years of operational life. The prices used for Fig. 6.1 are taken from such tenders and from the European bus project material, assuming that the cell price constitutes about half of the total system price (the estimated cost of a *Citaro F* bus relative to a similar bus with a conventional diesel traction system). The power coefficient $c_p$ is taken as 0.12, corresponding to 5000 h in operation over 5 years. This is probably a high estimate, as the vehicles would not be operating all their driving time at full fuel cell rated power (except possibly for hybrid cars charging batteries with excess power).

The future reduction in PEM fuel cell prices is indicated in Fig. 6.1 in terms of two curves corresponding to learning rates of 10-20% corresponding to the limiting cases of photovoltaic modules and wind turbines over the recent decade. Even with the lower curve, break-even with current vehicles would not be reached until the cumulated production reached about 500 GW. However, difficulties with the market for oil products (peaking production, instability of major supplier countries) may make PEM fuel cell vehicles competitive at higher prices than the currently seen break-even price.

This discussion has shown that many factors influence price development, in addition to industrial "learning". One should also be careful with methods used in the literature, which are often very different from one another, e.g. using running instead of as here inflation-corrected prices or comparing prices and cumulative installations only for a single country (Ibenholt, 2002; Junginger *et al.*, 2004). The learning of manufacturers in a single country may be a reasonable indicator, as technological progress spreads rapidly across regions when not prevented by patent issues, but it is not always the case. Prices, on the other hand, may vary geographically due to a new kind of industry not possessing the sales and maintenance infrastructure world-wide, which is why costs for a particular country have been used in Fig. 6.1. Another reminder is that use of double-logarithmic plots can make just about anything look linear, especially if several decades are included for both abscissa and ordinate. Care has been taken in Fig. 6.1 not to encourage this feature, which is striking in a large part of the literature on the subject of learning curves.

A final warning regards comparing prices of energy-producing equipment with different lifetimes. Wind turbines have an established lifetime of around 25 years. The lifetime of photovoltaic panels is believed to be similar or even longer, at least for those based upon crystalline and multicrystalline cells. In contrast, the 5-year lifetime currently aimed at for both automotive and stationary fuel cells should be considered in comparing technologies.

One may thus argue that the fuel cell curves in Fig. 6.1 should be moved upwards by a factor of five, if the purpose is to compare different technologies. The break-even of fuel cell vehicles is not affected, because it is already based on the assumed equipment life. Only in case the lifetime goal cannot be reached (or is surpassed) will the assessment have to be modified.

The time required to meet the target costs for both stationary and mobile PEM fuel cell systems has been explored by the Delphi method (interviewing a number of experts) (Kosugi *et al.*, 2004). The result was a distribution centred around 17 years for both technologies. It matches very well the assumptions of Tsuchiya *et al.* (2004) mentioned above and is consistent with the goals of 40-200 US$/kW$_{rated}$ expressed as needed to break-even in regional fuel cell programmes in Europe, Japan and the USA (European Commission, 1998; Iwai, 2004; USDoE, 2002a; Chalk *et al.*, 2004). The fact that the United States estimate of the break-even price is at the lower end of the range is due to the low price (ignoring externalities) of current subsidised vehicle fuels in that country. The reason that Tsuchiya and Kobayashi (2004) arrive at a lower set of curves as compared with Fig. 6.1 is that they use a lower present cost of PEM fuel cells as their starting point for learning (1833 US$/kW$_{rated}$ or about 15000 euro/kW$_{av}$).

## 6.1.3 Hydrogen storage costs

Hydrogen storage costs comprise capital costs for the equipment used and operational costs, such as power for compression or liquefaction. Shayegan *et al.* (2004) quote the additional cost of hydrogen recovered from a liquefied store as about 5 US$ kg$^{-1}$ for small units, decreasing to around 1 US$ kg$^{-1}$ for high compressor/liquefier rating. The additional H$_2$ cost going through a compressed hydrogen store in containers is around 0.4 US$ kg$^{-1}$ for short-term storage, but rises with length of storage (Amos, 1998; Padró and Putsche, 1999). The large-scale underground hydrogen storage in caverns, abandoned natural gas wells, aquifers or salt domes mentioned in Chapter 2, section 2.4 and Chapter 5, section 5.1 has much lower costs (capital cost of establishing the store at 3-20 US$ kg$^{-1}$, leading to a storage cost of one order of magnitude below that of liquefied H$_2$ storage and two orders of magnitude below that of compressed storage) and constitutes a natural choice for centralised storage of hydrogen.

For decentralised stationary storage, metal hydride stores or one of the similar concepts described in Chapter 2, section 2.4 may be more attractive than compressed gas flasks, but the cost is difficult to estimate because the final designs are not available (e.g., geometrical layout allowing sufficiently rapid extraction). Capital costs for metal hydride stores have been estimated as in the range of 2 000-80 000 2004-US$ per kg of H$_2$ capacity (Amos, 1998).

The storage cycle costs are estimated at 0.4-25 US$ kg$^{-1}$ (Padró and Putsche, 1999). If successful, such stores may also be considered for automotive applications, although the weight will be a problem except for some non-metal chemical or carbon storage types. Best current guesses are for a cost that is a few times higher than that of compressed storage containers.

## 6.1.4 Infrastructure costs

The cost of hydrogen transmission by pipeline depends on the pipe diameter and hydrogen flow rate. By increasing the pressure difference through the pipeline, cost can be reduced more than the additional cost of compressors. Amos (1998) arrives at a cost of about 5 US$ GJ$^{-1}$ for an optimised flow rate of $10^6$ kg day$^{-1}$ through a 160-km pipeline. The cost of liquid hydrogen transport by road is estimated as lower, using current levels of transportation fuel (diesel) costs, and in any case the additional cost of liquefying the hydrogen makes the total price unattractive except for the very longest transportation distances (such as intercontinental hydrogen transport by ship).

Converting filling stations for road vehicles to dispense compressed hydrogen may add some 0.1 US$ kg$^{-1}$ to the hydrogen price, but alternatively, the hydrogen production may take place at the filling station site using any of the methods available. The necessary filling station conversion to hydrogen cost is less than one year's maintenance costs for the current system (cf. review by Padró and Putsche, 1999; Campbell, 2004b).

In-building generation of hydrogen (by electrolysis) and dispensing to garage-parked vehicles ("one-car filling station") are likely to double the price of hydrogen production and filling, respectively (Padró and Putsche, 1999).

In case hydrogen is produced from natural gas, it may be considered to recover $CO_2$ (which is already separated in most current steam reforming plants) and store it away from the atmosphere, e.g., in abandoned wells or as carbonate on ocean floors. The additional cost is estimated at 0.05-0.1 US$ per kg of hydrogen (Padró and Putsche, 1999).

## 6.1.5 System costs

The cost of fuel cell systems is partly the cost of vehicles and building-based systems and in a wider context the total cost of a hydrogen economy with production, various types of usage and infrastructure such as storage and transmission, distribution and filling outlets.

One reason that system costs may reveal things not possible to derive from the component costs is that each system component has an efficiency characteristic that often differs from that of the equivalent component (if there is one) in the current energy system.

Looking first at the cost of PEM fuel cell vehicles, the studies aimed at probing into future fuel cell costs have also addressed the total system cost of fuel cell, hydrogen storage, handling, batteries in case of hybrid vehicles, and power control, in order to arrive at a total price development of car manufacture. Figure 6.2 shows the result of a Japanese study, with the assumption of a fuel cell cost declining to 40 US\$ kW$^{-1}$ by 2020 (one of several scenarios) and a corresponding increase in the stock of fuel cell vehicles (passenger cars, lorries, buses, etc.) to 5 million by 2020 and 15 million by 2030. One might have expected the demand for fuel cell vehicles to rise as the price comes down and not only after it has come down, but the scenario is just an indication for three selected years and does not pretend to be dynamic. By 2020, the fuel cell vehicle cost (15 788 US\$ in constant \$s) is nearly as low as that of current gasoline cars (13 136 US\$ assumed in the study).

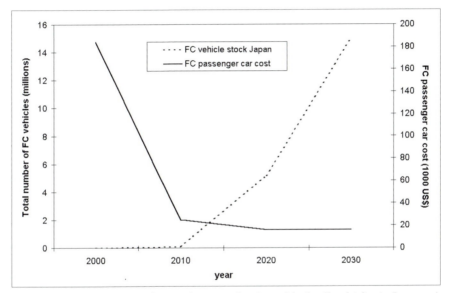

*Figure 6.2.* Scenario results for market introduction of fuel cell vehicles in Japan, giving the stock of fuel cell vehicles as a function of time, along with the decline in cost of passenger fuel cell cars (the stock comprises other types of fuel cell vehicles as well) (based on Tsuchiya *et al.*, 2004).

Figure 6.3 shows the amounts of hydrogen required in the Japanese scenario, along with the associated cost. Figure 6.4 shows the number of hydrogen filling stations required and the annual cost of constructing them.

The scenario assumptions made by Tsuchiya *et al.* (2004) comprise a slightly declining population in Japan, a modestly rising gross national product (GNP) and an unchanged demand for energy (achieved by the introduction of more efficient energy using equipment, not only in the transportation sector). The production of fuel cell vehicles rises from 50 000 in 2010 to 1.3 million in 2020 to 3.1 million per year in 2030, by which year the annual sales revenue is $59 \times 10^9$ US$ (production cost plus a 15% mark-up). The hydrogen activities constitute 1% of total Japanese GNP by the year 2030.

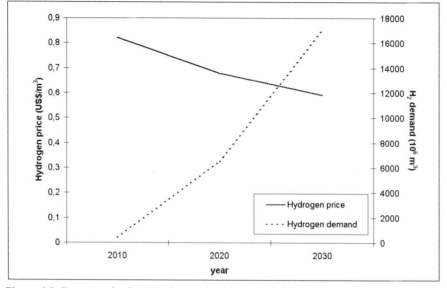

*Figure 6.3.* Japanese hydrogen demand and price delivered to the customer in the scenario of Fig. 6.2 (based on Tsuchiya *et al.*, 2004).

The cost estimates of key hydrogen handling equipment could be used to expand cost scenarios such as the Japanese one for road transportation to other sectors, e.g., the building-integrated use of fuel cells envisaged in the decentralised Danish scenario described in Chapter 5, section 5.5. In any case, such cost projections are highly uncertain, due to the wide range of future costs that may materialise for some of the most important components in a hydrogen-based energy system (cf. Fig. 6.1).

One criticism sometimes launched against the hydrogen infrastructure system is that the many conversion steps involved will make the overall efficiency of the system low (Bossel, 2004). From wind electricity through electrolysers to hydrogen, with transportation and storage losses, to finally regenerate electricity in buildings or fuel cell vehicles entails an overall efficiency of around 25%, which is seen as low compared with the present efficiency. The latter is between 13 and 30%, considering refinery losses (10%) and either vehicle combustion engines (average efficiency about 15%) or power plants (average efficiency at or below 35%). I believe the criticism is ill-placed, because the whole point is that the present fossil fuel system is drawing near its end and has to be replaced by another energy system. Since renewable energy offers the only sustainable future energy system, the handling of variable primary energy streams is a must. Handling such fluctuations requires storage of energy, and hydrogen seems to be the cheapest way of accomplishing this. Clearly, converting back and forth involves additional losses, but that cannot be helped, and a fair comparison of hydrogen systems would be with other systems capable of handling the large fluctuations of solar and wind energy, not with the obsolete system on its way out. Such a fair comparison will most likely designate hydrogen as the most viable energy storage medium, with underground hydrogen storage as very attractive compared with batteries, pumped hydro and other storage possibilities.

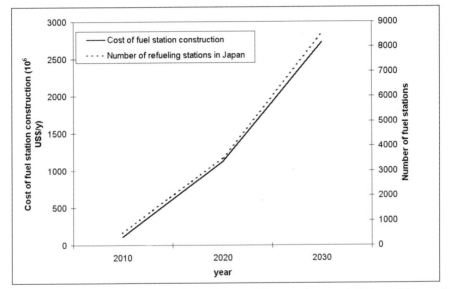

*Figure 6.4.* Japanese requirement for hydrogen filling stations and their cost (annual expenditure) for the scenario of Fig. 6.2 (based on Tsuchiya et al., 2004).

# 6.2 Life-cycle analysis of environmental and social impacts

The life-cycle assessment of hydrogen production is, in the cases where conventional fuels are converted into hydrogen, similar to many existing studies, in that major impacts are found to derive from air pollution and global warming issues (cf. Kuemmel *et al.*, 1997). I shall first, in section 6.2.1, give the definition of life-cycle analysis and assessment used here and subsequently a number of examples of the application of life-cycle analysis to hydrogen systems. This also brings about a discussion of environmental impacts from each component of the hydrogen systems, as well as other impacts that have been identified.

Among the examples of life-cycle analysis given will be one of direct conversion of biomass to hydrogen, one purpose of which is to clarify the claim that such hydrogen production pathways will reduce impacts. Although the technologies are emerging and a complete description of the industrial processing is not possible at the moment, there will be important implications from the generic assessment that can be made at present.

For fuel cell conversion of hydrogen, the industrial progress is such that meaningful life-cycle studies can start to be made on a fairly detailed basis. A concrete example of this will be given.

Finally, a life-cycle analysis of an entire system is described, choosing a passenger car as an important example. This involves both the analysis of the car manufacture, including the specific additions to traditional cars necessary for fuel cell operation, and also the infrastructure impacts and contributions from the fuel provision and from the final disposal of the product.

The short summary of the life-cycle method preceding the actual analyses is necessary, because life-cycle analysis is not a standard calculation. In the literature, one finds various definitions ranging from restricted net energy analyses over environmental impact studies to the full consideration of both environmental and social impacts. It is the latter methodology, more fully described in Sørensen (2004a), which will be employed here.

## 6.2.1 Purpose and methodology of life-cycle analysis

Technology assessment is a natural tool to use for evaluating both individual products and also more comprehensive systems aimed at handling general sectors such as that of energy supply. The purpose of the assessment may be licensing or establishing health and environmental requirements for a product or a system. To the individual customers, technology assessment is a tool for choosing between alternative ways of satisfying a given need, and for decision-makers, it may serve as a tool for evaluating industrial policy and allocating public investments.

Side effects of a particular technology may show up during its use, but might as well be associated with its production or procurement or with the disposal or recycling of the technology after use. The technology may also, during any of its life phases, require inputs of materials and services from other sectors of society, and they may again have negative effects associated with either of the life-cycle phases. It is therefore necessary to perform what is termed an entire "life-cycle analysis" (LCA), including such imbedded impacts, in order to be sure to present a fair comparison between different technologies, which may have their problem areas anywhere during the life-cycle outlined. It would certainly be wrong to compare energy technologies only on the basis of how they are produced and equally wrong to compare them only on their behaviour during use. For example, fossil fuels have their most severe impacts from emissions in case they are burned, whereas solar of wind energy have little impacts during the use phase. Nuclear energy has impacts (associated with radioactive waste) long after its usage phase, so, quite generally, fair comparison must be based upon a life-cycle assessment, following the gathering of necessary information which is the life-cycle analysis (Kuemmel *et al.*, 1997; Sørensen, 2004a).

The assessment problem is further aggravated by the fact that there is damage not being paid, neither by the manufacturer, by society nor by the customer. In such cases the damage will be left to deteriorate the collective value of society. Such impacts that escape the economic handling of technology may still be important for any comparative evaluation of alternatives. Costs not reflected in the marked prices associated with a technology are called "externalities". A more detailed discussion of the purpose and scope of life-cycle analysis, including ways to use an LCA, how to handle geographic distance and time displacement between occurrences of benefits and damage, and methodology for aggregation of data and for seeking common measures (such as monetary units) for presenting damage, may be found in Sørensen (2004a). Here, only a few remarks will be made, with the purpose to facilitate understanding the case studies presented below.

The actual work required for making a life-cycle analysis and assessment of a technology such as fuel cells may be summarised in the following way:

- Make an inventory of substances or processes with potential harm, from production through the use phase to final decommissioning,

- Perform an impact analysis including environmental effects, health effects and other less tangible social effects,

- Conduct a damage assessment in terms of actual physical damage, or

- Make an assessment in comparative terms, weighing the different damage contributions on a common scale as far as possible, and eventually

- Suggest alternative procedures for production, materials choice, conditions of use and decommissioning that will reduce those identified impacts deemed critical, and possibly formulate norms and regulation, if it is not preferred to impose taxation or otherwise create economic incentives or disincentives in order to guide the use of technology of a particular society in the way indicated by the life-cycle assessment made.

A fairly comprehensive list of items to include in a life-cycle analysis mention the following, without claiming the list to be complete:

- *Economic impacts,* as seen from the point of view of the owner (private economy) or from society (public economy, including considerations of employment and balance of foreign payments).

- *Environmental impacts,* e.g., land use, noise, visual impact, pollution of air, water, soil and biota, on a local, regional or global scale, including climate change induced by emissions of greenhouse gases or ozone depleting substances.

- *Social impacts,* including health impacts, accident risks, effect on work environment and satisfaction of human needs.

- *Security impacts,* including terror actions, misuse, as well as supply security issues.

- *Resilience,* i.e., sensitivity to system failures, planning uncertainty and future changes in value and impact assessments.

- *Impact on development,* i.e., furthering or countering the development goals of society (assuming that such exist, as they usually do at least for currently less developed nations).

- *Political impacts,* including requirements for control, regulation and centralisation of decision-making.

In order to analyse some (or all) of these impacts, including upstream and downstream components relative to the system or device of primary interest, it is as mentioned a good idea to establish an inventory of substances and processes of relevance and to assess their individual impacts, first in physical terms (emissions, etc.) and then in damage terms (injuries, disease, death, etc.). This list can in some cases be reused for assessing other technological items, provided that the inventory is not site- or time-specific. The outcome will be in different units, or in some cases non-quantifiable, and must be submitted to an assessment by decision-makers or by public debate. Only in some cases may it be meaningful to convert different impacts into common units (such as $\epsilon$ = *euro* or $ = *dollar* or "environmental points"). This involves many difficult issues, as damage often occurs at other places or times than the benefits of using the technology. For instance, one would have to estimate the cost to society of accidentally losing one member (the "statistical value of a life"), which has spurred discussion on whether to use the insurance "cost" of a life in Western Europe to assess a life lost in a Columbian

coal mine, rather than a Columbian value, perhaps corrected by use of purchase parity conversion. If the impacts are kept in different units, the assessment is said to be "multivariate".

| Impact category | Physical amount g/kWh of $H_2$ | Monetised value euro-c/kWh of $H_2$ | Uncertainty (range) |
|---|---|---|---|
| **Environment:** | Emissions: | | |
| Plant operation: $CO_2$ | 320 | 12.1 | (8–30) |
| $SO_x$ | 0.29 | 0.17 | high |
| $NO_x$ | 0.38 | 0.23 | high |
| $CH_4$ | 4.4 | 2.0 | (1–4) |
| $C_6H_6$ | 0.042 | NQ | |
| CO | 0.18 | NQ | |
| $N_2O$ | 0.0012 | – | |
| non-$C_6H_6$ hydrocarbons | 0.79 | NQ | |
| particulates | 0.06 | 0.04 | high |
| Ni catalyst material | NA | | |
| Plant construction/decommissioning | NA | | |
| **Occupational:** | Number: | | |
| Industrial disease and accident | 0.5 major injury/TWh | 0.0004 | low |
| **Economic:** | | | |
| Direct economy (production costs) | | 3–6 | |
| Resource usage | serious in long run | NQ | |
| Labour needs for manufacture | 5 person-years/MW | NQ | |
| Import fraction | NA | | |
| Benefits (value of product) | | 6–12 | |
| **Other:** | | | |
| Supply security | low to fair | NQ | |
| Robustness | medium | NQ | |
| Geopolitical | competition | NQ | |

*Table 6.1.* Life-cycle impacts from $H_2$ production by steam reforming of natural gas.
Mortality from stack emissions of ($SO_2$, $NO_x$) aerosols and particulate matter is taken
as $2 \times 10^{-9}$ per g, and morbidity is taken as $935 \times 10^{-6}$ workdays lost per g, corresponding to European population densities. Global warming cost is 0.38 euro/kg
$CO_2$-eq. NA/NQ: not analysed/not quantified (with use of data from Spath and
Mann (2001), Sørensen (2004a)).

## 6.2.2 Life-cycle analysis of hydrogen production

### Conventional production by steam reforming

Production of hydrogen by steam reforming of natural gas gives rise to a
number of impacts mediated by emissions to the atmosphere (Table 6.1), as
well as impacts from the equipment production and disposal stream and
from materials used (such as Ni catalysts) without complete recycling. The

impacts caused by the emissions include global warming (from emission of $CO_2$, $CH_4$, $N_2O$, etc.), waterway acidification (from $SO_x$) and eutrophication (from N and P) and human respiratory diseases (from $SO_x$, $NO_x$, benzene and particles, as well as from soot and winter smog involving C formed by the side reaction given by Chapter 2, Eq. (2.5); Koroneos *et al.*, 2004).

It is seen from Table 6.1 that the externality cost of hydrogen produced on the basis of natural gas is very high, due to the global warming caused by emission of greenhouse gases. This is also true if the lower estimates of global warming impacts are used. The issues related to these figures are discussed in Sørensen (2004a). The main question is related to the fact that negative consequences of warming mainly occur in the less developed countries near the Equator, and one has to decide whether to value lives lost in these regions at the same level as lives lost in Europe or the USA, as it is done in the central estimate, based on insurance payments for premature deaths in Europe.

As stated in section 6.1.4, the cost of removing $CO_2$ from the steam reforming plant appears to be much lower than the global warming externality associated with not handling the $CO_2$ issue.

## Production by electrolysis

If hydrogen production is based on surplus renewable energy such as wind power (as in the scenario in Chapter 5, section 5.5), the impacts are dramatically reduced. Only the occupational impacts are larger, at least at present, due to their roughly scaling with the cost of the conversion equipment, whether conventional electrolysers or reversely operated fuel cells.

## Direct bio-production of hydrogen from cyanobacteria or algae

The potential advantage of bio-hydrogen is to be able to use a production device or system, where major components are already part of the natural solar radiation disposition system operating on our planet to produce biomass. The same philosophy is behind agriculture, where a low conversion efficiency is accepted in return for obtaining a free "production system" requiring as its only input some seeds and some work (such as weeding and harvesting), and based on these inputs the system produces food and related biomass products.

However, both for agriculture and for bio-hydrogen production, it is basically wrong to consider plant or bacterial material "taken" from nature as free. There may be substantial life-cycle impacts and associated costs involved. The area used for plant growth or ocean farming is affected by the formation of large mono-cultures, and the ecological implications of residues left over, of water usage, and so on may be significant. It is therefore necessary to go through all the steps involved in bio-hydrogen production, including the farming itself, collection of crops, treatment made at any stage of the cultivation, residue handling and waste disposition, in order to make an inventory of impacts and the subsequent assessment in monetary or other terms.

The bio-reactor construction must cover the area (presumably of water, since land use would entail the higher land costs) from which solar radiation is to be collected and must, from a closed construction capable of keeping the hydrogen produced from escaping, pipe it to the shore. The cost can therefore be expected to be similar to that of thermal solar collectors. It follows from the discussion in Chapter 2 that the efficiency defined as the energy value of the hydrogen produced divided by the incoming solar radiation on the reactor surface will be considerably below 1%. Table 6.2 assumes for illustration an efficiency of 0.1% and estimates cost on the basis of the similarity to solar collectors.

| Impact category | Physical impacts | Monetised value euro-c/kWh of $H_2$ | Uncertainty, assumptions |
|---|---|---|---|
| **Environment:** | | | |
| Plant construction/decommissioning | NA | | |
| Land or ocean use | large | NQ | |
| Use of genetic engineering | problematic | NQ | |
| Hydrogen cleaning | NA | | |
| **Occupational:** | | | |
| Industrial disease and accident | NA | | |
| **Economic:** | | | |
| Direct economy (production costs) | | > 40 | efficiency 0.1% |
| Resource usage | area covered | NQ | |
| Labour needs for manufacture | 5 person-years/MW | NQ | |
| Import fraction | NA | | |
| Benefits (value of product) | | 6–12 | |
| **Other:** | | | |
| Supply security | good | NQ | |
| Robustness | medium | NQ | |
| Geopolitical | positive | NQ | |

*Table 6.2.* Life-cycle impacts from photo-induced $H_2$ production using cyanobacteria, as yet poorly quantified. See caption to Table 6.1 (with use of Sørensen, 2004c).

The implication of these estimates is that the productivity of diverting hydrogen from (possibly genetically modified) cyanobacteria or algae, which is only a fraction of the total energy turnover of the organism, must be substantially lower than the biomass productivity in agriculture, which on average is about 0.2% (a maximum for special plants being 4%; cf. Sørensen, 2004a). Even if the techniques suggested in Chapter 2 could be realised, it would therefore seem that the prospects for viable hydrogen production by direct photosynthesis are meagre.

*Impacts from use of genetically engineered organisms*
Although the form of genetic engineering has changed, the concept itself has been employed from the earliest periods of systematic crop farming, some $10^4$ years ago. Over the course of time, cereal species were improved to give higher yields, by sorting and cross-breeding of species or sub-species. Only recently, a few hundred years ago, did a theoretical understanding of the mechanisms underlying the processes of genetic modification start to emerge, and the process remained a very slow way of improving crops until some 50 years ago, when more systematic genetic experiments began to be performed with mutations induced e.g. by nuclear radiation. Today, the sequencing of a number of genes with identification of groups of code responsible for particular properties of the plant or organism has widely increased the options for modifying biological systems, such as changing a $CO_2$ or nitrogen fixing organism to become a hydrogen-producing organism, by suppression of the unwanted pathways of energy transfer. Still, the present knowledge of genetic functionality is very limited, and genetic engineering is not based on precise knowledge of what the smallest individual coding sequences do exactly, but rather on identifying the gross responsibilities of more substantial chunks of code. For this reason, genetic engineering is still to a considerable extent a game of trial and error, requiring monitoring and tests of the behaviour of the manipulated organisms, often during extended periods of time and with possible surprises in the form of unexpected side effects occurring only after delays and in particular circumstances.

The impacts of genetic manipulations could have been large, even during the Stone Age. We have no precise records to support this view, but it is likely, that agricultural species obtained by cross-breeding would sometimes turn out not only to improve yields or quality of the parts of the plant used for food, but also to make the plant more susceptible to infections or predator organisms, in which case the result could have been famine and an associated decimation of populations. Exactly the same problems could arise with present genetically modified species, which are often employed on a large scale after testing that must necessarily be incomplete and restricted to periods of time and diversity of environmental factors insufficient to rule out a number of adverse long-time effects. The increase in the industrial use of genetic manipulation techniques has rather had the effect that screening, especially for long-term effects, has become more superficial and less systematic. As it is required by current legislation, testing is often left to the industries themselves, as public control institutions with often limited budgets can only overcome a limited amount of independent verification (sometimes only by spot tests). This is like the situation for use of chemicals in industrial products, where the rate of new chemicals introduced is much higher than what is possible to do long-time screening of.

A profound moral (as well as practical) issue is the juxtaposition of benefits and risks associated with the introduction of genetically engineered crops and products. We have recently seen crops genetically engineered to accept the use of pesticides not compatible with previous semi-wild species, marketed by the same company that sells the pesticide in question. Would we accept crops engineered to accept the pesticide of one company, but made incompatible with pesticides of competing companies? Should we accept at all genetic engineering aimed at improving tolerance to pesticides that due to their possible wider environmental impacts would be better banned in favour of ecological methods of cultivation? Clearly, the questions raised make it very difficult to calculate any precise quantitative value for the externality associated with genetically engineered crop modification, whether for food or for hydrogen production.

| Impact category | Physical impacts | Monetised value euro-c/kWh of $H_2$ | Uncertainty, assumptions |
|---|---|---|---|
| **Environment:** | | | |
| Plant construction/decommissioning | NA | | |
| Agricultural production | substantial | 3 | high |
| Land or ocean use | large | NQ | |
| Transportation of feedstock | emissions, accidents | 40 | high |
| Bio-reactor operation | NA | | |
| Hydrogen cleaning | NA | | |
| **Occupational:** | | | |
| Industrial disease and accident | NA | | |
| **Economic:** | | | |
| Direct economy (production costs) | | > 10 | efficiency 30% |
| Resource usage | agricultural area | NQ | |
| Labour for manufacture and operation | 10 person-years/MW | NQ | |
| Import fraction | low | | |
| Benefits (value of product) | | 6–12 | |
| **Other:** | | | |
| Supply security | good | NQ | |
| Geopolitical | positive | NQ | |

Table 6.3. Life-cycle impacts from $H_2$ production by fermentation of agricultural waste, as yet poorly quantified. Transportation impacts include greenhouse gas warming, air pollution and traffic accidents, assuming current vehicles such as diesel trucks. See caption to Table 6.1 (with use of Sørensen, 2004c).

## Hydrogen from fermentation of biomass

The alternative route from photosynthesis to hydrogen mentioned in Chapter 2 separates the biomass production from the hydrogen extraction, proposing fermentation for the latter, in analogy to biogas (methane plus $CO_2$) production.

This technology has the potential for producing hydrogen at social costs that are acceptable. However, Table 6.3 indicates that a likely condition will be the use of less polluting vehicles for the substantial requirement for transport of biomass feedstock, which is here considered to be by road using the current variety of trucks. The transportation externality estimates in Table 6.3 are derived from impacts from driving passenger cars, and the agricultural externalities are from a similar study of biogas production (Sørensen, 2004a).

The sparsely populated Tables 6.2 and 6.3 should be regarded as indications of work to be done. A number of studies have identified environmental impacts for a range of hydrogen production pathways, ready for incorporation in life-cycle analyses (GM, 2001; Wang, 2002; Wurster, 2003; EC, 2004).

## 6.2.3. Life-cycle analysis of fuel cells

The further away from commercialisation a technology is, the more difficult is it to perform an accurate life-cycle analysis (LCA). On the other hand, precisely at an early stage, the LCA may be most useful. It may identify traits of a technology discouraging further development, and it may point to the most proper path for the development of a given technology, allowing important environmentally or socially relevant choices to be made at an early time during the development process. In many cases a broad-brush LCA may identify the most important externality cost items in the design and may allow them to be optimised or replaced by alternatives at a time when it is still possible to make such changes. Thus, it should not be necessary to make excuses for a life-cycle assessment being attempted at an early time, but of course its outcome must be viewed with appropriate caution.

### SOFCs and MCFCs

For the solid oxide fuel cells (SOFCs), a number of environmentally critical items have been identified (Zapp, 1996). The carrier sheet electrolyte may be produced from yttrium-stabilised zirconium oxide with added electrodes made of, e.g., LaSrMn-perovskite and NiO-cermet. Nitrates of these substances are used in manufacturing, and metal contamination of wastewater is a concern. The high temperature of operation makes the assembly very difficult to disassemble for decommissioning, and no process for recovering yttrium from the YSZ electrolyte material is currently known.

For molten carbonate fuel cells (MCFCs), a full life-cycle analysis has been attempted (Lunghi and Bove, 2003). Both electrodes and the electrolyte matrix are manufactured by mixing powdered constituents with binders and solvents to form a sheet after casting and drying. Some constituents of the slurry preparation have been omitted from the LCA due to industrial secrecy. The results of the analysis comprise resource usage and emissions to air, wastewater and soil. A critical resource may be Ni, for which the largest supplier is Cuba. The main life-cycle impacts for the areas mentioned are summarised in physical terms in Table 6.4.

| Life-cycle impact | Negative electrode | Positive electrode | Electrolyte matrix | Bipolar plate | Total | Unit |
|---|---|---|---|---|---|---|
| Electric energy input | 153.2 | 82.6 | 73.03 | 5.47 | 314 | kWh |
| $CO_2$ | 508 | 214 | 127 | 8.03 | 857 | kg |
| $CH_4$ | 423 | 131 | 36.2 | 0.502 | 591 | g |
| $N_2O$ | 12.2 | 3.8 | 3.5 | 0.0014 | 19.4 | g |
| $SO_2$ | 10.9 | 6.67 | 1.5 | 0.26 | 19.4 | kg |
| $SO_x$ (as $SO_2$) | 2.01 | 0.61 | 0.08 | – | 2.7 | kg |
| CO | 121 | 45.4 | 25.2 | 37.6 | 229 | g |
| $NO_2$ | 366 | 224 | – | 14.3 | 604 | g |
| $NO_x$ (as $NO_2$) | 697 | 214 | 27.8 | – | 939 | g |
| Non methane VOC | 420 | 129 | 16.8 | 0.03 | 566 | g |
| VOC | – | – | – | 15.5 | 15.5 | mg |
| Benzene | 0.895 | 0.31 | 0.031 | 0.01 | 1.3 | g |

Table 6.4. Contributions to life-cycle impacts from different components of a $1\text{-}m^2$ unit single molten carbonate fuel cell (based on data from Lunghi and Bove, 2003).

| Impact category | Physical amount | Monetised value euro-c/$m^2$ of cell | Uncertainty (range) |
|---|---|---|---|
| **Environment:** | Emissions: | | |
| Manufacture (global warming) | Table 6.4 | 301 | (200–400) |
| Manufacture (health effects) | Table 6.4 | 13 467 | high |
| Operation | NA | | |
| **Occupational:** | | | |
| Health effects (included above) | NA | | |
| **Economic:** | | | |
| Direct economy (production costs) | | NQ | |
| Resource usage | Ni availability | NQ | |
| Labour needs for manufacture | 5 person-years/MW | NQ | |
| Benefits (value of product) | | 6–12 | |

Table 6.5. Monetised life-cycle impacts corresponding to the physical impacts given in Table 6.4, using the same translation as in Table 6.1 The impacts are given for a $1\text{-}m^2$ cell, as the lifetime power production of the cell is not available.

It is seen from Table 6.4 that the largest life-cycle impacts come from the negative electrode manufacture. The attempt made in Table 6.5 to monetise these impacts shows a global warming impact that is dominated by the emission of methane, and health impacts that are dominated by the very large emissions of sulphur oxides indicated in Table 6.4. Presumably these values, derived from an Italian pilot production, would be substantially reduced for a realistic commercial production of MCFCs.

### PEMFCs

Figure 6.5 shows the main steps in the manufacture of a proton exchange membrane fuel cell (PEMFC) stack for automotive purposes. Several components in the fuel cell assembly are not significantly different from components found in other industrial products (metals, carbon as fibre or graphite, plastics), and their life-cycle impacts may be taken from generic studies. However, there are notable exceptions. First, the use of a polymer membrane (e.g., perfluorinated ionomers (Barbi *et al.*, 2003), hydrocarbon-based (Kreuer, 2003) or using organic materials (Evans *et al.*, 2003)) in each cell unit may cause special concerns regarding decommissioning or recycling (Handley *et al.*, 2002). Usually, recycling is difficult and incineration is recommended, although in some cases care should be taken to separate difficult materials such as Pd used with organic membranes before burning. Second, while the steel and carbon contents of the bipolar plates can easily be reused or in the case of carbon incinerated, there are again small amounts of special materials requiring care. The most important among these is Pt used as a catalyst at each electrode, possibly as a compound with other metals. Pt causes strongly negative impacts due to the emissions during extraction and purifying, particularly in Third World plants with modest environmental experience, such as seen in the case study of South African platinum performed by Pehnt (2001).

The effect of these impacts can be greatly reduced if Pt is recovered and reused, as all heavy metals should be. A strategy for supply of Pt to the fuel cell industry is discussed by Jaffray and Hards (2003). In terms of weight, the stack breakdown on components is shown in Fig. 6.6 for two conventional material choices for the bipolar plates, graphite or aluminium. Recently, bipolar plates made of conducting polymers have been developed (Middleman *et al.*, 2003), with thickness and weight reduction as a consequence.

If the PEM fuel cell is preceded by a reformer, there will be additional impacts, depending on the fuel and equipment used. The same is true for direct methanol fuel cells, with a dependence on the way methanol is produced, based on natural gas or bioenergy.

The short lifetime of present PEM fuel cells will in general terms increase life-cycle impacts relative to technologies with a long operational life. Current cells degrade significantly after less than 2000 h of operation (Ahn *et al.*,

2002), and even with the target 5-year life, replacement will be much more frequent than for most other energy technologies.

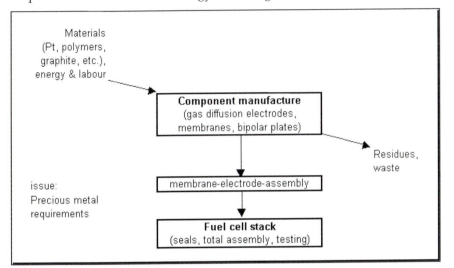

Figure 6.5. Life-cycle flow path for the industrial manufacture of PEM fuel cell stacks (Sørensen, 2004d).

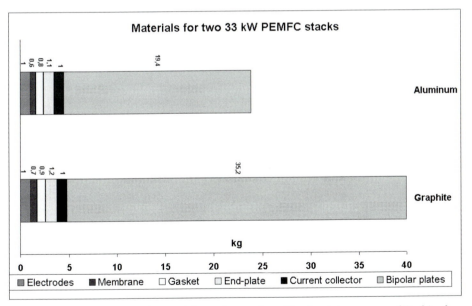

Figure 6.6. Distribution of material use for stacks with aluminium or graphite bipolar plates (Mepsted and Moore, 2003; Sørensen, 2004d).

## 6.2.4. Life-cycle comparison of conventional passenger car and passenger car with fuel cells

The passenger cars selected for this LCA are characterised by the features listed in Table 6.6. The DaimlerChrysler *f-cell* is the first fuel cell passenger car to enter the stage of limited series production (estimated at 60-80 units) for demonstration in Japan (Tokyo Gas Co., 2003) and subsequently in Europe and North America. It follows the *Citaro F* fuel cell bus entering a similar phase in 2003 (demonstration in Europe of a small series of about 30 units). The *f-cell* car is based on a slightly longer version of the commercial A2 series of Mercedes-Benz gasoline and diesel fuel cars, and Table 6.6 reflects the limited data available at the time of calculation. The two non-fuel cell cars studied for comparison are a Toyota *Camry* gasoline/Otto engine car used as a typical US year-2000 vehicle in a previous life-cycle study (Weiss *et al.*, 2000, 2003), and the *Lupo 3L TDI* Diesel car topping the European list for mixed driving efficiency (VW, 2002, 2003). Table 6.6 gives a gross material usage survey, as well as the weight and fuel consumption details to be used in the life-cycle analysis. These are also summarised in Figs. 6.7 and 6.8.

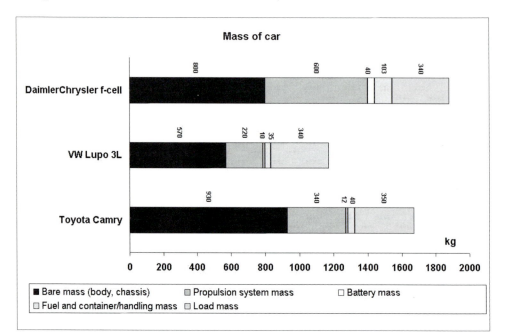

*Figure 6.7.* Comparison of mass distribution for the three passenger cars included in the LCA analysis (Sørensen, 2004d).

| Passenger car (1-5 persons plus luggage) Description | average USA, 2000 Otto engine Toyota Camry | best Europe, 2000 common-rail Diesel VW Lupo 3L | 35 MPa $H_2$ fuel PEMFC/elec. motor DaimlerChrysler f-cell | Unit | Reference |
|---|---|---|---|---|---|
| Bare mass (body, chassis) | 930 | 570 | 800 | kg | 3,4,7,est. |
| Propulsion system mass | 340 | 220 | 600 | kg | 3,est. |
| Battery mass | 12 | 10 | 40 | kg | 3,4, est. |
| Fuel and container/handling mass | <40 | <35 | 3+100 | kg | 3,4, est. |
| Proper mass (unloaded) | 1300 | 825 | 1589 | kg | 4,5,6 |
| Mass of steel | | 410 | | kg | 4 |
| Mass of plastics, rubber | | 130 | | kg | 4 |
| Mass of light metals | | 130 | | kg | 4 |
| Load mass | <350 | <340 | <340 | kg | 3,4 |
| Total mass (occupancy: 2, 0.67 full tank) | 1440 | 980 | 1725 | kg | 3,4 |
| Coefficient of rolling resistance | 0.009 | 0.0068 | 0.0068 | | 3,est. |
| Drag coefficient | 0.33 | 0.25 | 0.25 | | 3,4 |
| Auxiliary power | 0.7 | 0.6 | 1 | kW | 3,est. |
| Engine/fuel cell rating | 109 | 45 | 69 | kW | 3,5,6 |
| Electric motor rating | | | 65 | kW | 5,6 |
| Battery rating | | 4/732 | 20/1400 | kW/Wh | 5,6 |
| Reformer efficiency (not applicable) | | | | | |
| Engine/fuel cell efficiency* | 0.38 | 0.52 | 0.68 | | 3,7,calc. |
| Gear and transmission efficiency* | 0.75 | 0.87 | 0.93 | | 3,est. |
| Electric motor efficiency | | | 0.8 | | 3 |
| Fuel use* | 2.73 | 1.08 | 0.8-1.44¤ | MJ/km | 3,7,calc. |
| Fuel use* | 12 | 33 | | km/l | 3,4,5 |
| Fuel to wheel efficiency* | 0.15 | 0.27 | 0.36 | | 3,calc. |

Table 6.6. Basic vehicle data used (Sørensen, 2004d).
* For standard mixed driving cycle. Fuel-to-wheel efficiency is the work performed by the car to overcome air and road friction, plus the net work performed against gravity and for acceleration/deceleration, all divided by the fuel input (note that this efficiency concept varies linearly with the combined drag and rolling resistance).
¤ 1.44 MJ/km for the first f-cell cars manufactured, with hope of lowering (DC, 2004)
Key to references: 3: Weiss et al., 2000; 4: VW, 2002; 5: VW, 2003; 6: Tokyo Gas Co., 2003; 7: Weiss et al., 2003.

Figure 6.7 shows that while the *Lupo* has diminished weight as compared to an average car through use of lightweight materials where possible (but still being in the top safety category according to crash tests), the *f-cell* car, although small of appearance, has a higher mass than even the conventional car, due to the heavy equipment associated with hydrogen management and conversion. Figure 6.8 compares the efficiencies of the three cars studied. In terms of energy content, the fuel use of the *f-cell* car is slightly below that of the *Lupo*, both being considerably below the current average car. The fuel-to-wheel efficiency improves considerably for the fuel cell vehicle, over the efficient diesel car and of course over the conventional gasoline car. Fuel use is obtained by simulation, using the new European driving cycle used for official rating of cars in Europe. The shape of this driving profile is shown in Fig. 6.9.

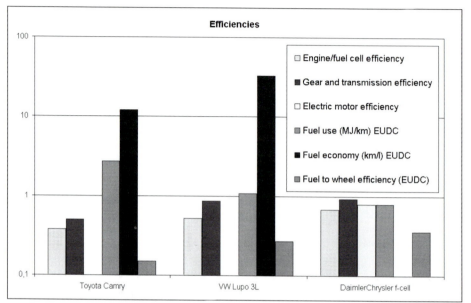

*Figure 6.8.* Comparison of contributions to and total efficiency of the three cars studied in the LCA (Sørensen, 2004d).

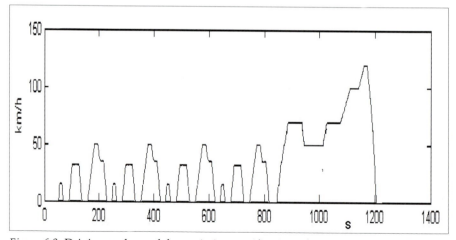

*Figure 6.9.* Driving cycle used for emission certification of passenger cars in Europe, with an urban stop-go section and at the end a continuous sequence at speeds up to 120 km/h (EC, 2001).

| Passenger car (1-5 persons & luggage) LCA environmental impacts Life-cycle emissions | Average USA, 2000 Otto engine Toyota Camry | Best Europe, 2000 common-rail Diesel VW Lupo 3L | $H_2$ from natural gas PEMFC/elec. motor DaimlerChrysler f-cell | $H_2$ from wind surplus PEMFC/elec. motor DaimlerChrysler f-cell | Unit | Ref. |
|---|---|---|---|---|---|---|
| **Car manufacture** | | Production+materials | Total/FC stack | | | |
| Energy use | 87 | 37+51 | 93/?, 178/80 § | 93/?, 178/80 § | GJ | 7,4,9 |
| Greenhouse gas emissions | 1.7 | 0.5+0.5 | 1.7/?, 2.8/1.4 § | 1.7/?, 2.8/1.4 § | $tC_{eq}$ | 7,4,9 |
| $SO_2$ emissions | | 1.6+10.0 | 36/14.5 § | 36/14.5 § | kg | 4,8 |
| CO emissions | | | ?/1.7 | ?/1.7 | kg | 8 |
| $NO_x$ emissions | | 1.8+4.6 | ?/14.5 | ?/14.5 | kg | 4,8 |
| Non-methane volatile organic compounds | | 2.0+1.3 | ?/1.7 | ?/1.7 | kg | 4,8 |
| Particulate matter emissions | | 0.3+4.0 | ?/2.6 | ?/2.6 | kg | 4,8 |
| Benzene | | | ?/2.3 | ?/2.3 | g | 8 |
| Benz(a)pyrine | | | ?/0.034 | ?/0.034 | g | 8 |
| **Fuel production (for 300 000 km)** | | | | | | |
| Energy use | 156 | 67 | 185 | 185 | GJ | 4,7 |
| Greenhouse gas emissions | 3.6 | 0.4 | 8.6 | 0 | $tC_{eq}$ | 4,7 |
| $SO_2$ | | 9 | | 0 | kg | 4 |
| $NO_x$ | | 40 | | 0 | kg | 4 |
| Non-methane volatile organic compounds | | 60 | | 0 | kg | 4 |
| Particulate matter | | 1 | | 0 | kg | 4 |
| Pd (if reformer is used) | | | | | | |
| **Lifetime operation (15 y, 300 000 km)** ¤ | | Incl. decomm. est.: | | | | |
| Energy use | 819 | 324 | 240 | 240 | GJ | 3,4 |
| Greenhouse gas emissions | 16.1 | 6.5 | 0 | 0 | $tC_{eq}$ | 3,4 |
| $SO_2$ | | 1.6 | 0 | 0 | kg | 4 |
| CO | | 30 | 0 | 0 | kg | 4 |
| $NO_x$ | | 75 | 0 | 0 | kg | 4 |
| Non-methane volatile organic compounds | | 2.7 | 0 | 0 | kg | 4 |
| Particulate matter | | 6 | 0 | 0 | kg | 4 |
| PAH | | 1.5 | 0 | 0 | kg | 4 |
| $N_2O$: effect on stratospheric ozone | 13 | 1 | ~0 | 0 | kg | 4,calc. |
| **Decommissioning** (not estimated separately) | | | | | | |
| **Totals** | | | | | | |
| Energy use | 1062 | 479 | 603 | 603 | GJ | 3,4,7,9 |
| Greenhouse gas emissions | 21.4 | 7.9 | 11.4 | 2.8 | $tC_{eq}$ | 3,4,7,9 |
| $SO_2$ | 61 | 22.2 | 36 | 36 | kg | 4,8 |
| CO | | 30 | | | kg | 4,8 |
| $NO_x$ | 70 | 121 | | | kg | 4,8 |
| Non-methane volatile organic compounds | | 66 | | | kg | 4,8 |
| Particulate matter | 12 | 11.3 | | | kg | 4,8 |

*Table 6.7.* Environmental life-cycle impacts (Sørensen, 2004d).
§ Pt manufacturing (assumed in South Africa) accounts for 30% of energy, 40% of greenhouse gases and 67% of acidification, with no recycling assumed (Ref. 8).
¤ Maintenance impacts not estimated.
Key to references: 3: Weiss *et al.*, 2000; 4: VW, 2002; 5: VW, 2003; 7: Weiss *et al.*, 2003; 8: Pehnt, 2001; 9: Pehnt, 2003.

*Environmental impact analysis*
Table 6.7 gives the environmental LCA data available for these cars, in terms of energy used and emissions occurring during the phases of the vehicle life-cycle, based on the studies mentioned with addition of my own calculations and estimates. The impacts are given in physical units as required for the LCA inventory and will subsequently be translated into concrete impacts on

health and environment, including global warming effects. Of particular interest are the impacts from the manufacture and use of the fuel cell component in the *f-cell* car, discussed in section 6.2.3.

The fuel cell vehicle considered in this study uses hydrogen directly. If reformation of methanol or gasoline was used, additional impacts would be derived from the reformer, including often quite large impacts from a catalyst such as Pd, which again should be recycled as fully as possible.

No separate data have been found for decommissioning impacts, although VW (2002) claims to have included them under "lifetime operation". In Denmark, cars delivered to a recycling station pay a fee of about 500 euro, assumed to cover the decommissioning costs minus income from selling extracted parts for reuse. European regulation is discussed, where decommissioning is part of the initial purchase price and the manufacturer is obliged to optimise assembly for decommissioning and to take the vehicle back at the end of service for maximum recycling.

The Volkswagen report (VW, 2002) is a detailed and site-specific LCA for the manufacturing plant at Wolfsburg, including materials and water delivered to or coming out of the plant. It is centred on the *Golf* cars, but the scaling for application to *Lupo* made here has in gross terms already been made in the VW environmental report (2002). The environmental impacts of Table 6.7 are summarised in Fig. 6.10.

| Passenger car (1-5 persons & luggage)# LCA social impacts and other environmental impacts | Average USA, 2000 Otto engine Toyota Camry | Best Europe, 2000 common-rail Diesel VW Lupo 3L | H₂ from natural gas PEMFC/elec. motor DaimlerChrysler f-cell | H₂ from wind surplus PEMFC/elec. motor DaimlerChrysler f-cell | Unit | Ref. |
|---|---|---|---|---|---|---|
| **Car manufacture/decommissioning** | | | | | | |
| Job creation | 0.3 | 0.3 | 1.8 | 1.8 | person-year | 1 |
| Occupational risk: death | 0.0001 | 0.0001 | 0.0005 | 0.0005 | | 1,12 |
| Occupational risk: severe injury | 0.003 | 0.002 | 0.015 | 0.015 | | 1,12 |
| Occupational risk: minor injury | 0.015 | 0.013 | 0.08 | 0.08 | | 1,12 |
| **Maintenance** | | | | | | |
| Job creation | 0.3 | 0.3 | | | | 1 |
| Occupat. risks (death/major/minor injury) | 0.0001/0.003/0.015 | 0.0001/0.002/0.013 | | | | 1,12 |
| **Driving** | | | | | | |
| Accidents (death/severe injury)£ | 0.005/0.050 | 0.005/0.050 | 0.005/0.050 | 0.005/0.050 | | 1 |
| Stress/inconveniences | some | some | some | some | | 1 |
| Mobility | advanced | advanced | advanced | advanced | | |
| Time use as social factor (different perception by individuals) | | | | | | |
| Noise (economic quantification in Table 6.10) some | | some | less | less | | 1 |
| Visual impacts (of cars in environment; different perception by individuals) | | | | | | |
| **Impacts from road infrastructure** (road construction, maintenance, visual impacts: estimated in monetary terms in Table 6.10) | | | | | | |
| **Impacts from car infrastructure** (service, repair, traffic police & courts, insurance: mostly included in cost given in Table 6.9) | | | | | | |

*Table 6.8.* Social life-cycle impacts (Sørensen, 2004d).
# All figures for service life of 15 years, 300 000 km.
£ Statistics for Denmark has been used.
Key to references: 1: Kuemmel *et al.*, 1997; 12: European Commission (1995).

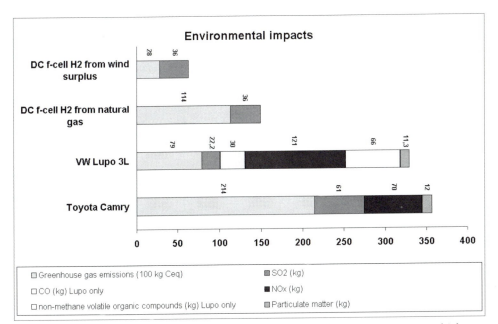

*Figure 6.10.* Comparison of environmental impacts from the three passenger vehicles considered in the life-cycle analysis, with hydrogen for the fuel cell car being derived from either natural gas or excess wind power (Sørensen, 2004d).

| Passenger car (1-5 persons & luggage) **LCA economic impacts** Life expectancy: 15 years, 300 000 km | Average USA, 2000 Otto engine Toyota Camry | Best Europe, 2000 common-rail Diesel VW Lupo 3L | H₂ from natural gas PEMFC/elec. motor DaimlerChrysler f-cell | H₂ from wind surplus PEMFC/elec. motor DaimlerChrysler f-cell | Unit | Ref. |
|---|---|---|---|---|---|---|
| **Direct economy** | | | | | | |
| Car (est. cost without taxes/subsidies) | 15 000 | 13 000 | 80 000¤ | 80 000¤ | euro | est. |
| Roads (monetary evaluation in Table 6.10) | | | | | | |
| Fuel cost (at filling station*, no tax) | 15 000 | 5 455 | 15 600 | 15 600 | euro | est. |
| Service and maintenance | 15 000 | 13 000 | 80 000 | 80 000 | euro | est. |
| Decommissioning (becoming included in purchase price) | | | | | | |
| Time use (of time that could have been paid, depends on individual case) | | | | | | |
| Reference cost of satisfying mobility needs | 35 000 | 35 000 | 35 000 | 35 000 | euro | § |
| **Ressource use** | | | | | | |
| See materials in Table 6.6, recycling will modify these) | | | | | | |
| **Balance of labour and trade** | | | | | | |
| Job intensity (near 50% local, even if no local car or fuel production) | | | | | | |
| Import and export fractions (varying between countries) | | | | | | |

*Table 6.9.* Economic life-cycle impacts (Sørensen, 2004d).
* Oil price staying at current level, hydrogen price dropping linearly from 100 to 30 euro/GJ (projected for 50 000 vehicle penetration; Jeong and Oh, 2002 ) during the 15-year period, initial cost of hydrogen filling stations not included.
¤ Small series cost is reflected; the current 85-kW PEMFC stack cost is about 10000 euro (with ~2500 euro projected for 2025) (Sørensen, 1998; Tsuchiya and Kobayashi, 2004).
§ Public transportation estimated cost.

*Social and economic impact analysis*

Table 6.8 gives occupational risks during the life-cycles of the vehicles, based on standard industrial data (i.e., the impacts are proportional to cost). The job content is based on statistics from the energy sector in Denmark (Kuemmel *et al.*, 1997). The rate of accidents on the road are taken from several Danish studies and is considerably higher in some parts of the world. Evaluation of the health and injury impacts are again based on several Danish studies (references in Kuemmel *et al.*, 1997), as are the less tangible visual and noise impacts (estimated by hedonic pricing) and the inconveniences such as children having to be supervised when near public roads or pedestrians in general having to use round-abouts to get to street crossings with traffic lights, where also the waiting time is valued.

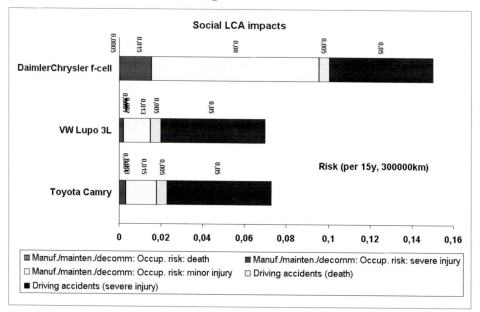

*Figure 6.11.* Comparison of social impacts from the three passenger vehicles considered in the life-cycle analysis (Sørensen, 2004d).

Cars need roads for driving, and the road infrastructure is thus an "externality" to vehicle LCA, which has to be evaluated along with the car operation infrastructure. This is done in monetary terms based on Kuemmel *et al.* (1997) and is included in Tables 6.9 and 6.10. Table 6.9 gives the direct costs involved (and for comparison the cost of public transportation), without including any of the substantial taxes and/or subsidies characterising the actual consumer costs in many countries. The cost of the *f-cell* car is not available at the present time, but has been taken as that of the corresponding

Mercedes-Benz (smallest car) plus a fuel cell stack taken as 100 euro/kW, and the other hydrogen handling and storage costs are assumed to be similar to that of the stack. Finally, a factor of two is applied due to the small series of production. This price distribution is similar to the one estimated for the *Citaro F* fuel cell bus (based on published Evobus EC project material). Maintenance costs are taken as a fixed fraction of capital cost and thus are large for the *f-cell* car (hardly unrealistic for a new construction). The hydrogen cost is that of production from natural gas, ramped down as a function of time. It does not include the initial high cost of establishing hydrogen filling stations. No separate estimate is made for the cost of producing hydrogen from wind, discussed in Sørensen *et al.* (2004). The fuel price for gasoline and diesel fuel has been taken at the current level, disregarding possible increases during the period of operating the vehicles. The social life-cycle impacts are summarised in Fig. 6.11.

| Passenger car (1-5 persons plus luggage) **Life-cycle assessment** externality monetising exercise | Average USA, 2000 Otto engine Toyota Camry | Best Europe, 2000 common-rail Diesel VW Lupo 3L | H₂ from natural gas PEMFC/elec. motor DaimlerChrysler f-cell | H₂ from wind surplus PEMFC/elec. motor DaimlerChrysler f-cell | Unit | Ref. |
|---|---|---|---|---|---|---|
| **Vehicle-related environmental emissions** (based on Table 6.7) | | | | | | |
| Human health impacts | 38 100 | 14 000-40 000# | 22 500§ | 22 500§ | euro | 1 |
| Global climate impacts* | 32 100 | 12 000 | 14 700 | 4 200 | euro | 1 |
| **Quantified social impacts** (based on Tables 6.8, 6.9) | | | | | | |
| Occupational health risks | 648 | 632 | 3 241 | 3 241 | euro | 1 |
| Traffic accidents, incl. rescue & hospital costs | 31 200¤ | 31 200¤ | 31 200¤ | 31 200¤ | euro | 1 |
| Traffic noise | 9 000 | 9 000 | 5 000 | 5 000 | euro | 1,est. |
| Road infrastructure (envir. & visual impacts) | 28 000 | 28 000 | 28 000 | 28 000 | euro | 1 |
| Inconvenience (to children, pedestrians etc.) | 30 000 | 30 000 | 30 000 | 30 000 | euro | 1 |

*Table 6.10.* Externality assessment (Sørensen, 2004d).
* Mainly caused by tropical diseases and with accidental deaths valued by European standards (3 Meuro/death), cf. discussion in Kuemmel *et al.* (1997).
# The upper estimate is due to possible increased impacts associated with $NO_x$ compared to earlier valuations (may be reduced by $NO_x$ exhaust cleaning).
§ Can be reduced by recycling of Pt (Pehnt, 2001).
¤ About half of this number is from the 3 Meuro valuation of accidental death.
Key to reference: 1: Kuemmel *et al.*, 1997.

*Overall assessment*
The total externality costs (i.e., those not reflected in direct consumer costs) are summarised in Table 6.10. This involves translating the impacts from physical units to common monetary units, with the problems inherent in such an approach, notably valuing the loss of a human life to society. The caveats are associated with the fact that impacts such as accidental deaths are not always occurring in the same society that harvests the benefits of car driving. These issues have been discussed, e.g., in Sørensen (2004a).

All monetised impacts are summarised in Fig. 6.12 for the three vehicles considered. A very large fraction of the impacts is derived from road infrastructure, traffic accidents and annoyance. These are identical for all vehicles, except for noise that is smaller for hydrogen vehicles. The other large contribution is from emissions of pollutants to the air. They are in part from manufacture and maintenance and, in the case of the gasoline and diesel cars, from emissions in breathing height, despite attempts of exhaust cleaning (much less efficient than for central power plants). This component is larger for the average car than for the *Lupo 3L*, as is the fuel cost. Regarding greenhouse gas emissions, the *f-cell* car using hydrogen from natural gas is no better than the *Lupo* car, but with hydrogen from renewable energy sources the advantage is substantial.

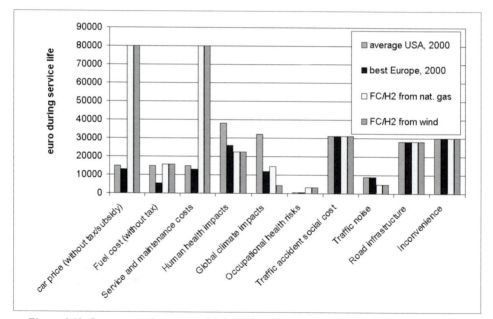

*Figure 6.12.* Summary of monetised life-cycle impacts from Tables 6.9 and 6. 10 (Sørensen, 2004d).

Concern over particulate air emissions involving small-diameter particles has made some countries prefer gasoline cars over diesel cars, except for trucks and buses where the higher efficiency has been the overriding factor. The mechanisms involved in the dispersion of such particles have been the subject of intense study (see, e.g., Kryukov *et al.*, 2004). The *Lupo* diesel car considered above has reduced the particulate emissions (Table 6.7) to levels comparable to those of gasoline cars, but newer European diesel cars, including both efficient passenger cars, buses and trucks, have electrostatic filters reducing the particle emissions by over 90%, which is better than the $SO_2$ removal by the small catalyst devices used in gasoline cars (but in both cases not as good as the exhaust cleaning at large, stationary power plants).

For fuel cell cars carrying methanol and using an on-board reformer, there is direct emission of greenhouse gases, as well as additional impacts from the manufacture of fuel and reformer, leading to an overall $CO_2$-equivalent contribution some 10% higher than for a corresponding car with a pure hydrogen fuel stream (Patyk and Höpfner, 1999; Pehnt, 2002; MacLean and Lave, 2003; Ogden *et al.*, 2004).

### 6.2.5 Life-cycle assessment of hydrogen storage and infrastructure

Most of the life-cycle assessments of hydrogen applications contain an analysis of the storage system used in that particular application. For several types of energy storage systems, the energy requirement for construction is a major concern. Table 6.11 gives the energy consumption for four current storage options for mobile applications, based on Neelis *et al.* (2004) and references quoted therein.

| Materials used in construction | Weight (kg) | Primary energy (MJ) |
|---|---|---|
| **Carbon reinforced epoxy resin container for compressed hydrogen storage** | | |
| Low-density polyethylene | 0.5 | 16 |
| Epoxy resin | 16.6 | 2 460 |
| Poly-acrylonitrile | 38.9 | 4 232 |
| Stainless steel | 2.8 | 125 |
| *Total* | *58.7* | *6 833* |
| **Container for liquid hydrogen** | | |
| Aluminium | 30.7 | 7 140 |
| Stainless steel | 5.0 | 223 |
| *Total* | *35.7* | *7 363* |
| **Low-temperature Fe-Ti hydride system** | | |
| Iron | 271.3 | 12 107 |
| Titanium | 232.7 | 99 945 |
| Aluminium | 88.0 | 20 467 |
| *Total* | *592* | *132 519* |
| **High-temperature Mg hydride system** | | |
| Magnesium | 119 | 35 486 |
| Aluminium | 84 | 19 536 |
| *Total* | *203* | *55 022* |

*Table 6.11.* Material and energy use for hydrogen storage containers (based on data from Neelis *et al.*, 2004).

For small containers suitable for portable applications, Paladini *et al.* (2003) give weights around 100 g and volumes around 20 cm³ for metal hydride stores using LaNi$_5$ or Mg$_2$Ni.

For fuel cell vehicles carrying reformers and using methanol as a fuel, additional life-cycle impacts from the methanol life-cycle have to be considered. A few of these have been analysed, e.g., by Pehnt (2003).

## 6.2.6 Life-cycle assessment of hydrogen systems

Individual hydrogen application systems include the passenger cars, for which life-cycle analysis is discussed in section 6.2.4. Similar investigations may be made for larger vehicles, such as buses (Tzeng *et al.*, 2004). Other transportation systems for which environmental assessments have been made include fuel cell ships (Altmann *et al.*, 2004). This European Commission study considers a 400-kW PEM fuel cell propulsion system based on compressed hydrogen gas for a small ferry and also a 2-MW auxiliary power system for a larger ferry on the route Oslo-Kiel. Here, a natural gas-based SOFC or MCFC system is compared to a liquid hydrogen-based PEMFC power unit. In the small ferry case, the base alternative used today is a diesel engine using light fuel oil, where the larger ship would use heavy fuel oil. The basic database used in the life-cycle study is said to be proprietary, but the numbers are very similar to those one would derive from publicly available emission and impact databases.

| Ship's energy system (engine or PEM fuel cell) | CO$_2$-eq. from fuel production (g/kWh$_{th}$) | CO$_2$-eq. from propulsion (t/y) | SO$_2$ from fuel production and use (t/y) |
|---|---|---|---|
| Diesel engine | 38 | 302 | 0.125 |
| H$_2$ from wind | 25 | 18 | 0.014 |
| H$_2$ from natural gas | 445 | 230 | 0.10 |
| H$_2$ from gasification of forest residues | 14 | 7 | small |
| H$_2$ from agricultural waste | 54 | 28 | small |

*Table 6.12.* Emissions of greenhouse gases and SO$_2$ from fuel production and use in propulsion for a small ferry, using different technologies (annual fuel consumption 991 MWh of diesel or 539 MWh of hydrogen). Based on Altmann *et al.*, 2004.

Table 6.12 gives rounded values found for the greenhouse emission associated with selected routes of hydrogen fuel production and usage for the small ship, for different energy systems and fuels, along with SO$_2$ emissions (omitting a discussion of a possible bonus for producing useful byproducts). The same investigation for the auxiliary power system of the large ship of course gives similar results for the hydrogen-based systems, but not for the diesel engine, where two conditions are different: the larger diesel engine for the large ship is considerably more efficient, giving lower per kWh emission, but the use of heavy fuel oil often entails higher emissions of $SO_2$, with details depending on the origin (and sulphur content) of the fuel.

From the same source are elements of a life-cycle analysis for a building-integrated fuel cell system, although somewhat larger than that of a detached residence (quoted in Fleischer and Oertel, 2003). Combined heat and power production is considered from various types of fuel cells and is compared to that using natural gas turbines. As always, the distribution of negative impacts on two types of energetic outputs, heat and power, makes the specific impact smaller per kWh of useful energy.

Turning from individual hydrogen systems to a complete hydrogen-based energy system, such as one of the scenarios considered in Chapter 5, one may ask if additional adverse impacts can be imagined. One suggestion put forward by Tromp *et al.* (2003) is that small amounts of unconverted hydrogen could escape from the transmission or conversion equipment and be released to the atmosphere as $H_2$ molecules and subsequently possibly diffuse to the stratosphere, where they might assist in forming additional water vapour. Such an increase could have substantial impacts on cloud cover, ozone chemistry and derived climatic changes. In a more detailed modelling work, Schultz *et al.* (2003) show that even with a larger than likely escape of hydrogen, the tropospheric burden of OH will decrease in a hydrogen-based energy system, due to reductions in $NO_x$ emissions. As a result, surface ozone concentrations become significantly increased, and the amounts reaching the stratosphere will more than compensate for the under 2% depletion that could be caused by added hydrogen molecule pollution.

# 6.3 Uncertainties

As a set of technologies in rapid development, hydrogen system components are difficult to evaluate, and the attempts described above, both as regards cost and life-cycle impacts, are necessarily incomplete and uncertain.

Most accuracy is expected with those hydrogen technologies that are similar to existing technologies, such as underground gas storage and container gas and liquid stores, and such as hydrogen production technologies based upon steam reforming and related technologies, or acid electrolysis.

The most uncertain technologies are the fuel cell technologies and new storage technologies such as hydrides and carbon materials. Although laboratory experiments have been carried through, and for a number of fuel cell types limited full-scale production, there are still major uncertainties in future cost and operational stability and durability of the equipment. The learning curves shown in Fig. 6.1 indicate the difference it would make, by shifting the curves upwards, if the life of fuel cells for both mobile and stationary applications turn out to be 4000 h (like present advanced batteries) rather than the 40 000 h hoped for or the 200 000 h characterising, e.g., wind turbines and being a lower limit for nearly all other current energy technology.

The largest technical problem remaining for PEM fuel cells seems to be water management. The cost of the membrane-electrode assembly is being decreased by use of graphite powder-polymer composites as separators, and the possibility of using new low-humidification concepts, where heat and water are simultaneously exchanged between oxygen/air flow channels, is under investigation, as shown in Fig. 6.13.

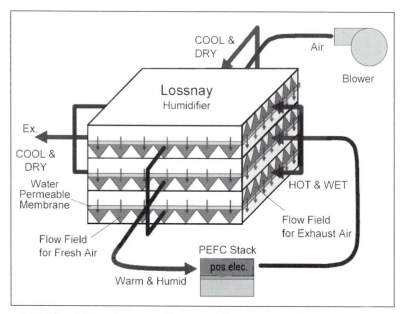

*Figure 6.13.* Humidifier (Lossnay™) concept for positive electrode area. (From K. Mitsuda, H. Maeda, T. Mitani, M. Matsumura, H. Urushibata, H. Yoshiyasu (2004). Technical issues of polymer electrolyte fuel cells. In 15th World Hydrogen Energy Conference, Yokohama 2004, Paper 01PL-02. Used by permission from Mitsubishi Electric Corp.)

Of course, many other factors will contribute to determining the fate of new energy technologies. Will the overall efficiency goals from primary energy through hydrogen and fuel cells to electric power be met? Probably yes as they are not too far away from the present status. Will the oil production peak over the next decade as predicted, and exactly what effect will that have on cost? How will the other fossil fuels fare – will the resources of natural gas be able to hold up to the demand, and will coal gasification and $CO_2$ removal make coal an acceptable fuel for future expansion? If fossil fuels are becoming increasingly problematic, is this the chance for hydrogen in the growing transportation sectors of China and India? Will these countries make two billion people take over the inefficient energy use habits of the USA, or will they look to Europe or even better develop a still higher energy efficiency, better allowing them to expand energy use as much as they wish for their populations? Current trends indicate that the emerging mega energy users will copy the USA, perhaps under influence of the multinational companies already establishing themselves in Asia. The questions are many, and the uncertainty regarding outcomes is large.

# 6.4 Problems and discussion topics

### 6.4.1

Try to extend the Japanese cost scenario for the transportation sector, described in section 6.1.5, to decentralised use of hydrogen and fuel cells in buildings, along the lines indicated in Chapter 5, Fig. 5.2.

### 6.4.2

Look at the production of hydrogen by fermentation and on the basis of biomass from waste or from dedicated crops. Estimate current and possibly reduced future costs of such a hydrogen production system, taking into account the cost of bio-feedstock, its transportation as well as the hydrogen-producing conversion equipment (cf. Chapter 2, section 2.1.5).

### 6.4.3

How would you suggest the introduction of hydrogen in less developed countries to proceed? Would the preferred systems be different from those envisaged in currently industrialised countries?

# CHAPTER 7

# CONCLUSION: A CONDITIONAL OUTCOME

The preceding chapters have considered the possible ingredients in a hydrogen system, i.e., an energy system where hydrogen plays a central role as energy carrier and possibly also as storage medium. The latter function is required for intermittent renewable energy primary sources, but is also useful for systems based on exhaustible resources such as coal or thorium. The story is still an unfinished one, not only because of challenging scientific and technical problems, but also due to the difficulties of economic systems such as current national entities to cope with radically new systems with a different cost structure from the ones that have prevailed for the last century. On the other hand, hydrogen technologies offer ways of handling some critical problems facing current societies, ways that could be preferable to other ways of coping with the problems. In this chapter, some of the main issues are recapitulated.

## 7.1 Opportunities

Looking broadly at the fuel cell technologies for converting fuels such as hydrogen into electric power, one striking feature is the fact that just about any primary source of energy can be used. Fuels can be fed into the high-temperature fuel cells or be converted into hydrogen for the low-temperature cells. Electricity can be transformed into hydrogen, either by fuel cells in reverse mode or by other electrolysers. This allows the flexibility of going back and forth between non-storable electricity and storable hydrogen. Of course there will be losses in these processes, but as ways of handling the fluctuating production from solar or wind power devices, there is no technology more efficient for matching supply and demand. Of course, only electricity that cannot be used when produced will have to be sent through the loss-prone conversions. This feedstock flexibility may prove invaluable in creating a smooth path for transition from the present to a hydrogen-based energy system.

The structure of the energy market may be influenced in a positive way by the hydrogen technologies. This is at least true for the PEM fuel cells, because they are modular and can be installed at nearly the same price per kW at any size, from family installations to central power plants. This means a large freedom for independent power producers who may avoid the restraints of conventional energy companies in regard to inertia in technologies used and services offered. Everyone can install a PEM fuel cell in the basement, or purchase a car with one, to produce power and associated heat for the building or for portable equipment on the basis of piped or locally stored hydrogen. Primary, intermittent power may come from rooftop photovoltaic panels or from a wind farm in which the user may possess shares. The implication is a decentralisation and restructuring of the energy industry that deserves the name much better than the current shifts in large-scale ownership.

In terms of environmental impacts, the hydrogen conversion and infrastructure technologies are generally benign. Some of the high-temperature fuel cells may produce nasty wastes that have to be dealt with delicately, but generally, from hydrogen through storage and converters to end-use energy, the negative impacts are small. The inefficient end-of-pipe remedies such as car exhaust catalysts used with current automotive fuels can totally be done away with for vehicles using hydrogen in fuel cells. This is not necessarily true for the production of hydrogen from primary energy sources or from on-board reformers in vehicles. The primary hydrogen production can be done with small as well as large negative impacts. Potentially large impacts arise from nuclear or fossil energy. In the nuclear case, remedies would require decades of research efforts in developing new, safe reactor types, just to partially solve the problems (cf. Chapter 5). In the fossil case, a possible remedy is offered by $CO_2$ sequestering and development of acceptable deposition and long-term $CO_2$-storage methods. Also production of hydrogen from biomass can have large negative impacts. A number of remedies appear possible, although few have been tried on a realistic scale as yet. Only hydrogen production from wind or solar power that is not used immediately seems to offer genuinely sustainable and ecologically acceptable avenues.

The decentralised potential of renewable and hydrogen technologies offers ways of delivering power to remote areas not yet deemed economic to reach by electricity transmission lines. This may turn out to be a very important feature for many developing regions of the world.

# 7.2 Obstacles

The obstacles to hydrogen and fuel cell development as a central energy carrier are both technical and social. The social obstacles are in part political and institutional, as they relate to the decision and economic structure prevailing in present societies. In the past, almost all new technology introduction was made by foresighted individuals (sometimes in government, sometimes in industry) and in plain disrespect for economic views at the time. Today, it seems that vested interests have become stronger, making transitions away from the way things happen to be at the moment more difficult. Furthermore, traditional economic thinking has managed to increase its influence on decisions made in private or public spheres. The introduction of the personal automobile 100 years ago would hardly have happened if economists had had an influence on the decision. Evaluating the cost of road infrastructure and the negative impacts of collisions between these poorly manoeuvrable vehicles, as well as the astronomical cost of producing them, plus considerations of the cost of extracting enough oil products to operate them, would certainly have discouraged any decision-maker listening to his economic advisors. As another example, electrification of the rural areas of currently industrialised regions by extension of grid lines to these areas would never have happened if city electricity users had not subsidised rural users through the uniform pricing policy of public power utilities. Had the electricity industry been privatised 100 years ago, no one would have dared to propose extending grid lines to thinly populated areas, as profits clearly were visible only in high power-use concentrations of people. Yet the comprehensive electrification is precisely what distinguishes the presently rich, industrialised countries from the less fortunate countries where rural life has hardly changed over the past several thousand years (except for a few transistor radios operated on expensive batteries).

The technical obstacles are more tangible. Many of the methods for producing hydrogen should be upgraded for supplying the amounts needed for energy supply rather than the current limited quantities for industrial feedstock. Hopefully, the price of traded hydrogen can be brought down along with the larger volumes. The alternative to such mass supply is decentralised production, which diminishes the need for transmission or transport. Although decentralised production is normally more expensive than mass production, its cost should also be compared with the higher price of energy currently delivered to the final customer.

The fuel cell technology is at its beginning, with a probably long learning process ahead. Even for the PEM technology currently receiving most attention, the cost is high, its performance has problems (such as water management), and the working life is too short. High-temperature fuel cells have some promising attributes for stationary uses, but are probably further away from the marketplace. Fuel cell conversion, including that of reversible cells, is probably the most important area for development and cost reduction efforts.

Hydrogen storage in containers (as compressed gas or liquefied) is an existing technology needing only modest refinement, unless radical changes such as ultra-high pressure is considered necessary. Also, underground storage is likely to be feasible as a bulk storage option at low cost for many geographical locations. Only the increased safety of metal or chemical hydride stores, or the possible advanced carbon storage forms, requires further research and development, both in terms of technology and cost, in order to become viable alternatives for storage. The acceptance of extra cost to an end-user liking to be in control of his energy supply is probably not stretching as far as accepting the high cost of metal hydride stores, in case cheap underground stores are centrally available, and provided that transmission costs are not excessive.

It is possible to consider a hydrogen "economy" based on hydrogen as the storage medium for intermittent energy sources, also in case the fuel cell conversion does not turn out to be technically or economically viable. The inexpensive cost of, e.g., the underground storage options will make this system feasible, even if its efficiency is lower if gas turbines have to be used for regenerating electricity. However, the efficiency of gas turbines is not much lower than that of fuel cells, and the factor of three penalty in volume relative to natural gas should make the cost at most a factor of three higher, which is quite acceptable in a stationary application because the turbine cost is low anyway.

By this line of argument is meant that as large penetration of intermittent renewable energy primary sources does require either storage or backup from fuels (which may not be possible in a not so distant future), then for the storage demanded, hydrogen storage may be cheaper or more convenient than other forms of storage, even if the roundtrip efficiency is slightly lower. However, many of the advantages of hydrogen as a storage medium are enhanced, if hydrogen can also be used in fuel cell systems, notably for the transportation sector. Therefore, the development of viable fuel cell systems remains a top priority.

Finally, the infrastructure required for hydrogen to be useful as a general energy carrier also needs further development, although both pipeline transportation, container transportation and dispensing fuel stations are all fairly well-known technologies needing only modest further development. The present uncertainty of which fuels may be used could cause some confusion, if filling stations should offer methanol, compressed hydrogen and gasoline. This is not impossible, but a likely situation will be one with different options offered in different locations. For example, in Europe this may imply incompatibility of filling station infrastructure in different countries, causing trouble for people driving their fuel cell vehicle from one country to another (just as present drivers of bio-diesel cars find that some countries do not offer this fuel but only truck diesels). The same is true if efforts to internationally standardise safety norms and equipment standards are not successful.

# 7.3 The way forward

The assessments made above suggests five key areas of hydrogen introduction, which could happen individually or in combination. They are

- Hydrogen as energy store for variable renewable energy systems.
- Applications of hydrogen and fuel cells in the transportation sector.
- Stationary applications of fuel cells in building environments.
- Portable applications of fuel cells.
- Stationary application of fuel cells in large power plants.

Below, each option if briefly assessed on the basis of the treatment made in previous sectors of this book. This constitutes a set of proposals for development efforts (in the US called "roadmaps", in Europe "action plans") in the hydrogen and fuel cell area.

## 7.3.1 Hydrogen storage in renewable energy systems

In order for fluctuating renewable energy sources such as wind and solar radiation to attain a substantial share in any energy system, storage must be part of the system. Only for small levels of penetration can demand-supply mismatch be handled by trading (e.g. in international power pools). Sudden large imports are costly and sudden spot-market exports must accept low prices. In any cases, this set-up only works if neighbouring systems are based on systems that can adjust their production arbitrarily (i.e. fuel-based or hydro systems). If the transition to variable primary sources is taking place generally, there is no alternative to energy storage. An assessment of possible energy storage technologies (Sørensen, 2004a) does not identify any other storage option more suitable than hydrogen underground storage. The options of aquifer or salt intrusion storage are natural extensions of their proven usefulness for natural gas storage and promise a very low storage cost. In areas without these options, rock cavern storage is technically feasible but somewhat more expensive.

If stored hydrogen is not used to regenerate electric power, it has to be forwarded to hydrogen users, by pipeline, truck, ship or other means of transport. The cost of such infrastructure is often high, and an optimisation of hydrogen transport requirements by considering locations of components such as filling stations for vehicles or industry using hydrogen should be made. The additional cost of alternative hydrogen storage nearer to the customer (at filling stations or completely decentralised to individual buildings) should be evaluated in relation to the total cost to the final consumer.

In any case, the discussion indicates that hydrogen as the storage medium to use in connection with variable renewable energy systems based on wind turbines or photovoltaic collectors is an option that can and very likely should be used, independent of the outcome of the development of fuel cell technologies towards a wider role for hydrogen in the energy system. Without fuel cells, conventional methods for regeneration of electricity from hydrogen (e-g. gas turbines, Sterling engines) have to be used, implying a lower efficiency than that of stationary fuel cells, but not in a dramatic way (40-50% rather than 50-65%).

In conclusion, the consideration of hydrogen stores should be part of any energy policy aiming to replace environmentally adverse and politically if not resource-wise uncertain reliance on fossil fuels by renewable energy technologies.

## 7.3.2 Fuel cell vehicles

The transportation sector is clearly the sector where a transition to sustainable energy sources is most difficult. For the past several decades, hopes were placed in battery-based electric vehicles, but it has taken long time to reach the technical goals for battery performance, and the economic goals that would make a purely battery-operated vehicle with suitable range become economically acceptable have not yet been reached. As a consequence, automobile manufacturers have lowered the expectations to battery-driven vehicles and instead hope that fuel cell vehicles will be able to meet both technical and cost goals. The analysis in Chapter 4 indicates that this may not be the best approach, because fuel cell-battery hybrids may well turn out to be a better solution, due to the complementary advantages of the two systems. The current emphasis on PEM fuel cells requires substantial further technological development of PEM fuel cells in regard to reliability and long service life, and particularly in regard to cost, particularly if the fuel cell has the total responsibility for powering the vehicle. The hybrid fuel cell-battery option may offer full performance with a lower sum rating of the two systems, as compared with both pure fuel cell vehicles or the pure battery-operated vehicles (and consequently a lower cost).

If the PEM cell development does not meet the set goals, a possible alternative would be acid polymer cells operating at temperatures around 200°C. However, the development stage of this concept is currently much less advanced, and a shift to this technology will likely have the effect of delaying the deployment of viable vehicle fuel cells in the general automobile manufacturing lines.

Assuming that the PEM fuel cells do reach the technical goals, their introduction will depends on the cost reduction linked to market penetration and hence accumulated mass production. This phase can be speeded up by creating infrastructure facilities at an early stage, and by market introduction initiatives involving a reward for the absence of pollution during operation of the fuel cell vehicles (and also during hydrogen production if this is based on renewable energy sources). The speed of the transition to hydrogen is in my view influenced by whether or not fuel cell-battery hybrids are made available at an early stage, because this would give each stepwise improvement in either technology the possibility of exerting a positive impact on cost. A catalytic role may be seen for the currently available gasoline-battery hybrids, which can help lower the cost of advanced batteries and make their deployment in fuel-cell-battery hybrids happen at an earlier stage.

Initial introduction of fuel cell vehicles may be in applications where the range requirements are modest, such as fixed route buses, delivery vehicles and trucks. Also the markets for fuel cell trains and ships seem more interesting that current activities indicate. The auxiliary system components should not be neglected, because as mentioned the cost of filling stations and transport/transmission systems for of delivering hydrogen to these (or producing it on-site) can substantially influence the overall attractiveness of the fuel cell system.

## 7.3.3 Building-integrated fuel cells

The PEM fuel cell is the leading alternative for building-integrated applications, where multi-functionality would allow current natural gas burners to be replaced by combined heat-and-power systems, possibly with the additional option of supplying hydrogen to a one-vehicle filling station. One may think that if the automobile industry is successful in developing a viable

PEM fuel cell for vehicles, then it will be directly applicable for stationary purposes. However, this is only partially true, as the service-life requirements are much higher for stationary uses. The aim at the current natural gas customers implies that a gas reformer needs to be integrated in the system. The customers that currently have access to piped natural gas is only a segment of the total market, with a share that varies between countries.

Looking a bit further ahead, an interesting system is that of a reversible PEM cell capable of converting excess electricity supplies (from renewables) to hydrogen for moderate time storage. A break-through in the efficiency of the PEM electrolysis operation of specially designed PEM cells (section 3.5.5) makes this an interesting solution (with parallels in the area of filling station on-site conversion), and the remaining problem is suitable hydrogen stores for safe operation in building environments (beyond what can be directly stored in vehicles parked at the building, cf. Chapter 5).

Overall, the cost and technical performance of PEM fuel cells exhibit the same problems for decentralised stationary as for mobile applications, with the mentioned qualification that durability requirements are substantially higher. Both applications are in areas, where consumers in most parts of the world are accustomed to paying a fairly high energy price, as compared with that characterising central power plant operators.

## 7.3.4 Fuel cells in portable equipment

The current generation of portable consumer products are in several areas close to the technical limits of battery technology. This is true for portable computers, where high performance requirements have led to considerable efforts to make energy use as efficient as possible. Flat screens of very low power consumption have been developed, and both central processors and peripheral equipment are rapidly approaching a very good energy performance. Driving forces in this direction include also the stationary computers, because damage caused by excess heat is a decisive determinant of life time and performance. For these reasons, the current autonomous operating time for laptops of some 4 hours between battery recharging is a major limitation both for present users and for further performance development.

This suggests that portable applications may offer a very attractive upstart niche market for fuel cells and small-scale hydrogen or methanol stores, just as they did some years ago for advanced battery types (first NiMH then Li ion batteries). The discussion in section 4.6 suggests that the most appropriate technology for this type of application may be a direct methanol fuel cell, due to the volume requirements of the fuel store. Increasing the computer's operational time between recharging or reloading to some 10-20 hours, this option offers advantages to users that they would likely be willing to pay a considerable price for, as they already do for advanced Li ion batteries typically consuming more than 200 US$ of the cost of a portable computer.

Also for the fuel cell development in general, the existence of such niche markets can have a positive effect, reminding of the success of Japanese manufacturers obtained by incorporating solar cells in consumer products such as watches and calculators and thereby earning a profit capable of covering the total Japanese development costs of solar cells, at least for an initial ten year period.

## 7.3.5 Fuel cells in centralised power production

Because the current bulk power production sector is dominated by coal-fired power plants, the fuel price issue may be seen as less critical than for oil in the transportation sector. However, considerations of greenhouse gas emissions put coal low on the list of acceptable fuels, and ways of de-carbonising coal are intensely discussed. Here, primary hydrogen conversion (if not coal to hydrogen replacement) is high on the agenda, because recovering carbon from exhaust gases is a fairly low-efficiency and energetically unfavourable options. In case the renewable energy transition is successful in the power industry, hydrogen will as mentioned in section 7.3.1 also have an important role already for central energy storage, even if the re-generation of electricity is not done by fuel cells.

If the higher efficiency of fuel cells warrants their higher cost, the most likely fuel cell type for the bulk power sector is the SOFC. However, there are still many technical issues to resolve, especially if fuels other than pure hydrogen are to be used (the poisoning problem from sulphur or nitrogen compounds as well as chlorides and other halogens mentioned in section 3.3, and the long cold-upstart times). Because the power sector is not pushing this development as strongly as the automobile sector is pushing the PEMFC development, it may take longer before a cost-effective SOFC is available, or the PEM cells make take over the utility market due to the lower cost achieved through its use in the transportation sector, despite slightly lower conversion efficiency.

The feeling of a lower urgency for introduction of fuel cells in the power sector may not be warranted, considering the possibility of an increased effort to combat the global warming problem by introduction of renewable energy, which has its primary market aim precisely in the area of power generation. The scenarios in Chapter 5 indicate centralised solutions that may be considerably lower in cost than the building-integrated solutions, due to the possibly lower cost of large fuel cell installations (e.g. due to balance of system costs, as the stacks themselves are modular and hence with little scale effect), including stores and taking advantage of lower-cost infrastructure for bringing hydrogen from stores to power plant sites and of the availability of existing electricity transmission networks, all in comparison with the duplication of many infrastructure components in a multitude of buildings.

## 7.3.6 Efficiency considerations

In Chapter 6, section 6.1, a criticism sometimes forwarded against fuel cells and hydrogen going through energy stores was briefly discussed. The claim is that the efficiency of hydrogen energy systems is unacceptably low when all stages of conversion is included in the considerations. My reply was that use of variable renewable energy sources in any case must involve some part of the energy going from the form of electricity through stores based on other energy forms and subsequently back to electricity, with the associated unavoidable losses. In order to put this argument in perspective, Figs. 7.1 and 7.2 shows the accumulated efficiencies through all conversion stages from renewable energy leading to the two specific end-uses of personal transportation and electric appliances. For comparison, the corresponding efficiency chains are shown for oil and coal as used in the current energy system. I shall now describe the individual steps involved in the calculations.

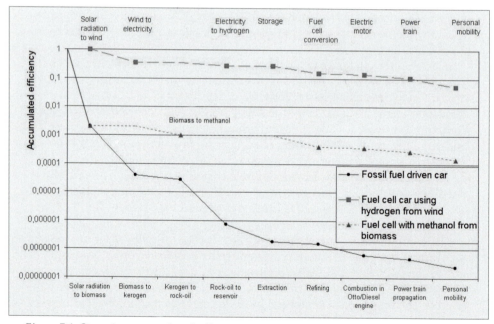

*Figure 7.1.* Stepwise accumulated efficiency of chains of energy conversion from primary sources to end-use, taken here as personal transportation. Current vehicles based on oil products are compared to fuel cell vehicles using methanol from biomass or hydrogen from wind (Sørensen, 2004g; see text for details).

Consider first the routes to personal transportation. For fossil and biomass routes, Fig. 7.1 starts with the production of biomass from solar radiation, at an average efficiency of 0.2% (Sørensen, 2004a). For the petroleum products, the efficiency of the (million year) fossilisation processes of kerogen formation (2%), oil formation in rocks (67%, the rest being natural gas, of which some escapes) and flow into exploitable reservoirs (2.8%) are taken from Dukes (2003). Average extraction efficiencies are estimated as 24.5% and refinery losses as 15%. The combustion efficiency in vehicle Otto or Diesel engines is about 40%, power train efficiency 75% and the end-use benefits of overcoming air resistance and friction or elevation losses 50% (cf. section 6.2.4). The overall accumulated fossil-fuel car efficiency is thus as low as 2.3 $\times 10^{-8}$.

For the methanol route the solar-to-biomass efficiency of 0.2% is followed by a methanol production efficiency (50%), a DMFC or reformer-PEMFC efficiency of 40%, an electric motor efficiency of 93% and the same power train and driving efficiencies as for the fossil fuel car.

Finally, for the hydrogen fuel cell car, the solar radiation to wind conversion-efficiency is taken as 100% (following arguments of Sørensen, 1996c), the wind turbine efficiency as 35%, the electrolysis efficiency as 80%, the fuel cell conversion efficiency as 55% and the rest as for methanol. The overall accumulated efficiencies are $1.4 \times 10^{-4}$ for the methanol route and 0.054 for the wind-hydrogen route.

Turning now to the routes leading to energy for electronic devices such as computers, consider first the coal route. In Fig. 7.2, following the solar energy to biomass efficiency (0.2%), the successive transformation into peat, highly volatile bituminous coal and finally hard coal (anthracite) carries efficiencies of 15%, 92.5% and 63% (Dukes, 2003). A typical efficiency of coal extraction (average of surface and deep mining) is 69%, that of refining and transport 90%. The steam power plant efficiency is taken as 42%, the electricity transmission and distribution efficiency as 94%. Finally, an average to high-end end-use efficiency of using electricity (say in a microelectronic device) is taken as 20%.

The first steps of the renewable routes are as for Fig. 7.1 (35% wind turbine), but as under half of the power produced is assumed to be used directly and the remaining going through a hydrogen store (as discussed in the Chapter 5 scenarios), an average storage cycle efficiency of 75% is assumed, covering electrolysis and hydrogen storage/transmission losses, but not electricity regeneration which is assumed to be 60% for the centralised SOFC route and 50% electric plus 40% heat efficiency for the decentralised PEM route. In the centralised case, a transmission and distribution loss of 6% is subtracted, while for the decentralised case, the in-house losses are confined to 1%. The electricity to microelectronic device performance efficiency is still taken as 20%, but a 40% in-house use of waste heat is added in the decentralised case.

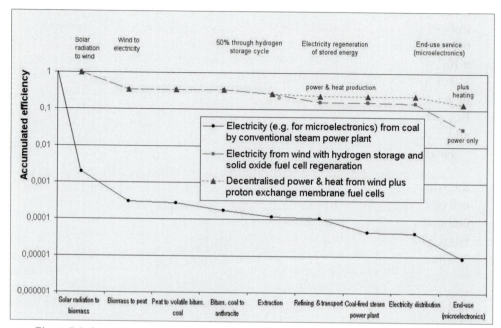

*Figure 7.2.* Stepwise accumulated efficiency of chains of energy conversion from primary sources to end-use, taken here as the operation of electronic equipment by electricity. Current electricity production based on coal is compared to renewable electricity from wind, of which some go through hydrogen stores and are converted back to electricity with and without utilisation of associated heat (Sørensen, 2004g; see text for details).

In the perspective created by these two examples, the renewable energy, hydrogen store and fuel cell routes are much more efficient than the current systems.

# 7.4 How much time do we have?

The time-frame for the necessary changes in the energy system and thus the urgency of the development efforts related to hydrogen, is illustrated by the prospects for continued reliance on oil and expectations of oil price developments. Figure 7.3 shows the decline in finding new exploitable oil fields. The energy contents of new finds topped before 1970, and in contrast to the erratic exploration efforts prior to 1950, the subsequent global efforts have been much more systematic and hence the probability of unexpected large finds in the future is considered low. Enhanced recovery techniques beyond the gas injection method already in widespread use could lead to an upgrading of existing reserve estimates. However, the proposed methods (such as *in situ* combustion, using bacteria to release oil from rock pores, high-pressure chemistry) do not have universal applicability and can at best release some 20% of the 60% of oil resources presently left trapped in the geological formations (cf. Giles, 2004). High investment costs are associated with increasing production above the present level, whether using new enhanced recovery methods or exploiting new reservoirs, such as Canadian tar sands or Venezuelan shales.

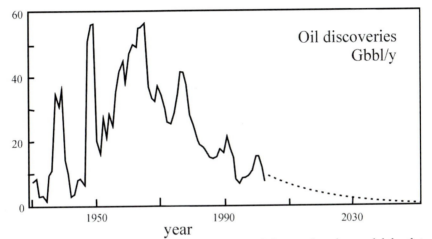

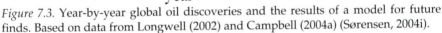

*Figure 7.3.* Year-by-year global oil discoveries and the results of a model for future finds. Based on data from Longwell (2002) and Campbell (2004a) (Sørensen, 2004i).

Figure 7.4 gives some model results for the near-term development of oil production and oil prices. On the basis of the data shown in Fig. 7.3 and the remarks on unconventional oil recovery techniques, it is assumed that half of the exploitable reserves have been used by 2010. The uncertainty of this estimate is low (of the order of ±10 years, unless quite unexpected new oil finds materialise) and is supported by many international investigations (Campbell, 2004a; PFC Energy, 2004). However, the fact that we are at the mid-point of reserves does not determine the rate of oil use during the next decades. High use will be followed by a more abrupt decline, and price excursions are likely to increase. Figure 7.4 shows the behaviour in 3 models, assuming that oil production is growing (a), is constant (b) or declines symmetrically with its historical growth (c). The latter possibility, the Hubbert bell shaped curve (Hubbert, 1962), is unlikely because it assumes readily available substitution of oil by other energy forms at similar prices. Case (b) of staying at current oil production level requires a partial substitution to be available for covering the demand growth. Candidates such as liquefaction of coal or biomass into oil-substituting fuels currently have price tags equivalent to around 100 US$/bbl of oil or higher, which is reflected in the lower part of Fig. 7.4. In this case, reserves will be exhausted around 2043 (consistent with the appraisal given in section 5.3.2). Finally case (a) assumes that production is increased to cater to the new consumer countries (China, India, etc.). Its 25% increase in oil usage from 2004 to 2030 assumes a 30% increase in world car ownership over the next 25 years, an assumption that few will find exaggerated. The reserves now last only to around year 2038. A recent IEA report (IEA, 2004) has an even higher 60% increase in energy use to 2030 in its reference scenario. The implication of such scenarios is a 60% increase in $CO_2$ emissions (IEA, 2004) and a 60% increase in OPEC oil production, while non-OPEC oil production will decline during the period (PFC Energy, 2004, in a study prepared for the Center for Strategic & International Studies (CSIS) in Washington, DC).

The geopolitical and supply-security implications of this development are huge. Both IEA and the CSIS study recommend energy policies reducing demand and speeding up a transition away from oil dependence. They differ in that IEA believes that increasing the oil production sufficiently is possible at least to 2030, while the CSIS predicts a demand-supply gap already during the period 2014-2020, for a corresponding global demand growth varying from 2.4 to 1.1% per year. A recent study by the oil industry supports the CSIS view (ExxonMobile, 2004), while the analysis in Fig. 7.4 is also consistent with IEA, by setting year 2038 as the limit for covering demand by increased oil production. In all cases, the price of oil is predicted to go up substantially. In the high production scenarios, the sudden onset of decline in supply is likely preceded by crisis conditions that could have been avoided by a planned winding-down of production. Short-term behaviour of the oil price is likely to exhibit the fluctuations characteristic for a nervous and hypersensitive market, perhaps including price decreases below the current on. However, the solidity of the reasoning behind expecting a phase-out of conventional cheaply produced oil is high.

The analysis above provides a clear indication of the importance of making viable energy alternatives, including hydrogen technologies, available as soon as possible, and emphasises the urgency of getting R&D started in order to achieve the technical maturity and economic acceptability levels aimed for. At the same time, it is clear that hydrogen technologies cannot reach a penetration in the marketplace large enough to solve the near-term problems. Other solutions are required for the short term, and as underlined by investigations such as the one made for passenger cars in section 6.2.4, efficient use of energy is the only technology that is ready and can be implemented in time (as also recommended by Sørensen, 2004a; IEA, 2004; PFC Energy, 2004). Hydrogen and fuel cell technologies may be the solution for the long-term energy transition, but short-term oil supply problems can only be handled by the improving energy efficiency.

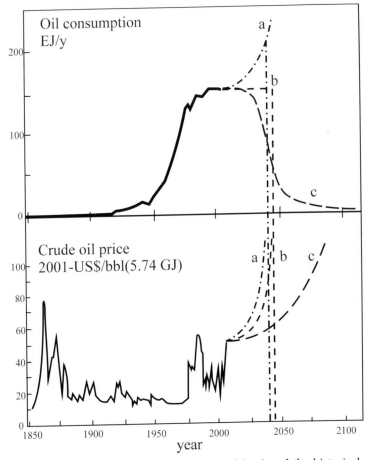

*Figure 7.4.* The historical global consumption of oil (top) and the historical price levels of crude oil (bottom), supplemented with 3 models for future behaviour: a extrapolates the expected growth in demand, notably carried by countries such as China and India, b assumes a constant usage of oil (i.e. all growth substituted) and c is a Hubbert (1962) model based on the assumption of generously available substitute fuels at prices similar to that of current oil. The rapid price increases in the models a and b simply reflect the lack of available substitutes for oil at equivalent prices under 100 US$/bbl. The figure uses historical consumption data from Fig. 5.6 and historical prices from Enquete Kommission des Deutschen Bundestages (1995) and Energy Information Administration of the US DoE (2004) (cf. Sørensen, 2004i).

# 7.5 The end, and a beginning

It is time to end this book, but not without a few concluding remarks signalling the need to create the conditions for a new beginning, based on a partnership between research and industry and between the voter and the parliamentarians, to make the conditions for a hydrogen society realistic. The motivation is the fact that there are so few alternatives that we better have to make the hydrogen route work out if we want to preserve the progression in wealth for both the rich and the presently poor.

The success of the hydrogen project is conditional on a number of things falling in place, notably solving the technical and economic problems of fuel cells. There are plenty of drivers to spur us in this direction: peaking and then declining fossil fuel production, political instability of the largest supplier countries and the quest for a transition to a sustainable energy system that most voters like to see based upon renewable energy sources. The hydrogen technologies have already brought a large number of new players into the field. But numbers do not guarantee success, and there are a range of "conventional" perceptions to question. Do hydrogen and fuel cell systems have to match the current low prices of conventional systems, without consideration of externality costs? I do not think that is possible, not for hydrogen and not for the primary sources of energy that may replace fossil fuels. This has further implications, e.g., for the transportation sector.

Should we encourage driving in armoured cars, like a certain Californian governor, or one person in a four-wheel-drive oversized van (the one day a year we need to drive off-road we could rent a four-wheel-driven car), or should we rather go for a super-efficient small car (cf. the assessment made in Chapter 6, section 6.2.4)? The first option would more than double the energy demand of existing industrialised societies, plus add the additional demand from developing economies. The second option could reduce existing demand by a factor of three, leaving more room for newcomers to the industrialised world. Some national fuel cell programmes appear to be construed in order to push the automobile problem 25 years into the future (probably the earliest time to have viable massive penetration of fuel cell vehicles in the marketplace), rather than use the already available low or no cost energy efficient car option capable of reducing the oil supply problem here and now by a factor of three or more.

The energy efficient automobile is also a perfect starting point for subsequent introduction of fuel cell drives, where one would no longer need to put a 65-100 kW fuel cell in to get a decent performance and hence could do with smaller on-board hydrogen storage pressure and get a safer system still with acceptable range (cf. section 4.1.3). Furthermore, future cars must have an electronic control system not only for optimum performance, but also for avoiding collisions and preventing speeding. Energy is not the only problem of the current personal transportation system, so solutions that solve more than one problem should be considered, and the possibly higher cost would then have a higher chance to become accepted by customers.

All this requires a new attitude to shaping our way of life. We want more self-control, and technologies such as the reversible fuel cells offer us just that. But being more in charge ourselves also implies a greater responsibility for how it all affects other people, which again should make us think of changing the priorities of what we want to spend money on. What do we do with people willing to pay more for a car capable of going three times the maximum speed limit, but unwilling to pay just a little bit more for an energy-efficient car. There must be a balance between increased freedom of choice and setting common rules for society to work decently. "Freedom of choice" often just means that the best advertiser wins, and the much-devalued concept of regulation may be unequivocally needed to make society an acceptable place to be for every member of it. Individualism is often to the detriment of the common goods or – environmentally speaking – of the commons.

> The law locks up the man or woman
> who steals the goose from off the common
> but lets the greater villain loose
> who steals the common from the goose
>> (English rhyme quoted from McMichael, 2001).

# REFERENCES

Abashar, M. (2004). Coupling of steam and dry reforming of methane in catalytic fluidized bed membrane reactors. *Int. J. Hydrogen Energy* **29**, 799-808.

Abe, A., Nakamura, M., Sato, I., Uetani, H., Fujitani, T. (1998). Studies of the large-scale sea transportation of liquid hydrogen. *Int. J. Hydrogen Energy* **23**, 115-121.

Adamo, C., Barone, V. (2002). Physically motivated density functionals with improved performances: the modified Perdew-Burke-Ernzerhof model. *J. Cham. Phys.*, **116**, 5933-5940.

Adamson, K-A., Pearson, P. (2000). Hydrogen and methanol: a comparison of safety, economics, efficiencies and emissions. *J. Power Sources* **86**, 548-555.

Agranat, V., Tchouvelev, A. (2004). CFD modelling of gas-liquid flows in water electrolysis units. In Proc. 15th World Hydrogen Energy Conf., Yokohama. 28C-05, CD Rom, Hydrogen Energy Soc. Japan.

Ahluwalia, R., Wang, X., Rousseau, A., Kumar, R. (2004). Fuel economy of hydrogen fuel cell vehicles. *J. Power Sources* **130**, 192-201.

Ahn, S-Y., Shin, S-J., Ha, H., Hong, S-A., Lee, Y-C, Lim, T., Oh, I-H. (2002). Performance and lifetime analysis of the kW-class PEMFC stack. *J. Power Sources* **106**, 295-303.

Akansu, S., Dulger, Z., Kahraman, N., Veziroglu, T. (2004). Internal combustion engines fueled by natural has-hydrogen mixtures. *Int. J. Hydrogen Energy* **29**, 1527-1539.

Aki, H., Yamamoto, S., Kondoh, J., Maeda, T., Yamaguchi, H., Murata, A., Ishii, I., Sugimoto, I. (2004). Fuel cells and hydrogen energy networks in urban residential buildings. In Proc. 15th World Hydrogen Energy Conf., Yokohama. 28I-05, CD Rom, Hydrogen Energy Soc. Japan.

Akkerman, I., Janssen, M., Rocha, J., Wijffels, R. (2002). Photobiological hydrogen production: photochemical efficiency and bioreactor design. *Int. J. Hydrogen Energy*, **27**, 1195-1208. The article is difficult to read because of editing errors.

Alcaide, F., Brillas, E., Cabot, P-L. (2004). Limiting behaviour during the hydroperoxide ion generation in a flow alkaline fuel cell. *J. Electroanalytical Chem.* **566**, 235-240.

Altmann, M., Weindorf, W., Wurster, R., Mostad, H., Weinberger, M., Filip, G. (2004). FCSHIP: environmental impacts and costs of hydrogen, natural gas and conventional fuels for fuel cell ships. In Proc. 15th World Hydrogen Energy Conf., Yokohama. 30A-05, CD Rom, Hydrogen Energy Soc. Japan.

Amos, W. (1998). Cost of storing and transporting hydrogen. Internal Report, Nat. Renewable Energy Lab., Golden, CO.

Andreassen, K. (1998). Hydrogen production by electrolysis,. In "Hydrogen Power: Theoretical And Engineering Solutions" (Sætre, T., ed.), p.91. Kluwer, Dordrecht.

Angrist, S. (1976). "Direct Energy Conversion", 3rd ed. Allyn and Bacon, Boston.

Antonkine, M., Jordan, P., Fromme, P., Krauss, N., Golbeck, J., Stehlik, D. (2003). Assembly of protein subunits within the stromal ridge of photosystem I: structural changes between unbound and sequentially PS I-bound popypeptides and correlated changes of the magnetic properties of the terminal iron sulphur clusters. *J. Mol. Biol.* **327**, 671-697.

Aroutiounian, V., Arakelyan, V., Shahnazaryan, G. (2004). Metal oxide photoelectrodes for hydrogen generation using solar radiation-driven water splitting. *Solar Energy* (in press).

Arroyo, M. y de Dompablo, Ceder, G. (2004). First principles investigations of complex hydrides $AMH_4$ and $A_3MH_6$ (A = Li, Na, K, M = B, Al, Ga) as hydrogen storage systems. *J. Alloys Compounds* **364**, 6-12.

Asada, Y., Koike, Y., Schnackenberg, J., Miyake, M., Uemura, I., Miyake, J. (2000). Heterologous expression of clostidial hydrogenase in the cyanobacterium *Synechococcus* PCC7942. *Biochim. Biophys. Acta* **1490**, 269-278.

Asakuma, Y., Miyauchi, S., Yamamoto, T., Aoki, H., Miura, T. (2004). Homogenization method for effective thermal conductivity of metal hydride bed. *Int. J. Hydrogen Energy* **29**, 209-216.

Au, S., Hemmes, K., Woudstra, N. (2003). Flowsheet calculation of a combined heat and power fuel cell plant with a conceptual molten carbonate fuel cell with separate $CO_2$ supply. *J. Power Sources* **122**, 19-27.

Ayabe, S., Omoto, H., Utaka, T., Kikuchi, R., Sasaki, K., Teraoka, Y., Eguchi, K. (2003). Catalytic autothermal reforming of methane and propane over supported metal catalysts. *Appl. Catalysis A: General* **241**, 261-269.

Azam, F., Worden, A. (2004). Microbes, molecules and marine ecosystems. *Science* **303**, 1622-1623.

Badsberg, U., Jørgensen, A., Gasmar, H., Led, J., Hammerstad, J., Jespersen, L., Ulstrup, J. (1996). Solution structure of reduced plastocyanin from the blue-green alga *Anabena variabilis. Biochemistry* **35**, 7021-7031.

Bahatyrova, S., Frese, R., Siebert, C., Olsen, J., Werf, K. van der, Grondelle, R. van, Niederman, R., Bullough, P., Otto, C, Hunter, C. (2004). The native architecture of a photosynthetic membrane. *Nature* **430**, 1058-1062.

Bain, A., Vorst, W. van (1999). The Hindenburg tragedy revisited: the fatal flaw found. *Int. J. Hydrogen Energy* **24**, 399-403.

Balachandran, U., Lee, T., Wang, S., Dorris, S. (2004). Use of mixed conducting membranes to produce hydrogen by water dissociation. *Int. J. Hydrogen Energy* **29**, 291-296.

Bao, D. (2001). A panoramic review of hydrogen energy activity in China. In "Hydrogen Energy Progress XIII, Proc. 13th World Energy Conf., Beijing 2000" (Mao, Z., Veziroglu, T., eds.), pp. 181-185. Int. Assoc. Hydrogen Energy, Beijing.

Barber, J. (2002). Photosystem II: a multisubunit membrane protein that oxidises water. *Current Opinion Structural Biology* **12**, 523-530.

Barbi, V., Funari, S., Gehrke, R., Scharnagl, N., Stribeck, N. (2003). Nanostructure of Nafion membrane material as a function of mechanical load studied by SAXS. *Polymer* **44**, 4853-4861.

Barbir, F. (2003). System design for stationary power generation. In "Handbook of Fuel Cells, Vol. 4" (Vielstich, W., Lamm, A., Gasteiger, H., eds.), Ch. 51. Wiley, Chichester.

Bard, A., Faulkner, L. (1998). "Electrochemical methods. 2nd ed." Wiley, New York.

Barone, V. (1996). Chapter in "Recent Advances In Density Functional Methods, Part I" (Chong, D., ed.). World Scientific Publ., Singapore.

Basile, A., Paturzo, L., Laganà, F. (2001). The partial oxidation of methane to syngas in a palladium membrane reactor: simulation and experimental studies. *Catalysis Today* **67**, 65-75.

Batra, V., Maudgal, S., Bali, S., Tewari, P. (2002). Development of aplha lithium aluminate matrix for molten carbonate fuel cell. *J. Power Sources* **112**, 322-325.

Bauer, F., Willert-Porada, M. (2004). Microstructural charaxterization of Zr-phosphate-Nafion membranes for direct methanol fuel cell (DMFC) application. *J. Membrane Science* **233**, 141-149.

Bauernschmitt, R., Ahlrichs, R. (1996). Treatment of electronic excitations within the adiabatic approximation of time dependent density functional theory. *Chem. Phys. Lett.* **256**, 454-464.

Becke, A. (1993). Density-functional thermochemistry. III: the role of exact exchange. *J. Chem. Phys.* **98**, 5648.

Béja, O., Aravind, L., Koonin, E., Suzuki, M., Hadd, A., Nguyen, L., Jovanovich, S., Gates, C., Feldman, R., Spudich, J., Spudich, E., DeLong, E. (2000). Bacterial Phodopsin: evidence for a new type of phototrophy in the sea. *Science* **289**, 1902-1906.

Bell, B. (1998). Looking beyond the internal combustion engine: the promises of methanol fuel cell vehicles. Paper presented at "Fuel Cell Technology Conference, London, September". IQPC Ltd., London.

Berman-Frank, I., Lundgren, P., Chen, Y., Küpper, H., Kolber, Z., Bergman, B., Falkowski, P. (2001). Segregation of nitrogen fixation and oxygenic photosynthesis in the marine cyanobacterium *Trichodesmium. Science* **294**, 1534-1537.

Berthold, O., Bünger, U., Niebauer, P., Schindler, P., Schurig, V., Weindorf, W. (1999). Analyse von Einsatzmöglichkeiten und Rahmenbedingungen von Brennstoffzellen in Haushalten und im Kleinverbrauch in Deutschland und Berlin. Ludwig-Bölkow Systemtechnik GmbH, Ottobrun.

Bird, R., Stewart, W., Lightfoot, E. (2001). "Transport phenomena", 2nd ed., John Wiley & Sons, New York.

Bischoff, M., Farooque, M., Satou, S., Torazza, A. (2003). MCFC fuel cell systems. In "Handbook of Fuel Cells, vol. 4" (Vielstich, W., Lamm, A., Gasteiger, H., eds.). Ch. 92. Wiley, Chichester.

Bitsche, O., Gutmann, G. (2004). Systems for hybrid cars. *J. Power Sources* **127**, 8-15.

BMW (2004). Model 745h, 750hL. Website: http://www.bmwworld.com/models

Bockris, J. (1975). "Energy: The Solar–hydrogen Alternative". Australia and New Zealand Book Co., Brookvale, Australia.

Bockris, J. (1972). A hydrogen economy. *Science* **176**, 1323.

Bockris, J., Despic, A. (2004). Principles of physical science: fields. "Encyclopædia Brittannica Library", CDROM Deluxe Ed., London.

Bockris, J., Reddy, A. (1998). "Modern Electrochemistry", 2nd ed., Vol. 1 (and Vol. 2B, 2000). Plenum Press, New York.

Bockris, J., Reddy, A., Gamboa-Aldeco, M. (2000). "Modern Electrochemistry", 2nd ed., Vol. 2A. Plenum Press, New York.

Bockris, J., Shrinivasan, S. (1969). "Fuel Cells: Their Electrochemistry". McGraw–Hill, New York.

Boettner, D., Moran, M. (2004). Proton exchange membrane (PEM) fuel cell-powered vehicle performance using direct-hydrogen fueling and on-board methanol reforming. *Energy* **29**, 2317-2330.

Bogdanovic, B., Brand, R., Marjanovic, A., Schwickardi, M., Tölle, J. (2000). Metal-doped sodium aluminium hydrides as potential new hydrogen storage materials. *J. Alloys Compounds* **302**, 36-58.

Bogdanovic, B., Schwickardi, M. (1997). Ti-doped alkali metal aluminium hydrides as potential novel reversible hydrogen storage materials. *J. Alloys Compounds* **253-204**, 1-9.

Bolton, J. (1996). Solar photoproduction of hydrogen: a review. *Solar Energy* **57**, 37-50.

Borglum, B. (2003). From cells to systems: Global Thermoelectric's critical path approach to planar SOFC development. Presentation at "8th Grove Fuel Cell Symposium 2003", http://www.globalte.com.

Borgwardt, R. (1998). Methanol production from biomass and natural gas as transportation fuel. *Industrial Eng. Chem. Res.* **37**, 3760-3767.

Bossel, U. (2004). Hydrogen: why its future in a sustainable energy economy will be bleak, not bright. *Renewable Energy World* **7**, No. 2, 155-159.

Bowman, C., Arthur, E., Heighway, E., Lisowski, P., Venneri, F., Wender, S. (1994). Accelerator-driven transmutation technology, Vol. 1, pp. 1-11. Los Alamos Laboratory, report LALP-94-59.

Boysen, D., Uda, T., Chisholm, C., Haile, S. (2004). High-performance solid acid fuel cells through humidity stabilization. *Science* **303**, 68-70.

Brehm, N., Mayinger, F. (1989). A contribution to the phenomenon of the transition from deflagration to detonation. VDI–Forschungsheft No. 653/1989, pp. 1-36. (website: http://www.thermo–a.mw.tu–muenchen.de/lehrstuhl/foschung/eder_gerlach.html).

Breitung, W., Bielen, U., Necker, G., Veser, A., Wetzel, F-J., Pehr, K. (2000). Numerical simulation and safety evaluation of tunnel accidents with a hydrogen powered vehicle. In "Proc. 13th World Energy Conf., Beijing 2000" (Mao, Z. and Veziroglu, T., eds.), pp. 1175-1181. Int. Assoc. Hydrogen Energy, Beijing.

Brettel, K., Leibl, W. (2001). Electron transfer in photosystem I. *Biochim. Biophys. Acta*, **1507**, 100-114.

Brown, S. (1998). The automakers' big-time bet on fuel cells. *Fortune Mag.*, 30 March, 12 pages (http://www.pathfinder.com/fortune/1998/980330).

Bruggeman, D. (1935). *Ann. Phys.* **24**, 636.

Buchmann, I. (1998). Understanding your batteries in a portable world. Cadex Inc., Canada, http://www.cadex.com/cfm.

Buono, S., Rubbia, C. (1996). Simulation of total loss of power accident in the energy amplifier. CERN/ET internal note 96-015, 10 pp.

Butler, D. (2004). Nuclear power's new dawn. *Nature* **429**, 238-240.

Callen, H. (1960). "Thermodynamics", John Wiley & Sons, New York.

Camara-Artigas, A., Williams, J., Allen, J. (2001). Structure of cytochrome $c_2$ from Rhodospirillum centenum. *Acta Cryst., Biol. Cryst.* **D57**, 1498-1505.

Campanari, S., Iora, P. (2004). Definition and sensitivity analysis of a finite volume SOFC model for a tubular cell geometry. *J. Power Sources* **132**, 113-126.

Campbell, C. (2004a). Oil and gas liquids 2004 scenario. http://www.peakoil.net/uhdsg/Default.htm (update of "ASPO Statistical Review of Oil and Gas, 2002" (Aleklett, K., Bentlay, R., Campbell, C., eds.).

Campbell, D. (2004b). Fuel cells international: PEM perspective. Oral presentation at Proc. 15th World Hydrogen Energy Conf., Yokohama.

Carcassi, M., Cerchiara, G., Marangon, A. (2004). Experimental studies of gas vented explosion in real type environment. In "Hydrogen Power – Theoretical and Engineering Solutions, Proc. Hypothesis V, Porto Conte 2003" (Marini, M., Spazzafumo, G., eds.), pp. 589-597. Servizi Grafici Editoriali, Padova.

Casida, M., Casida, K., Jamorski, C., Salahub, D. (1998). *J. Chem. Phys.* **108**, 4439.

Ceder, G., Chiang, Y–M., Sadoway, D., Aydinol, M., Jang, Y–I., Huang, B. (1998). Identification of cathode materials for lithium batteries guided by first–principles calculations. *Nature* **392**, 694-696.

Center for the Evaluation of Risks to Human Reproduction (2004). NTP-CERHR expert panel report on the reproductive and developmental toxicity of methanol. *Reproductive Toxicology* **18**, 303-390.

Ceyer, S. (1990). New mechanisms for chemistry at surfaces. *Science* **249**, 133-139.

CFCL (2004). Ceramic Fuel Cells Ltd. Distributed Generation Product Concept, web: http://www.cfcl.com.au.

Chahine, R. (2003). Review of progress in $H_2$ storage technologies. In "Proc. 1st European Hydrogen Energy Conf., Grenoble 2003", CDROM produced by Association Francaise de l'Hydrogène, Paris.

Chalk, S., Devlin, P., Gronich, S., Milliken, J., Sverdrup, G. (2004). The United States' FreedomCAR and hydrogen fuel initiative. In Proc. 15th World Hydrogen Energy Conf., Yokohama. 28A-09, CD Rom, Hydrogen Energy Soc. Japan.

Chan, S.H., Abou-Ellail, M., Yan, T. (2004). Prediction and measurement of fuel cell flammability limits. *Int. J. Green Energy* **1**, 101-114.

Chang, F., Lin, C. (2004). Biohydrogen production using an up-flow anaerobic sludge blanket reactor. *Int. J. Hydrogen Energy*, **29** 33-39.

Chartier, P., Meriaux, S. (1980). *Recherche* **11**, 766–776.

Chaudhuri, S., Lovley, D. (2003). Electricity generation by direct oxidation of glucose in mediatorless microbial fuel cells. *Nature Biotechnology* **21**, 1229-1232.

Cheng, H., Scott, K., Ramshaw, C. (2002). Intensification of water electrolysis in a centrifugal field. *J. Electrochem. Soc.* **149**, D172-D177.

Chibing, S., Qinwu, L., Chunlin, J., Jin, Z., Zhenguo, W. (2001). Combustion performance of $H_2/O_2$/hydrocarbon tripropellant engine operating in dual mode. In "Hydrogen Energy Progress XIII, Proc. 13th World Energy Conf., Beijing 2000" (Mao, Z., Veziroglu, T., eds.), pp. 677-683. Int. Assoc. Hydrogen Energy, Beijing.

Chitnis, P. (2001). Photosystem I: function and physiology. *Ann. Rev. Plant Physiol. Mol. Biol.* **52**, 593-626.

Chitose, K., Takeno, K., Kouchi, A., Yamada, Y., Okabayashi, K. (2004). Activities on hydrogen safety for hydrogen refueling stations – experiment and simulation of gaseous hydrogen dispersion. In Proc. 15th World Hydrogen Energy Conf., Yokohama. CD Rom, Hydrogen Energy Soc. Japan.

Choudhary, V., Banerjee, S., Rajput, A. (2002). Hydrogen from step-wise steam reforming of methane over $Ni/ZrO_2$: factors affecting catalytic methane decomposition and gasification by steam of carbon formed on the catalyst. *Appl. Catalysis A: General* **234**, 259-270.

Chung, T. (1997). The role of thorium in nuclear energy. US Dept. Energy Info.

Agency, http://www.eia.doe.gov/cneaf/nuclear/uia/thorium/thorium.html.

Cifrain, M., Kordesch, K. (2004). Advances, aging mechanism and lifetime in AFCs with circulating electrolytes. *J. Power Sources* **127**, 234-242.

Ciureanu, M., Miklailenko, S., Kaliaguine, S. (2003). PEM fuel cell as membrane reactors: kinetic analysis by impedance spectroscopy. *Catalysis Today* **82**, 195-206.

Clarke, S., Dicks, A., Pointon, K., Smith, T., Swann, A. (1997). Catalytic aspects of the steam reforming of hydrocarbons in internal reformiing fuel cells. *Catalysis Today* **38**, 411-423.

Clayton, R. (1965). "Molecular Physics in Photosynthesis". Blaisdell, New York.

Consoli, F. *et al.* (eds.) (1993). "Guidelines for Life–cycle Assessment: A Code of Practice." Society of Environmental Toxicology and Chemistry (SETAC).

Cortright, R., Davda, R., Dumesic, J. (2002). Hydrogen from catalytic reforming of biomass-derived hydrocarbons in liquid water. *Nature* **418**, 964-967.

Costamagna, P., Selimovic, A., Borghi, M., Agnew, G. (2004). Electrochemical model of the integrated planar solid oxide fuel cell (IP-SOFC). *Chem. Eng. J.* **102**, 61-69.

Courson, C., Makaga, E., Petit, C., Kiennemann, A. (2000). Development of Ni catalysts for gas production from biomass gasification. Reactivity in steam- and dry-reforming. *Catalysis Today* **63**, 427-437.

CRC (1973). "Handbook of Chemistry and Physics" (Weast, R., ed.). The Chemical Rubber Co., Cleveland, OH.

Cuddy, M. (1998). Volkswagen gearbox description for ADVISOR software. File notes, National Renewable Energy Lab., Golden, CO.

Culley, A., Lang, A., Suttle, C. (2003). High diversity of unknown picorna-like virusses in the sea. *Nature*, **424**, 1054-1057.

Dabrock, B., Bahl, H., Gottschal, G. (1992). Parameters affecting solvent production by *Clostridium pasteurianum*. *Appl. Environm. Microbiol.* **58**, 1233-1239.

Dahl, J., Buechler, K., Weimer, A., Lewandowski, A., Bingham, C. (2004). Solar-thermal dissociation of methane in a fluid-wall aerosol flow reactor. *Int. J. Hydrogen Energy* **29**, 725-736.

DaimlerChrysler (2001). Accident-free driving – a vision. Strategies for safety, Hightech Report, 18-23.

DaimlerChrysler-Ballard (last accessed 1998). Fuel-cell development programme. http://www.daimler–benz.com/research/specials/necar/necar_e.htm.

Damen, K., Faaij, A., Walter, A., Souza, M. (2002). Future prospects for biofuel production in Brazil. In "12th European Biomass Conf.", Vol. 2, pp. 1166–1169. ETA Firenze & WIP Munich.

Danish DoE (1998). Energy-2100. Plan scenario and Action plan with Update (1999). Danish Department of Energy and Environment, Copenhagen.

Danish Hydrogen Committee (1998). Brint – et dansk energi perspektiv (Sørensen, B., ed.). Danish Energy Agency, Copenhagen.

Danish National Agency for Enterprise and Construction (1999). Building Registry (extract). Description: http://www.ebst.dk/Publikationer/0/10 (2003).

Danish Power Utilities (1997). Action plan for off-shore wind parks (in Danish). SEAS Wind Dept.

Danish Transport Council (1993). Externaliteter i transportsektoren. Report 93–01, Copenhagen.

Danish Windpower Industry Association (2003). Web: http://www.windpower.dk.

Dante, R. (2004). Hypotheses for direct PEM fuel cells applications of photobioproduced hydrogen by *Chlamydomonas reinhardtii. Int. J. Hydrogen Energy* (in press).

Dapprich, S., Komáromi, I., Byun, K., Morokuma, K., Frisch, M. (1999). A new ONIOM implementation in Gaussian98. Part I. The calculation of energies, gradients, vibrational frequencies and electric field derivatives. *J. Molec. Structure (Theochem)* **461-462**, 1-21.

Dayton, D., Ratcliff, M., Bain, R. (2001). Fuel cell integration – a study of the impacts of gas quality and impurities. Report NREL/MP-510-30298, National Renewable Energy Lab., Golden CO.

DC (2004). F-Cell brochure (Japanese/English). http://www.daimlerchrysler.co.jp

Debe, M. (2003). Novel catalysts, catalysts support and catalysts coated membrane methods. In "Handbook of Fuel Cells - Fundamentals, Technology and Applications (Vielstich, W., Gasteiger, H., Lamm, A., eds.), Vol. 3, Ch. 45, pp. 576-589. John Wiley & Sons, New York.

Dell, R., Bridger, N. (1975). Hydrogen – the ultimate fuel. *Appl. Energy* **1**, 279-292.

deMouy, L. (1998). Projected world uranium requirements, reference case. USDoE Energy Information Administration: Int. Energy Information Report, website: http://www.eia.doe.gov/cneaf/nuclear/n_pwr_fc/apenf1.html.

Desrosiers, R. (1981). In "Biomass Gasification" (Reed, T., ed.), pp. 119-153. Noyes Data Corp., Park Ridge, NJ

Dey, R. (2004). Facilitating commercialization of hydrogen technologies through the activities of ISO/TC 197. In Proc. 15th World Hydrogen Energy Conf., Yokohama. CD Rom, Hydrogen Energy Soc. Japan.

Dirac. P. (1930). *Proc. Cambridge Phil. Soc.* **27**, 240.

Doctor, R., Wade, D., Mendelsohn, M. (2002). STAR-H2: a calcium-bromine hydrogen cycle using nuclear heat. Paper for "American Inst. Chem. Eng. Spring Meeting", New Orleans, http://www.eere.energy.gov/hydrogenandfuelcells.

DONG (2003). Gas stores. Danish Oil and Gas Co. http://www.dong.dk/dk/publikationer/lagerbrochure/(last assessed 2003).

Drift, A. (2002). An overview of innovative biomass gasification concepts. In "Proc. PV in Europe Conf.". WIP, Munich & ETA, Florence.

DTI (2000). 1996 Danish Reference Year, obtained from Danish Technological Institute, Tåstrup.

Duffie, J., Beckman, W. (1991). "Solar Energy Thermal Processes", 2nd ed., Wiley, New York.

Dukes, J. (2003). Burning buried sunshine: Human consumption of ancient solar energy. *Climate Change* **61**, 31-44.

Dutta, S., Morehouse, J., Khan, J. (1997). Numerical analysis of laminar flow and heat transfer in a high temperature electrolyzer. *Int. J. Hydrogen Energy* **22**, 883-895.

EC (1994). "Biofuels" (M. Ruiz–Altisent, ed.), DG XII Report EUR 15647 EN, European Commissior., Brussels.

EC (2001). The ECE-EUDC driving cycle. European Commission Report 90/C81/01, Brussels.

EC (2004). Well-to-wheels analysis of future automotive fuels and powertrains in the European context. Joint study of the European Council for Automotive R&D, European Oil Companies' Association for environment, health and safety in refining and distribution (CONCAWA), the Institute for Environment and

Sustainability of the European Commission's Joint Research Centre, L-B Systemtechnik and Institut Francais de Pétrole. WTW Report 220104. CORDIS.

EC-ATLAS (2003). European Commission DG Energy: ATLAS programme, http://europa.eu.int/energy_transport/atlas/htmlu/lbpot2.html.

EEA (2002). Size of vehicle fleet. Indicator fact sheet TERM 32AC, European Environmental Agency, Copenhagen.

Eguchi, K., Fujihara, T., Shinozaki, N., Okaya, S. (2004). Current work on solar RFC technology for SPF airship. In Proc. 15th World Hydrogen Energy Conf., Yokohama. 30A-07, CD Rom, Hydrogen Energy Soc. Japan.

Eichler, A., Hafner, J. (1997). Molecular precursors in the dissociative adsorption of $O_2$ on Pt(111). *Phys. Rev. Lett.* **79**, 4481-4484.

Einsle, O., Tezcan, F., Andrade, S., Schmid, B., Yoshida, M., Howard, J., Rees, D. (2002). Nitrogenase MoFe-protein at 1.16Å resolution: a central ligand in the FeMo-cofactor. *Science* **297**, 1696-1700.

Elliott, J., Hanna, S., Elliott, A., Cooley, G. (2000). Interpretation of the small-angle X-ray scattering from swollen and oriented perfluorinated ionomer membranes. *Macromolecules* **33**, 4161-4171.

Eltra/Elkraft (2001). Time series of Danish wind power production 2000, available at websites http://www.eltra.dk and http://www.elkraft-system.dk.

Eltra (2003). Søkabel, Danish power utility webpage http://www.eltra.dk.

EnBW (2004). Sulzer-Hexis SOFC-field test. Energie Baden-Würtemburg AG. Website: http://www.enbw.com.

Energi- og Miljødata (2002). Danish wind turbine price lists. Previously published by Danish Energy Agency and then Renewable Energy Information Secretariat, beginning 1981. http://www.emd.dk.

Energy Information Administration of the US DoE (2004). Annual Energy Review 2002. Washington DC; contains price data to 2003 and is supplemented by 2004 data from newscasts.

Enquete Kommission des Deutschen Bundestages (1995). Mehr Zukunft für die Erde. Economica Verlag, Bonn.

Erdmann, G. (2003). Future economies of the fuel cell housing market. *Int. J. Hydrogen Energy* **28**, 685-694.

Eroglu, E., Gündüz, U., Yücel, M., Türker, L., Eroglu, I. (2004). Photobiological hydrogen production by using olive mill wastewater as a sole substrate source. *Int. J. Hydrogen Energy* **29**, 163-171.

Escudero, M., Rodrigo, T., Soler, J., Daza, L. (2003). Electrochemical behaviour of lithium-nickel oxides in molten carbonate. *J. Power Sources* **118**, 23-34.

European Commission (1995). ExternE: externalities of Energy, Vols. 1-6. Reports EUR 16520-16525 EN, DGXII, Luxembourg.

European Commission (1997). Energy in Europe, 1997 – Annual Energy Review, DGXII: Science, Research & Development, Special Issue, 179 pp.

European Commission (1998). A fuel cell RDD strategy for Europe to 2005. DGXIIF, Brussels.

European Platform for Hydrogen and Fuel Cell Technologies (2004). Strategic Research Agenda and Deployment Strategy. Draft report available on website: http://www.hfpeurope.org/

Evans, B., O'Neill, H., Malyvanh, V., Lee, I., Woodward, J. (2003). Palladium-bacterial cellulose membranes for fuel cells. *Biosensors Bioelectronics* **18**, 917-923.

ExxonMobil (2004). A report on energy trends, greenhouse gas emissions and alternative energy, Houston, TX.

Faaij, A., Hamelinck, C. (2002). Long term perspectives for production of fuels from biomass; integrated assessment and R&D priorities. In "12th European Biomass Conf." Vol. 2, pp. 1110-1113. ETA Firenze & WIP Munich.

Fan, Y., Li, C., Lay, J., Hou, H., Zhang, G. (2004). Optimization of initial substrate and pH levels for germination of sporing hydrogen-producing anaerobes in cow dung compost. *Bioresource Technology,* **91** 189-193.

Fang, H., Liu, H., Zhang, T. (2002). Characterization of a hydrogen-producing granular sludge. *Biotechnology Bioeng.* **78**, 44-52.

Farooque, M., Ghezel-Ayagh, H. (2003). System design. In "Handbook of Fuel Cells, Vol. 4" (Vielstich, W., Lamm, A., Gasteiger, H., eds.), Ch. 68. Wiley, Chichester.

Feng, W., Wang, S., Ni, W., Chen, C. (2004). The future of hydrogen infrastructure for fuel cell vehicles in China and a case of application in Beijing. *Int. J. Hydrogen Energy* **29**, 355-367.

Fermi, E. (1928). *Zeitschrift f. Physik* **48**, 73.

Fernandez, R., Mandrillon, P., Rubbia, C., Rubio, J. (1996). A preliminary estimate of the economic impacts of the energy amplifier. Report CERN/LHC/96-01(EET), 75 pp.

Ferreira, K., Iverson, T. Maghlaoui, K., Barber, J., Iwata, S. (2004). Architecture of the photosynthetic oxygen-evolving center. *Science* **303**, 1831-1835.

Fingersh, L. (2003). Optimized hydrogen and electricity generation from wind. National Renewable Energy Lab., Report NREL/TP-500-34364, Golden, CO.

Finneran, K., Johnsen, C., Lovley, D. (2003). *Rhodoferax ferrireducens* sp. nov., a psychrotolerant, facultatively anaerobic bacterium that oxidizes acetate with the reduction of Fe(III). *Int. J. Systematic Evolutionary Microbiology* **53**, 669-673.

Fischer, G., Schnagl, J., Sarre, C., Lechner, W. (2003). Function of the liquid hydrogen fuel system for the new BMW 7 series. In "Proc. 1st European Hydrogen Energy Conf.", Grenoble 2003. CDROM published by Association Francaise de l'Hydrogène, Paris, 6 pp.

Fleischer, T., Oertel, D. (2003). Fuel cells - impact and consequences of fuel cell technology on sustainable development. Report EUR 20681 EN, European Commission JRC, Sevilla.

Fock, V. (1930). *Zeitschrift f. Physik* **61**, 126.

Folkesson, A., Andersson, C., Alvfors, P., Aläküla, M., Overgaard, L. (2003). Real life testing of a hybrid PEM fuel cell bus. *J. Power Sources* **118**, 349-357.

Fontana, G., Galloni, E., Jannelli, E., Minutillo, M. (2004). Different technologies for hydrogen engine fuelling. In "Hydrogen Power: Theoretical and Engineering Solutions, Proc. Hypothesis V Conf., Porto Conte 2003" (Marini, M., Spazzafumo, G., eds.), pp. 917-927. Servizi Grafici Editoriali, Padova.

Fontell, E., Kivisaari, T., Christiansen, N., Hansen, J-B., Pålsson, J. (2004). Conceptual study of a 250 kW planar SOFC system for CHP application. *J. Power Sources* **131**, 49-56.

Foresman, J., Head-Gordon, M., Pople, J., Frisch, M. (1992). Towards a systematic molecular orbital theory for excited states. *J. Phys. Chem.*, **96**, 135-149.

Fowler, M., Mann, R., Amphlett, J., Peppley, B., Roberge, P. (2002). Incorporation of voltage degradation into a generalised steady state electrochemical model for a PEM fuel cell. *J. Power Sources* **106**, 274-283.

Frangini, S., Masci, A. (2004). Intermetallic FeAl based coatings deposited by the electrospark technique: corrosion behavior in molten (Li+K) carbonate. *Surface Coatings Tech.* **184**, 31-39.

Freni, S., Barone, F., Puglisi, M. (1998). The dissolution process of the NiO cathodes for molten carbonate fuel cells: state-of-the-art. *Int. J. Energy Res.* **22**, 17-31.

Freni, S., Passalacqua, E., Barone, F. (1997). The influence of low operating temperature on molten carbonate fuel cells decay processes. *Int. J. Energy Res.* **21**, 1061-1070.

Friedlmeier, G., Friedrich, J., Panik, F. (2001). Test experiences with the Daimler-Chrysler fuel cell electric vehicle NECAR 4. *Fuel Cells* **1**, 92-96.

Frisch, M. J., G. W. Trucks, H. B. Schlegel, G. E. Scuseria, M. A. Robb, J. R. Cheeseman, J. A. Montgomery, Jr., T. Vreven, K. N. Kudin, J. C. Burant, J. M. Millam, S. S. Iyengar, J. Tomasi, V. Barone, B. Mennucci, M. Cossi, G. Scalmani, N. Rega, G. A. Petersson, H. Nakatsuji, M. Hada, M. Ehara, K. Toyota, R. Fukuda, J. Hasegawa, M. Ishida, T. Nakajima, Y. Honda, O. Kitao, H. Nakai, M. Klene, X. Li, J. E. Knox, H. P. Hratchian, J. B. Cross, C. Adamo, J. Jaramillo, R. Gomperts, R. E. Stratmann, O. Yazyev, A. J. Austin, R. Cammi, C. Pomelli, J. W. Ochterski, P. Y. Ayala, K. Morokuma, G. A. Voth, P. Salvador, J. J. Dannenberg, V. G. Zakrzewski, S. Dapprich, A. D. Daniels, M. C. Strain, O. Farkas, D. K. Malick, A. D. Rabuck, K. Raghavachari, J. B. Foresman, J. V. Ortiz, Q. Cui, A. G. Baboul, S. Clifford, J. Cioslowski, B. B. Stefanov, G. Liu, A. Liashenko, P. Piskorz, I. Komaromi, R. L. Martin, D. J. Fox, T. Keith, M. A. Al-Laham, C. Y. Peng, A. Nanayakkara, M. Challacombe, P. M. W. Gill, B. Johnson, W. Chen, M. W. Wong, C. Gonzalez, and J. A. Pople, (2003). Gaussian 03 software, Revision B.02. Gaussian, Inc., Pittsburgh PA (use of this reference format is part of user licence).

Fromme, P., Jordan, P., Krauss, N. (2001). Structure of photosystem I. *Biochim. Biophys. Acta* **1507**, 5-31.

Fromme, P., Melkozernov, A., Jordan, P., Krauss, N. (2003). Structure and function of photosystem I: interaction with its soluble electron carriers and external antenna systems. *FEBS Lett.* **555**, 40-44.

Fuhrmann, J., Gärtner, K. (2003). A detailed numerical model for DMFC: discretization and solution methods. Paper for "Computaional fuel cell dynamics II", Banff Int. Res. Center, USA.

Fujioka, Y., Ozaki, M., Takeuchi, K., Shindo, Y., Herzog, H. (1997). Cost comparison of various $CO_2$ ocean dispersal options. *Energy Conversion & Management* **38**, S273-S277.

Fujishima, A., Honda, K. (1972). Electrochemical photolysis of water at a semiconductor electrode. *Nature* **283**, 37.

Fukada, S., Nakamura, N., Monden, J. (2004). Effects of temperature, oxygen-to-methane molar ratio and superficial gas velocity on partial oxydation of methane for hydrogen production. *Int. J. Hydrogen Energy* **29**, 619-625.

Fukushima, Y., Shimada, M., Kraines, S., Hirao, M., Koyama, M. (2004). *J. Power Sources* **131**, 327-339.

Futerko, P., Hsing, I-M. (2000). Two-dimensional finite-element method study of the resistance of membranes in polymer electrolyte fuel cells. *Electrochimica Acta* **45**,

1741-1751.

FZK (1999). LH2 release in tunnel. Hydrogen research at Forschungszentrum Karlsruhe. In "Proc. Hydrogen Workshop at European Commission DG XII", website http://www.eihp.org/eihp1/workshop

Gallucci, F., Paturzo, L., Basile, A. (2004). A simulation study of the steam reforming of methane in a dense tubular membrane reactor. *Int. J. Hydrogen Energy* **29**, 611-617.

Gambardella, P., Sljivancanin, Z., Hammer, B., Blanc, M., Kuhnke, K., Kern, K. (2001). Oxygen dissociation at Pt steps. *Phys. Rev. Lett.* **87**, 056103.1-4.

Gates, D. (1966). *Science* **151**, 523–529.

Gaussian (2003). Software package, see Frisch *et al.* (2003).

Ghenciu, A. (2002). Review of fuel processing catalysts for hydrogen production in PEM fuel cell systems. *Current Opinion Solid State Material Sci.* **6**, 389-399.

Ghosh, D. (2003). Development of stationary solid oxide fuel-cells at Global Thermoelectric Inc. In "14th World Hydrogen Energy Conference", Montreal 2002, File B001g, 5 pp. CD published by CogniScience Publ. for l'Association Canadienne de l'Hydrogène, revised CD issued 2003.

Gierke, T., Hsu, W. (1982). The cluster-network model of ion clustering in perfluorosulfonated membranes. In "Perfluorinated Ionomer Membranes", American Chemical Society Symp. Series **180**, Washington, DC.

Gigliucci, G., Petruzzi, L., Cerelli, E., Garzisi, A., LaMendola, A. (2004). Demonstration of a residential CHP system based on PEM fuel cells. *J. Power Sources* **131**, 62-68.

Gil, M., Ji, X., Li, X., Na, H., Hampsay, J., Lu, Y. (2004). Direct synthesis of sulfonated aromatic poly(ether ether ketone) proton exchange membranes for fuel cell applications. *J. Membrane Sci.* **234**, 75-81.

Giles, J. (2004). Every last drop. *Nature* **429**, 694-695.

GM (2001). Well-to-wheel energy use and greenhouse gas emissions of advanced fuel/vehicle systems – North American analysis. Report from General Motors Corp., Argonne Nat. Lab., BP, ExxonMobil and Shell. (For adjacent European study see Wurster, 2003).

Gøbel, B., Bentzen, J., Hindsgaul, C., Henriksen, U., Ahrenfeldt, J., Houbak, N., Qvale, B. (2002). High performance gasification with the two-stage gasifier, In "12th European Biomass Conf.", pp. 289-395. ETA Firenze & WIP Munich.

González, A., McKeogh, E., Gallachóir, B. (2003). The role of hydrogen in high wind energy penetration electricity systems: the Irish case. *Renewable Energy* **29**, 471-489.

Gouérec, P., Poletto, L., Denizot, J., Sanchez-Cortezon, E., Miners, J. (2004). The evolution of the performance of alkaline fuel cells with circulating electrolytes. *J. Power Sources* **129**, 193-204.

Grand, P. (1979). *Nature* **278**, 693-696.

Griffith, D. (1995). "Introduction to quantum mechanics". Prentice Hall, New Jersey.

Grochala, W., Edwards, P. (2004). Thermal decomposition of the non-interstitial hydrides for the storage and production of hydrogen. *Chem. Rev.* **104**, 1283-1315.

Groenestijn, J. v., Hazewinkel, J., Nienroord, M., Bussmann, P. (2002). Energy aspects of biological hydrogen production in high rate bioreactors operated in the thermophilic temperature range. *Int. J. Hydrogen Energy* **27**, 1141-117.

Güllü, D., Demirbas, A. (2001). Biomass to methanol via pyrolysis process. *Energy Conversion Management* **42**, 1349-1356.

Gülzow, E. (1996). Alkaline fuel cells: a critical view. *J Power Sources* **61**, 99–104.

Gülzow, E., Schulze, M. (2004). Long-term operation of AFC electrodes with $CO_2$ containing gases. *J. Power Sources* **127**, 243-251.

Güther, V., Otto, A. (1999). Recent developments in hydrogen storage applications based on metal hydrides. *J. Alloys Compounds* **293-295**, 889-892.

Ha, S., Adams, B., Masel, R. (2004). A miniature air breathing direct formic acid fuel cell. *J. Power Sources* **128**, 119-124.

Hahn, R.,Wagner, S., Schmitz, A., Reichl, H. (2004). Development of a planar micro fuel cell with thin film and micro patterning technologies. *J. Power Sources* **131**, 73-78.

Haile, S., Boysen, D., Chisholm, C., Merle, R. (2001). Solid acids as fuel cell electrolytes. *Nature* **410**, 910-913.

Hamann, C., Hammett, A., Vielstich, W. (1998). "Electrochemistry". Wiley-VCH, Weinheim.

Hamilton Sundstrand Inc. (2003). Water electrolysis. Website http://xnwp021.utc.com/ssi/ssi/Applications/Echem/Background/waterelec.html.

Hammer, B., Nørskov, J. (1995). Why gold is the noblest of all the metals. *Nature* **376**, 238-240.

Hammer, B., Nørskov, J. (1997). Adsorbate reorganization at steps: NO on Pd(211). *Phys. Rev. Lett.* **79**, 4441-4444.

Hamnett, A. (2003). Direct methanol fuel cells (DMFC). In "Handbook of Fuel Cells Vol. 1" (Vielstich, W., Lamm, A., Gasteiger, H., eds.), Ch. 18. Wiley, Chichester.

Han, S., Shin, H. (2004). Biohydrogen production by anaerobic fermentation of food waste. *Int. J. Hydrogen Energy* **29**, 569-577.

Handley, C., Brandon, N., van der Vorst, R. (2002). Impact of the European vehicle waste directive on end-of-life options for polymer electrolyte fuel cells. *J. Power Sources* **106**, 344-352.

Hanneman, R., Vakil, H., Wentorf Jr., R. (1974). Closed loop chemical systems for energy transmission. In "Proc. 9th Intersociety Energy Conversion Engineering Conf.". American Society of Mechanical Engineers, New York.

Hannerz, K. (1983). *Nuclear Engineering International* Dec., p. 41

Happe, T., Schütz, K., Böhme, H. (1999). Transcriptional and mutational analysis of the uptake hydrogenase of the filamentous cyanobacterium *Anabaena variabilis* ATCC 29413. *J. Bacteriology* **182**, 1624-1631.

Harrington, D., Conway, B. (1987). *Electrochimica Acta* **32**, 1703.

Harth, R., Range, J., Boltendahl, U. (1981). EVA-ADAM system, a method of energy transportation by reversible chemical reactions. In "Energy storage and transportation" (Beghi, G., ed.), pp. 358-374. Reidel, Dordrecht.

Hartree, D. (1928). *Proc. Cambridge Phil. Soc.* **24**, 89.

Hasunuma, T., Fukusaki, E., Kobayashi, A. (2003). Methanol production is enhanced by expression of an *Aspergillus niger* pectin methylesterase in tobacco cells. *J. Biotechnology* **106**, 45-52.

Haugen, H., Eide, L. (1996). $CO_2$ capture and disposal: the realism of large scale scenarios. *Energy Conversion Management* **37**, 1061-1066.

Hawkes, F., Dinsdale, R., Hawkes, D., Hussy, I. (2002). Sustainable fermentative hydrogen production: challenges for process optimisation. *Int. J. Hydrogen Energy,* **27**, 1339-1347.

Hayashi, T., Watanabe, S. (2004). Hydrogen safety for fuel cell vehicles. In Proc. 15th World Hydrogen Energy Conf., Yokohama. CD Rom, Hydrogen Energy Soc. Japan.

Hedrick, J. (1998). Thorium. US Geological Survey, Mineral commodity summary & Yearbook. http://minerals.er.usgs.gov/minerals/pubs/commodity/thorium.

Herrmann, M., Meusinger, J. (2003). Hydrogen storage systems for mobile applications. In "Proc. 1st European Hydrogen Energy Conf., Grenoble 2003", CDROM produced by Association Francaise de l'Hydrogène, Paris.

Herrmann, A., Schimmele, L., Mössinger, J., Hirscher, M., Kronmüller, H. (2001). Diffusion of hydrogen in heterogeneous systems. *Appl. Phys.* **A72**, 197-208.

Herzog, H., Adams, E., Auerbach, D., Caulfield, J. (1996). Environmental impacts of ocean disposal of $CO_2$, *Energy Conversion Management* **37**, 999-1005.

Hibino, T., Hashimoto, A., Yano, M., Suzuki, M., Yoshia, S., Sano, M. (2002). *J. Electrochem. Soc.* **149**, A133.

Hibino, T., Hashimoto, A., Yano, M., Suzuki, M., Sano, M. (2003). Ru-catalyzed anode materials for direct hydrocarbon SOFCs. *Electrochimica Acta* **48**, 2531-2537.

Hirsch, R., Gallagher, J., Lessard, R., Wesselhoft, R. (1982). *Science* **183**, 909-915.

Hirsch, D., Steinfeld, A. (2004). Solar hydrogen production by thermal decomposition of natural gas using a vortex-flow reactor. *Int. J. Hydrogen Energy* **29**, 47-55.

Hodges, H. (1970). "Technology in the ancient world". Barnes & Nobles, New York.

Hodoshima, S., Arai, H., Saito, Y. (2001). Liquid-film type catalytic decalin dehydrogeno-aromatization for mobile storage of hydrogen. In "Hydrogen Energy Progress XIII, Proc. 13th World Energy Conf., Beijing 2000" (Mao, Z., Veziroglu, T., eds.), pp. 504-509. Int. Assoc. Hydrogen Energy, Beijing.

Hoffmann, P. (1998). ZEVCO unveils fuel cell taxi. *Hydrogen and Fuel Cell Letter*, feature article, August (http://www.mhv.net/~hfcletter/letter).

Hoffmann, J., Yuh, C-Y., Jopek, A. (2003). Electrolyte and material challenges. In "Handbook of Fuel Cells, Vol. 4" (Vielstich, W., Lamm, A., Gasteiger, H., eds.), Ch. 67. Wiley, Chichester.

Hoganson, C., Babcock, G. (1997). A metalloradical mechanism for the generation of oxygen from water in photosynthesis. *Science* **277**, 1953-1956.

Hohenberg, P., Kohn, W. (1964). Inhomogeneous electron gas. *Phys. Rev.* **136**, B864-B871.

Holladay, J., Wainright, J., Jones, E., Gano, S. (2004). Power generation using a mesoscale fuel cell integrated with a microscale fuel processor. *J. Power Souces* **130**, 111-118.

Honda (2004). Honda's vision of future home life. http://www.honda.com.

Horch, S., Lorensen, H., Helweg, S., Lægsgaard, E., Stensgaard, I., Jacobsen, K., Nørskov, J., Besenbacher, F. (1999). Enhancement of surface self-diffusion of platinum atoms by adsorbed hydrogen. *Nature* **398**, 134-136.

Horny, C., Kiwi-Minsker, L., Renken, A. (2004). Micro-structured string-reactor for autothermal production of hydrogen. *Chem. Eng. J.* **101**, 3-9.

Hu, M., Zhu, X., Wang, M., Gu, A., Yu, L. (2004). Three dimensional, two phase flow mathematical model for PEM fuel cell: Parts I and II. Analysis and discussion of the internal transport mechanism. *Energy Conversion & Management* **45**, 1861-

1882 and 1883-1916.

Hubbert, M. (1962). Energy resources, a report to the Committee on Natural Resources. Nat. Acad. Sci., Publ. 1000D.

Ibenholt, K. (2002). Explaining learning curves for wind power. *Energy Policy* **30**, 1181-1189.

IEA (2002). Key world energy statistics, International Energy Agency, Paris, Available at http://www.iea.org

IEA (2004). World Energy Outlook 2004. Executive Summary. OECD/IEA, Paris.

IEA-PVPS (2003). Trends in photovoltaic applications. Report IEA-PVPS T1-12:2003. International Energy Agency, Paris.

Industry Canada (2003). Canadian fuel cell commercialization roadmap. Industry Canada, Government Canada, Vancouver, BC.

Ioroi, T., Yasuda, K., Siroma, Z., Fujiwara, N., Miyazaki, Y. (2002). Thin film electrocatalyst layer for unitized regenerative polymer electrolyte fuel cells. *J. Power Sources* **112**, 583-587.

Ioroi, T., Yasuda, K., Miyazaki, Y. (2004). Polymer electrolyte-type unitized regenerative fuel cells. In "15th World Hydrogen Energy Conference, Yokohama 2004". Paper P09-09. Hydrogen Energy Systems Soc. of Japan (CDROM).

IPCC (1996). "Climate Change 1995: Impacts, Adaptation and Mitigation of Climate Change: Scientific-Technical Analysis. Contribution of WGII" (Watson *et al.*, eds.), 572 pp. Cambridge University Press, Cambridge.

Ishihara, T., Shibayama, T., Ishikawa, S., Hosoi, K., Nishiguchi, H., Takita, Y. (2004). Novel fast oxide ion conductor and application for the electrolyte of solid oxide fuel cell. *J. European Ceramic Soc.* **24**, 1329-1335.

Itoh, N., Kaneko, Y., Igarashi, A. (2002). Efficient hydrogen production via methenol steam reforming by preventing back-permeation of hydrogen in a palladium membrane reactor. *Industrial Eng. Chem. Res.* **41**, 4702-4706.

Iwai, Y. (2004). Japan's approach to commercialization of fuel cell/hydrogen technology. In Proc. 15th World Hydrogen Energy Conf., Yokohama. 28PL-02, CD Rom, Hydrogen Energy Soc. Japan.

Iwata, S., Lee, J., Okada, K., Lee, J., Iwata, M., Rasmussen, B., Link, T., Ramaswamy, S., Jap, B. (1998). Complete structure of the 11-subunit bovine mitochondrial cytochrome $bc_1$ complex. *Science* **281**, 64-71.

Jaffray, C., Hards, G. (2003). Precious metal supply requirements. In "Handbook of Fuel Cells - Fundamentals, Technology and Applications", Vol 3, Ch. 41 (Vielstich, W., Gasteiger, H., Lamm, A., eds.), pp. 509-513. John Wiley & Sons, New York.

Jang, S., Molinero, V., Cagin, T., Goddard, W. III (2004). Nanophase-segregation and transport in Nafion 117 from molecular dynamics simulations: effect of monomeric sequence. *J. Phys. Chem.* **B108**, 3149-3157.

Janssen, H., Bringmann, J., Emonts, B., Schroeder, V. (2004). Safety-related studies on hydrogen production in high-pressure electrolysers. *Int. J. Hydrogen Energy* **29**, 759-770.

Jensen, J., Li, Q., He, R., Xiao, G., Gao, J-A., Bjerrum, N. (2004). High temperature polymer fuel cells and their interplay with fuel processing systems. In "Hydrogen Power – Theoretical and Engineering Solutions, Proc. Hypothesis V, Porto Conte 2003" (Marini, M., Spazzafumo, G., eds.), pp. 675-683. Servizi Grafici

Editoriali, Padova.

Jensen, J., Sørensen, B. (1984). "Fundamentals of Energy Storage". Wiley, New York, 345 pp.

Jeong, K., Oh, B. (2002). Fuel economy and life-cycle cost analysis of a fuel cell hybrid vehicle. *J. Power Sources* **105**, 58-65.

Jo, J-H., Yi, S-C. (1999). A computational simulation of an alkaline fuel cell. *J. Power Sources* **84**, 87-106.

Joensen, F., Rostrup-Nielsen, J. (2002). Conversion of hydrocarbons and alcohols for fuel cells. *J. Power Sources* **105**, 195-201.

Jordan, P., Fromme, P., Witt, H., Klukas, O., Saenger, W., Krauss, N. (2001). Three-dimensional structure of cyanobacterial photosystem I at 2.5 Å resolution, *Nature* **411**, 909-917.

Jun, J., Jun, J., Kim, K. (2002). Degradation behaviour of Al-Fe coatings in wet-seal area of molten carbonate fuel cells. *J. Power Sources* **112**, 153-161.

Junginger, M., Faaij, A., Turkenburg, W. (2004). Global experience curves for wind farms. *Energy Policy* **33**, 133-150.

Jurewicz, K., Frackowiak, E., Béguin, F. (2004). Towards the mechanism of electro-chemical hydrogen storage in nanustructured carbon materials. *Appl. Phys.* **A78**, 981-987.

Kahn, J. (1996). Fuel cell breakthrough doubles performance, reduces cost. Berkeley Lab. Research News, 29. May (http://www.lbl.gov/science–articles/archive/fuel–cells.html).

Kalkstein, L. (1993). Health and climate change: direct impacts in cities. *The Lancet* **342**, 1397-1399.

Kalkstein, L., Smoyer, K. (1993). The impact of climate change on human health: some international implications. *Experientia* **49**, 969–979.

Kalnay, E., Kanamitsu, M., Kistler, R., Collins, W., Deaven, D., Gandin, L., Iredell, M., Saha, S., White, G., Woollen, J., Zhu, Y., Leetmaa, A., Reynolds, R., Chelliah, M., Ebisuzaki, W., Higgins, W., Janowiak, J., Mo, K., Ropelewski, C., Wang, J., Jenne, R., Joseph, D. (1996). The NCEP/NCAR 40-year reanalysis project. *Bull. Am. Met. Soc.* (March).

Kaltschmitt, M., Reinhardt, G., Stelzer, T. (1996). LCA of biofuels under different environmental aspects, In "Biomass for Energy and the Environment" (Chartier, P., Ferrero, G., Henius, U., Hultberg, S., Sachau, J., Wiinblad, M., eds.), Vol. 1, pp. 369-386. Pergamon/Elsevier, Oxford.

Kamiya, N., Shen, J-R. (2003). Crystal structure of oxigen-evolving photosystem II from *Thermosynechococcus vulcanus* at 3.7-Å resolution. *Proc. Nat. Acad. Sci. (US)* **100**, 98-103.

Karim, G., Wierzba, A. (2004). The lean flammability and operational mixture limits of gaseous fuel mixtures containing hydrogen in air. In "Hydrogen Power: Theoretical and Engineering Solutions, Proc. Hypothesis V Conf., Porto Conte 2003" (Marini, M., Spazzafumo, G., eds.), pp. 839-845. Servizi Grafici Editoriali, Padova.

Karmazyn, A., Fiorin, V., Jenkins, S., King, D. (2003). First-principles theory and mi-crocalorimetry of CO adsorption on the {211} surfaces of Pt and Ni. *Surface Sci.* **538**, 171-183.

Kato, T., Suzuoki, Y. (2004). Energy saving potential of home co-generation system using PEFC in both individual household and overall energy system. In Proc. 15th World Hydrogen Energy Conf., Yokohama. P12-02, CD Rom, Hydrogen Energy Soc. Japan.

Katz, E., Shipway, A., Willner, I. (2003). Biochemical fuel cells. In "Handbook of Fuel Cells, Vol. 1" (Vielstich, W., Lamm, A., Gasteiger, H., eds.). Wiley, Chichester.

Kawada, T., Mizusaki, J. (2003). Current electrolytes and catalysts. Ch. 70 in "Handbook of fuel cells vol. 4" (Vielstich, W., Lamm, A., Gasteiger, H., eds.), Ch. 21. Wiley, Chichester.

Kendrich, D., Herding, G., Scouflaire, P., Rolon, C., Candel, S. (1999). Effects of a recess on cryogenic flame stabilization. *Combustion Flame* **118**, 327-339.

Key, T., Sitzlar, H., Geist, T. (2003). Fast response, load-matching hybrid fuel cell. Report NREL/SR-560-32743, Nat. Renewable Energy Lab., Golden, CO.

Khaselev, O., Turner, J. (1998). A monolithic photovoltaic-photoelectrochemical device for hydrogen production via water splitting. *Science* **280**, 425-427.

Kikuchi, E., Menoto, Y., Kajiwara, M., Uemiya, S., Kojima, T. (2000). Steam reforming of methane in membrane reactors: comparison of electroless-plating and CVD membranes and catalyst packing methods. *Catalysis Today* **56**, 75-81.

Kikuzawa, H., Ohmura, T., Yamaguchi, R., Ohtuka, M., Sawada, Y., Tomihara, I. (2004). Japanese national project for establishment of codes and standards for stationary PEM fuel cell system. In Proc. 15th World Hydrogen Energy Conf., Yokohama. CD Rom, Hydrogen Energy Soc. Japan.

Kim, J-D., Honma, I. (2004). Synthesis and proton conducting properties of zirconia bridged hydrocarbon/phosphotungstic acid hybrid materials. *Electrochimica Acta* **49**, 3179-3183.

Kim, W., Voiti, T., Rodriguez-Rivera, Dumesic, J. (2004). Powering fuel cells with CO via aqueous polyoxometalates and gold catalysts. *Science* **305**, 1280-1283.

King, J., McDonald, B. (2003). Experience with 200 kW PC25 fuel cell power plant. In "Handbook of Fuel Cells, Vol. 4" (Vielstich, W., Lamm, A., Gasteiger, H., eds.), Ch. 61. Wiley, Chichester.

Kittel, C. (1971). "Introduction to Solid State Physics". Wiley, New York.

Klueh, P. (1986). *New Scientist* 3. April, 41-45.

Klug, H., Faass, R. (2001). Cryoplane: hydrogen fuelled aircraft – status and challenges. *Air & Space Europe* **3**, 252-254.

Knorr, H., Held, W., Prümm, W., Rüdiger, H. (1998). The MAN hydrogen propulsion system for city bus. *Int. J. Hydrogen Energy* **23**, 201-208.

Knudsen, M. (1934). "The kinetic theory of gases". Methuen, London.

Kocha, S., Turner, J., Nozik, A. (1991). *J. Electroanalyt. Chem.* **367**, 27.

Kohn, W., Sham, L. (1965). Self-consistent equations including exchange and correlation effects. *Phys. Rev.* **140** (1965) A1133.

Koide, H., Shindo, Y., Tazaki, Y., Iijima, M., Ito, K., Kimura, N., Omata, K. (1997). Deep sub-seabed disposal of $CO_2$ – the most protective storage. *Energy Conversion Management* **38**, S253-S258.

Kok, B., Forbush, B., McGloin, M. (1970). *Photochem. Photobiol.* **11**, 457-475.

Kondo, T., Arakawa, M., Wakayama, T., Miyake, J. (2002). Hydrogen production by combining two types of photosynthetic bacteria with different characteristics. *Int. J. Hydrogen Energy* **27**, 1303-1308.

Kordesch, K., Simader, G. (1996). "Fuel cells and their applications". VCH Verlag,

Weinheim.

Koroneos, C., Dompros, A., Roumbas, G., Moussiopoulos, N. (2004). Life cycle assessment of hydrogen fuel production processes. *Int. J. Hydrogen Energy* **29**, 1443-1450.

Kosugi, T., Hayashi, A., Tokimatsu, K. (2004). Forecasting development of elemental technologies and efficiency of R&D investments for polymer electrolyte fuel cells in Japan. *Int. J. Hydrogen Energy* **29**, 337-346.

Kratzer, P., Pehlke, E., Scheffler, M., Raschke, M., Höfer, U. (1998). Highly site-specific $H_2$ adsorption on vicinal Si(001) surfaces. *Phys. Rev. Lett.* **81**, 5596-5599.

Kreuer, K. (2001). On the development of proton conducting polymer membranes for hydrogen and methanol fuel cells. *J. Membrane Sci.* **185**, 29-39.

Kreuer, K. (2003). Hydrocarbon membranes. In "Handbook of Fuel Cells - Fundamentals, Technology and Applications", Vol. 3 (Vielstich, W., Gasteiger, H., Lamm, A., eds.), ch. 33. John Wiley & Sons, Chichester.

Kryukov, A., Levashov, V., Sazhin, S. (2004). Evaporation of diesel fuel droplets: kinetic versus hydrodynamic models. *Int. J. Heat Mass Transfer* **47**, 2541-2549.

Kubo, Y. (2004). Micro fuel cells for portable electronics. In Proc. 15th World Hydrogen Energy Conf., Yokohama. CD Rom, Hydrogen Energy Soc. Japan.

Kudin, K., Scuseria, G. (1998). A fast multipole algorithm for the efficient treatment of the Coulomb problem in electronic structure calculations of periodic systems with Gaussian orbitals. *Chem. Phys. Lett.* **289**, 611-616.

Kudin, K., Scuseria, G. (2000). Linear-scaling density-functional theory with Gaussian orbitals and periodic boundary conditions: efficient evaluation of energy and forces via the fast multipole method. *Phys. Rev.* **B61**, 16443.

Kudin, K., Scuseria, G., Martin, R. (2002). Hybrid density-functional theory and the insulating gap of $UO_2$. *Phys. Rev. Lett.* **89**, 266402.

Kuemmel, B., Nielsen, S., Sørensen, B. (1997). "Life–cycle Analysis of Energy Systems". Roskilde University Press, Copenhagen, 216 pp.

Kulinovsky, A., Scharmenn, H., Wippermann, K. (2004). Dynamics of fuel cell performance degradation. *Electrochem. Comm.* **6**, 75-82.

Kümmel, S., Perdew, J. (2003). Simple iterative construction of the optimized effective potential for orbital functionals, including exact exchange. *Phys. Rev. Lett.* **90**, 043004.

Kurchatov Institute (1997). Hypertext data base: Chernobyl and its consequences, Website http://polyn.net.kiae.su/polyn/manifest.html (last accessed 1999).

Kurisu, G., Zhang, H., Smith, J., Cramer, W. (2003). Structure of the cytochrome $b_6f$ complex of oxygenic photosynthesis: tuning the cavity. *Science* **302**, 1009-1014.

Kussmaul, K., Deimel, P. (1995). Materialverhalten in $H_2$-Hochdrucksystemen. *VDI Berichte* **1201**, pp. 87-101.

LaConti, A., Hamdan, M., McDonald, R. (2003). Mechanisms of membrane degradation. In "Handbook of Fuel Cells", Vol. 3 (Vielstich, W., Lamm, A., Gasteiger, H., eds.), Ch. 49. Wiley, Chichester.

Ladebeck, J., Wagner, J. (2003). Catalyst development for water-gas shift. In "Handbook of Fuel Cells" (Vielstich, W., Lamm, A., Gasteiger, H., eds.), Ch. 16. Wiley, Chichester.

Lamm, A., Müller, J. (2003). System design for transport applications. In "Handbook of Fuel Cells", Vol. 4 (Vielstich, W., Lamm, A., Gasteiger, H., eds.), Ch. 64.

Wiley, Chichester.

Lange, J-P. (1997). Perspectives for manufacturing methanol at fuel value. *Industrial Eng. Chem. Res.* **36**, 4282-4290.

Lauritsen, A., Svendsen, T., Sørensen, B. (1996). A study of the integration of wind energy into the national energy systems of Denmark, Wales and Germany as illustrations of success stories for renewable energy. Wind Power in Denmark, EC project report RENA.CT94-0012 (106 pp). IMFUFA, Roskilde University.

Lay, J. (2000). Biohydrogen generation by mesophilic anaerobic fermentation of microcrystalline cellulose. *Biotechnology Bioeng.* **74**, 280-287.

Lecocq, A., Furukawa, K. (1994). Accelerator molten salt breeder. In "Procedings 8$^{th}$ Journées Saturne, Saclay", pp. 191-192. Website (last accessed 1999): http://db.nea.fr/html/trw/docs/saturne8/.

Lee, C., Yang, W., Parr, R. (1988). Development of the Colle-Salvetti correlation-energy formula into a functional of the electron density. *Phys. Rev.*, **B37**, 785.

Lee, H., Hong, H., Kim, Y-M., Choi, S., Hong, M., Lee, H., Kim, K. (2004). Preparation and evaluation of sulphonated-fluorinated poly(arylene ether)s membranes for a proton exchange membrane fule cell (PEMFC). *Electrochimica Acta* **49**, 2315-2323.

Lee, S., Mukerjee, S., McBreen, J., Rho, Y., Kho, Y., Lee, T. (1998). Effects of Nafion impregnation on performances of PEMFC electrodes. *Electrochimica Acta* **43**, 3693-3701.

Lennard-Jones, J. (1932). *Trans. Faraday Soc.* **28**, 333.

Levin, D., Pitt, L., Love, M. (2004). Biohydrogen production: prospects and limitations to practical application. *Int. J. Hydrogen Energy* **29**, 173-185.

Li, P-W. and M. Chyu, M. (2003). Simulation of the chemical/electrochemical reactions and heat/mass transfer for a tubular SOFC in a stack. *J. Power Sources* **124**, 487-498.

Li, Q., Jensen, J., He, R., Bjerrum, N. (2004). New polymer electrolyte membranes based on acid doped PBI for fuel cells operating above 100°C. In "Hydrogen Power – Theoretical and Engineering Solutions, Proc. Hypothesis V, Porto Conte 2003" (Marini, M., Spazzafumo, G., eds.), pp. 685-696. Servizi Grafici Editoriali, Padova.

Liang, G. (2003a). Magnesium-based alloys for hydrogen storage. In "Hydrogen and Fuel Cells Conference. Towards a greener world", Vancouver June, CDROM published by Canadian Hydrogen Association and Fuel Cells Canada, Vancouver.

Liang, J. (2003b). Theoretical insight on tailoring energetics of Mg hydrogen absorption/desorption through nano-engineeering. *Appl. Phys.* **A** DOI:10.1007/s00339-003-2383-3.

Lim, C., Wang, C-Y. (2004). Effects of hydrophobic polymer content in GDL on power performance of a PEM fuel cell. *Electrochimica Acta* **49**, 4149-4156.

Lin, C., Lay, C. (2004). Carbon/nitrogen-ratio effect on fermentative hydrogen production by mixed microflora. *Int. J. Hydrogen Energy* (in press).

Lin, Y.-M., Rei, M.-H. (2000). Process development for generating high purity hydrogen by using supported palladium membrane reactor as steam reformer. *Int. J. Hydrogen Energy* **25**, 211-219.

Linden, S. van der (2003). The commercial world of energy storage: a review of operating facilities. Presentation for "1$^{st}$ Ann. Conf. Energy Storage Council", Houston, Texas.

Lipman, T., Edwards, J., Kammen, D. (2004). Fuel cell system economics: comparing the costs of generating power with stationary and motor vehicle PEM fuel cell systems. *Energy Policy* **32**, 101-125.

Lister, S., McLean, G. (2004). PEM fuel cell electrodes. *J. Power Sources* **130**, 61-76.

Liu, P., Nørskov, J. (2001). Kinetics of the anode processes in PEM fuel cells – the promoting effect of Ru in PtRu anodes. *Fuel Cells* **1**, 192-201.

Liu, S., Takahashi, K., Ayabe, M. (2003). Hydrogen production by oxidative methanol reforming on Pd/ZnO catalyst: effects of Pd loading. *Catalysis Today* **87**, 247-253.

Longenbach, J., Rutkowski, M., Klett, M., White, J., Schoff, R., Buchanan, T. (2002). "Hydrogen Production Facilities, Plant Performance and Cost Comparisons". US DoE, Nat. Energy Technology Lab., Reading, PA.

Longwell, H. (2002). The future of the oil and gas industry: Past approaches, new challenges, *World Energy* (Houston) **5**, 100-104.

Losciale, M. (2002). Technical experiences and conclusions from introduction of biogas as a vehicle fuel in Sweden. In "12$^{th}$ European Biomass Conf.", vol. 2, pp. 1124-1127. ETA Firenze & WIP Munich.

Lostao, A., Daoudi, F., Irún, M., Ramón, Á., Fernández-Cabrera, C., Romero, A., Sancho, J. (2003). How FMN binds to *Anabaena* apoflavodoxin. *J. Biol. Chem.* **278**, 24053-24061.

Løvvik, O. (2004). Adsorption of Ti on LiAlH$_4$ surfaces studied by band structure calculations. *J. Alloys Compounds* **373**, 28-32.

Lu, G., Wang, C-Y. (2004). Electrochemical and flow characterization of a direct methanol fuel cell. *J. Power Sources* **134**, 33-40.

Lu, G., Wang, C., Yen, T., Zhang, X. (2004). Development and characterization of a silicon-based micro direct methanol fuel cell. *Electrochimica Acta* **49**, 821-828.

Lung, M. (1997). Reactors coupled with accelerators. Joint Research Center (ISPRA) seminar paper (5 pp., revision 12.3.1997), Website (last accessed 1999): http://itumagill.fzk.de/ADS/mlungACC.htm.

Lunghi, P., Bove, R. (2003). Life cycle assessment of a molten carbonate fuel cell stack. *Fuel Cells* **3**, 224-230.

Lusardi, M., Bosio, B., Arato, E. (2004). An example of innovative application in fuel cell system development: CO$_2$ segregation using molten carbonate fuel cell. *J. Power Sources* **131**, 351-360.

Lutz, A., Bradshaw, R., Bromberg, L., Rabinovich, A. (2004). Thermodynamic analysis of hydrogen production by partial oxidation reforming. *Int. J. Hydrogen Energy* **29**, 809-816.

MacLean, H., Lave, L. (2003). Evaluating automobile fuel/propulsion system technologies. *Progress Energy Combustion Sci.* **29**, 1-69.

Madsen, B. (2002). International wind energy development. Annual reports, BTM consult. http://www.btm.dk.

Magazu, V., Freni, A., Cacciola, G. (2003). Hydrogen storage: strategic fields and comparison of different technologies. In "Hydrogen Power – Theoretical and Engineering Solutions, Proc. Hypothesis V, Porto Conte 2003" (Marini, M., Spazzafumo, G., eds.), pp. 371-386. Servizi Grafici Editoriali, Padova.

Magill, J., O'Carroll, C., Gerontopoulos, P., Richter, K., van Geel, J. (1995). Advantages and limitations of thorium fuelled energy amplifiers. In "Proc. Unconventional Options for Plutonium Dispositions, Obninsk", Int. Atomic Energy Agency TECDOC-840, pp. 81-86.

MAN (2004). Nutzfahrzeuge. http://www.brennstoffzellenbus.de/bus/bus.html.

Mao, W., Mao, H., Goncharov, A., Struzhkin, V., Guo, Q., Hu, J., Shu, J., Hemley, R., Somayazulu, M., Zhao, Y. (2002). Hydrogen clusters in clathrate hydrate. *Science* **297**, 2247-2249.

MAPINFO (1997). Professional GIS Software v 4.5, country boundaries. Troy, NY.

Marchetti, C. (1973). *Chem. Econ. & Eng. Rev.* **5**, 7.

Markel, T., Brooker, A., Hendricks, T., Johnson, V., Kelly, K., Kramer, B., O'Keefe, M., Sprik, S., Wipke, K. (2002). ADVISOR: a systems analysis tool for advanced vehicle modeling. *J. Power Sources* **110**, 255-266. A commercial version of the software is under preparation by AVL. Graz, Austria; http://www.avl.com

Maron, S., Prutton, C. (1959). "Principles of Physical Chemistry". Macmillan, New York.

Masukawa, H., Nakamura, K., Mochimaru, M., Sakurai, H. (2001). Photohydrogen production and nitrogenase activity in some heterocystous cyanobacteria. In "BioHydrogen II" (Miyake, J., Matsunaga, T., Pietro, A., eds.), pp. 63-66.

Masukawa, H., Mochimaru, M., Sakurai, H. (2002). Hydrogenases and photobiological hydrogen production utilizing nitrogenase system in cyanobacteria. *Int. J. Hydrogen Energy* **27**, 1471-1474.

Matsumoto, H., Okubo, M., Hamajina, S., Katahira, K., Iwahara, H. (2002). Extraction and production of hydrogen using high-temperature proton conductor. *Solid State Ionics* **152-3**, 715-720.

Matsumura, Y., Minowa, T. (2004). Fundamental design of a continuous biomass gasification process using a supercritical water fluidized bed. *Int. J. Hydrogen Energy* **29**, 701-707.

Matsuo, Y., Saito, K., Kawashima, H., Ikehata, S. (2004). Novel solid acid fuel cell based on a superprotonic conductor $Tl_3H(SO_4)_2$. *Solid State Comm.* **130**, 411-414.

Matter, P., Braden, D., Ozkan, U. (2004). Steam reforming of methanol to $H_2$ over nonreducing Zr-containing CuO/ZnO catalysts. *J. Catalysis* **223**, 340-351.

Maxoulis, C., Tsinoglou, D., Koltsakis, G. (2004). Modeling of automotive fuel cell operation in driving cycles. *Energy Conversion & Management* **45**, 559-573.

McLean, G., Niet, T., Prince-Richard, S., Djilali, N. (2002). An assessment of alkaline fuel cell technology. *Int. J. Hydrogen Energy* **27**, 507-526.

McMichael, T. (2001). "Human frontiers, environments and disease". Cambridge University Press, Cambridge.

Meibom, P., Svendsen, T., Sørensen, B. (1999). Trading wind in a hydro-dominated power pool system. *Int. J. Sustainable Development* **2**, 458-483.

Meisen, A., Shuai, X. (1997). Research and development issues in $CO_2$ capture. *Energy Conversion Management* **38**, S37-S42.

Mendoza, L., Baddour-Hadjean, R., Cassir, M., Pereira-Ramos, J. (2004). Raman evidence of the formation of $LT-LiCoO_2$ thin layers on NiO in molten carbonate at 650°C. *Appl. Surface Sci.* **225**, 356-361.

Mepsted, G., Moore, J. (2003). Performance and durability of bipolar plate. In "Handbook of Fuel Cells - Fundamentals, Technology and Applications", Vol. 3

(Vielstich, W., Gasteiger, H., Lamm, A., eds.), Ch. 23. John Wiley & Sons, Chichester.

Mercedes-Benz (2004). Fuel cells in field trials. http://www.mercedes-benz.com under "citarofcell".

Mérida, W., Maness, P., Brown, R., Levin, D. (2004). Enhanced hydrogen production from indirectly heated, gasified biomass, and removal of carbon gas emissions using a novel biological gas reformer. *Int. J. Hydrogen Energy* **29**, 283-290.

Meyers, J., Maynard, H. (2002). Design considerations for miniaturized PEM fuel cells. *J. Power Sources* **109**, 76-88.

Middleman, E., Kout, W., Vogelaar, B., Lenssen, J., Waal, E. de (2003). Bipolar plates for PEM fuel cells. *J. Power Sources* **118**, 44-46.

Michel, F., Fieseler, H., Meyer, G., Theissen, F. (1998). On-board equipment for liquid hydrogen vehicles. *Int. J. Hydrogen Energy* **23**, 191-199.

Milczarek, G., Kasuya, A., Mamykin, S., Arai, T., Shinoda, K., Tohji, K. (2003). Optimization of a two-compartment photoelectrochemical cell for solar hydrogen production. *Int. J. Hydrogen Energy* **28**, 919-926.

Mimura, T., Simayoshi, H., Suda, T., Iijima, M., Mituoka, S. (1997). Development of energy saving technology for flue gas carbon dioxide recovery in power plant by chemical absorption method and steam system. *Energy Conversion Management* **38**, S57-S62.

Minami, E., Kawamoto, H., Saka, S. (2002). Reactivity of lignin in supercritical methanol studied with some lignin model compounds. In "12$^{th}$ European Biomass Conf.", pp. 785—788. ETA Firenze & WIP Munich.

Minkevich, I., Laurinavichene, T., Tsygankov, A. (2004). Theoretical and experimental quantum efficiencies of the growth of anoxygenic phototrophic bacteria. *Process Biochem.* **39**, 939-949.

Mirenowicz, J. (1997). Le CERN reste silencieux face au nucléaire "propre" proposé par Carlo Rubbia, Journal de Géneve, 21. June, p. 13.

Mishra, P., Shukla, P., Singh, A., Srivastava, O. (2003). Investigation and optimization of nanostructures $TiO_2$ photoelectrode in regard to hydrogen production through photoelectrochemical process. *Int. J. Hydrogen Energy* **28**, 1089-1094.

Mitsuda, K., Maeda, H., Mitani, T., Matsumura, M., Urushibata, H., Yoshiyasu, H. (2004). In Proc. 15$^{th}$ World Hydrogen Energy Conf., Yokohama. 01PL-02, CD Rom, Hydrogen Energy Soc. Japan.

Miyake, J., Miyake, M., Asada, Y. (1999). Biotechnological hydrogen production: research for efficient light energy conversion. *J. Biotechnology* **70**, 89-101.

Mohitpour, M., Golshan, H., Murray, A. (2000). "Pipeline Design & Construction". ASME Press, New York.

Molina-Heredia, F., Wastl, J., Navarro, J., Bendall, D., Hervás, M., Howe, C., Rosa, M. (2003). A new function for an old cytochrome? *Nature* **424**, 33-34.

Møller, C., Plesset, M. (1934). Note on an approximation treatment for many-body systems. *Phys. Rev.* **46**, 618.

Morales, R., Chron, M., Hudry-Clegeon, G., Pélillot, Y., Nørager, S., Medina, M., Frey, M. (1999). Refined X-ray structures of the oxidized, at 1.3 Å, and reduced, at 1.17 Å, [2Fe-2S] ferredoxin from the cyanobacterium *Anabaene* PCC7119 show redox-linked conformation changes. *Biochemistry*, **38**, 16764-15773.

Morinaga, M., Yukawa, H. (2002). Nature of chemical bond and phase stability of hydrogen storage compounds. *Materials Science Eng.* **A329**, 268-275.

Morse, P. (1964). "Thermal Physics". W. A. Benjamin, New York.

MPS (2004). Hot module MCFCs. Modern Power Systems Inc. Website: http://www. connectingpower.com.

MSI (2000). WebLab ViewerLite software, v3.7. Molecular Simulations Inc.

Mueller, J., Urban, P. (1998). Characterization of direct methanol fuel cells by ac impedance spectroscopy. *J. Power Sources* **75**, 139-143.

Mugikura, Y. (2003). Stack material and stack design. In "Handbook of Fuel Cells", Vol. 4 (Vielstich, W., Lamm, A., Gasteiger, H., eds.), Ch. 66. Wiley, Chichester.

Müller, J., Frank, G., Colbow, K., Wilkinson, D. (2003). Transport/kinetic limitations and efficiency losses. Ch. 62 in "Handbook of fuel cells vol. 4" (Vielstich, W., Lamm, A., Gasteiger, H., eds.). Wiley, Chichester.

Müller, J., Urban, P., Hölderich, W. (1999). Impedance studies on direct methanol fuel cell anodes. *J. Power Sources* **84**, 157-160.

Murakami, M., Hirose, K., Kawamura, K., Sata, N., Ohishi, Y. (2004). Post-perovskite phase transition in MgSiO$_3$. *Science* **304**, 855-858.

Nakamori, Y., Orimo, S. (2004). Destabilization of Li-based complex hydrides. *J. Alloys Compounds* **370**, 271-275.

Nakao, M., Yoshitake, M. (2003). Composite perfluorinate membranes. In "Handbook of Fuel Cells", Vol. 3 (Vielstich, W., Lamm, A., Gasteiger, H., eds.), Ch. 32. Wiley, Chichester.

Nakicenovic, N., Grübler, A., Ishitani, H., Johansson, T., Marland, G., Moreira, J., Rogner, H-H. (1996). Energy Primer, pp. 75-92 in IPCC (1996).

Narayanasamy, J., Anderson, A. (2003). Mechanism for the electrooxidation of carbon monoxide on platinum by H$_2$O. Density functional theory calculation. *J. Electroanalytical Chem.* **554-555**, 35-40.

NASA (1971). Report No. R-351 and SP-8005, May.

National Academy of Sciences (US) (1994). "Management and Disposition of Excess Weapons Plutonium". National Academy Press, Washington, DC.

Nazeeruddin, M., *et al.* (2001). Engineering of efficient panchromatic sensitizers for nanocrystalline TiO$_2$-based solar cells. *J. Am. Chem. Soc.* **123**, 1613-1624.

NCAR (1997). The NCAR Community Climate Model CCM3 with NCAR/CSM Sea Ice Model. University Corporation for Atmospheric Research, National Center for Atmospheric Research, and Climate and Global Dynamics Division, http://www.cgd.ucar.edu:80/ccr/bettge/ice.

NCEP–NCAR (1998). The NOAA NCEP–NCAR Climate Data Assimilation System I, described in Kalnay *et al.* (1996), data available from Univ. Columbia at http://ingrid.ldgo.columbia.edu.

Neelis, M., Kooi, H. van der, Geerlings, J. (2004). Exergetic life cycle analysis of hydrogen production and storage systems for automotive applications. *Int. J. Hydrogen Energy* **29**, 537-545.

Neij, L., Andersen, P., Durstewitz, M., Helby, P., Hoppe-Kilpper, M., Morthorst, P. (2003). Experience curves: a tool for energy policy assessment. EC Extool project report ENG1-CT2000.00116. Lund University.

NFC (2000). Brintbil med forbrændingsmotor – et pilot projekt. Report 1763/99-003 to Danish Energy Agency, Nordvestjysk Folkecenter for Vedvarende Energi, Hurup.

Nguyen, P., Berning, T., Djilali, N. (2004). Computational model of a PEM fuel cell with serpentine gas flow channels. *J. Power Sources* **130**, 149-157.

Nguyen, T., Knobbe, M. (2003). A liquid water management strategy for PEM fuel cell stacks. *J. Power Sources* **114**, 70-79.

Nielsen, S., Sørensen, B. (1996). Long-term planning for energy efficiency and renewable energy. Paper presented at "Renewable Energy Conference, Cairo April 1996"; revised as: Interregional power transmission: a component in planning for renewable energy technologies, *Int. J. Global Energy Issues* **13**, No. 1-3 (2000) 170-180.

Nielsen, S., Sørensen, B. (1998). A fair market scenario for the European energy system. In "Long-Term Integration of Renewable Energy Sources into the European Energy System" (LTI-research group, ed.), pp. 127-186. Physica-Verlag, Heidelberg.

Nijkamp, M., Raaymakers, J., Dillen, A. van, Jong, K. de (2001). Hydrogen storage using physisorption: materials demands. *Appl. Phys.* **A72**, 619-623

Ochmann, F., Fürst, S., Müller, C. (2004). Industrialization of automotive hydrogen technology. In Proc. 15th World Hydrogen Energy Conf., Yokohama. 01PL-26, CD Rom, Hydrogen Energy Soc. Japan.

OECD (1994). Overview of Physics Aspects of Different Transmutation Concepts, 118 pp. Nuclear Energy Agency, Paris, Report New/Nsc/Doc(94)11.

OECD (1996). Energy Balances and Statistics of OECD and Non-OECD Countries. Annual Publications, Paris.

OECD and IAEA (1993). Uranium: Resources, production and demand. Nuclear Energy Agency, Paris.

OECD/IEA (2002a). Energy Balances of OECD Countries, 1999-2000. International Energy Agency, Paris.

OECD/IEA (2002b). Energy Balances of Non-OECD Countries, 1999-2000. International Energy Agency, Paris.

Ogata, H., Mizoguchi, Y., Mizuno, N., Miki, K., Adachi, S., Yasuoka, N., Yagi, T., Yamauchi, O., Hirota, S., Higuchi, Y. (2002). Structural studies of the carbon monoxide complex of [NiFe] hydrogenase from *Desulfovibrio vulgaris* Miyazaki F: suggestions for the initial activation site for dihydrogen. *J. Am. Chem. Soc.* **124**, 11628-11635.

Ogden, J. (1999). Developing an infrastructure for hydrogen vehicles: a Southern California case study. *Int. J. Hydrogen Energy* **24**, 709-730.

Ogden, J., Williams, R., Larson, E. (2004). Societal lifecycle costs of cars with alternative fuels/engines. *Energy Policy* **32**, 7-27.

O'Hayre, R., Lee, S., Cha, S., Prinz, F. (2002). A sharp peak in the performance of sputtered platinum fuel cells at ultra-low platinum loading. *J. Power Sources* **109**, 483-493.

Oh, Y., Roh, H., Jun, K., Baek, Y. (2003). A highly active catalyst, Ni/Ce-ZrO$_2$/θ-Al$_2$O$_3$, for on-site H$_2$ generation by steam methane reforming: pretreatment effect. *Int. J. Hydrogen Energy* **28**, 1387-1392.

Ohi, J., Rossmeissl, N. (2004). Hydrogen codes and standards: an overview of US DoE Activities. In Proc. 15th World Hydrogen Energy Conf., Yokohama. CD Rom, Hydrogen Energy Soc. Japan.

Oi, T., Wada, K. (2004). Feasibility study on hydrogen refueling infrastructure for fuel cell vehicles using off-peak power in Japan. *Int. J. Hydrogen Energy* **29**, 347-354.

Okada, T. (2003). Effect of ionic contaminants. In "Handbook of Fuel Cells",V vol. 3 (Vielstich, W., Lamm, A., Gasteiger, H., eds.), Ch. 48. Wiley, Chichester.

Onsager, L. (1938). *J. Am. Chem. Soc.*, **58**, 1486.

Orimo, S., Fujii, H. (2001). Materials science of Mg-Ni-based new hydrides. *Appl. Phys.* **A72**, 167-186.

Osaka Gas Co. (2004). Super compact on-site hydrogen production unit: Hyserve-30. http://www.osakagas.co.jp.

Ovesen, C., Clausen, B., Hammershøi, B., Steffensen, G., Askgaard, T., Chorkendorff, I., Nørskov, J., Rasmussen, P., Stoltze, P., Taylor, P. (1996). A microkinetic analysis of the water-gas shift reaction under industrial conditions, *J. Catalysis* **158**, 170-180.

Pacheco, M., Sira, J., Kopasz, J. (2003). Reaction kinetics and reactor modelling for fuel processing of liquid hydrocarbons to produce hydrogen: isooctane reforming. *Appl. Catalysis A: General* **250**, 161-175.

Paddison, S. (2001)., *J. New Materials Electrochem. Sys.* **4**, 197.

Padró, C., Putche, V. (1999). Survey of the economics of hydrogen technologies. US National Renewable Energy Lab. Report NREL/TP-570-27079. Golden, CO.

Paladini, V., Miotti, P., Manzoni, G., Ozebec, J. (2003). Conception of modular hydrogen storage systems for portable applications. In "Hydrogen and Fuel Cells Conference. Towards a Greener World", Vancouver June, CDROM, 12 pp. Published by Canadian Hydrogen Association and Fuel Cells Canada, Vancouver.

Palenik, B., Brahamsha, B., Larimer, F., Land, M., Hauser, L., Chain, P., Lamerdin, J., Regala, W., Allen, E., McCarren, J., Paulsen, I., Dufresne, A., Partensky, F., Webb, E., Waterbury, J. (2003). The genome of a motile marine Synechococcus. *Nature* **424**, 1037-1042.

Pallassana, V., Neurock, M., Hansen, L., Hammer, B., Nørskov, J. (1999). Theoretical analysis of hydrogen chemisorption on Pd(111), Re(0001) and $Pd_{ML}$/Re(0001), $Re_{ML}$/Pd(111) pseudomorphic overlayers. *Phys. Rev.* **B60**, 6146-6134.

Palo, D., Holladay, J., Rozmiarek, R., Guzman-Leong, C., Wang, Y., Hu, J., Chin, Y.-H., Dagle, R., Baker, E. (2002). Development of a soldier-portable fuel cell power system. Part I: a bread-board methanol fuel processor. *J. Power Sources* **108**, 28-34.

Patil, P. (1998). The US DoE fuel cell program. Investing in clean transportation. Paper presented at "Fuel Cell Technology Conference, London, September", IQPC Ltd, London.

Patyk, A., Höpfner, U. (1999). Ökologischer Vergleich von Kraftfarhzeugen mit verschiedenen Antriebsenergien unter besonderer Berücksichtigung der Brennstoffzelle. IFEU, Heidelberg.

Pehnt, M. (2001). Life-cycle assessment of fuel cell stacks. *Int. J. Hydrogen Energy* **26**, 91-101.

Pehnt, M. (2002). Ganzheitliche Bilanzierung von Brennstoffzellen in der Energie- und Verkehrstechnik. Dissertation, *Fortschrittsberichte* **6**, No. 476. VDI-Verlag Dusseldorf.

Pehnt, M. (2003). Life-cycle analysis of fuel cell system components. In "Handbook of Fuel Cells - Fundamentals, Technology and Applications, Vol. 4 (Vielstich, W.,

Gasteiger, H., Lamm, A., eds.), ch. 94. John Wiley & Sons, Chichester.

Penev, E., Kratzer, P., Scheffler, M. (1999). Effect of the cluster size in modelling the $H_2$ desorption and dissociative adsorption on Si(001). *J. Chem. Phys.* **110**, 3986-3994.

Perdew, J., Burke, K., Ernzerhof, M. (1996). Generalized gradient approximation made simple. *Phys. Rev. Lett.* **77**, 3865-3868. Erratum: **78**, 1396-1397.

Perdew, J., Kurth, S., Zupan, A., Blaha, P. (1999). Accurate density functional with correct formal properties: a step beyond the generalised gradient approximation. *Phys. Rev. Lett.* **82**, 2544-2547. Erratum: 5179.

Perednis, D., Gauckler, L. (2004). Solid oxide fuel cells with electrolytes prepared via spray pyrolysis. *Solid State Ionics* **166**, 229-239.

Perrette, L., Chelhaoui, S., Corgier, D. (2003). Safety evaluation of a PEMFC bus. In "Hydrogen Power – Theoretical and Engineering Solutions, Proc. Hypothesis V, Porto Conte 2003" (Marini, M., Spazzafumo, G., eds.), pp. 599-610. Servizi Grafici Editoriali, Padova.

Perrin, G. (1981). *Verkehr und Technik*, issue no. 9.

Peters, J., Lanzilotta, W., Lemon, B., Seefeldt, L. (1998). X-ray crystal structure of the Fe-only hydrogenase (Cpl) from *Clostridium pasteurianum* to 1.8 Ångström resolution. *Science* **282**, 1853-1858.

PFC Energy (2004). Global crude oil and natural gas liquids supply forecast. Presentation at Center for Strategic & Int. Studies, Washington DC, http://www.csis.org/energy/040908_presentation.pdf

Pigford, T. (1991). In Transmutation as a waste management tool, pp. 97-99. Unpublished Conf. Proc.

Pinto, F., Troshina, O., Lindblad, P. (2002). A brief look at three decades of research on cyanobacterial hydrogen evolution. *Int. J. Hydrogen Energy* **27**, 1209-1215.

Pohl, H., Malychev, V. (1997). Hydrogen in future civil aviation. *Int. J. Hydrogen Energy* **22**, 1061-1069.

Polle, J., Kanakagiri, S., Jin, E., Masuda, T., Melis, A. (2002). Truncated chlorophyll antenna size of the photosystems – a practical method to improve microalgal productivity and hydrogen production in mass culture. *Int. J. Hydrogen Energy* **27**, 1257-1264.

Pooley, D. (chairman) (1997). Opinion of the scientific and technical committee on a nuclear energy amplifier. European Commission, Nuclear Science and Technology Report EUR 17616 EN 1996; UKAEA Government Division, Harwell, UK, assessed 1997 at http://itumagill.fzk.de/ADS/pooley.html.

Pöpperling, R., Schwenk, W., Venkateswarlu, J. (1982). Abschätzung der Korrosionsgefärdung von Behältern und Rohrleitungen aus Stahl für Speicherung von Wasserstoff und wasserstofhältigen Gasen unter hohen Drücken. VDI Zeitschriften Reihe 5, No. 62.

Presting, H., Konle, J., Starkov, V., Vyatkin, A., König, U. (2004). Porous silicon for micro-sized fuel cell reformer units. *Materials Sci. Eng.* **B108**, 162-165.

Proton Energy Systems (2003). Unigen. Website http://www.protonenergy.com.

Qi, Z., Kaufman, A. (2003). Low Pt loading high performance cathodes for PEM fuel cells. *J. Power Sources* **113**, 37-43.

Qian, D., Nakamura, C., Wenk, S., Wakayama, T., Zorin, N., Miyake, J. (2003). Electrochemical hydrogen evolution by use of a glass carbon electrode sandwiched with clay, poly(butylviologen) and hydrogenase. *Materials Lett.* **57**, 1130-1134.

Qing, H., Chengzhong, Y. (2001). Application of liquid hydrogen in hypersonic aero-engine. In "Hydrogen Energy Progress XIII, Proc. 13th World Energy Conf., Beijing 2000" (Mao, Z., Veziroglu, T., eds.), pp. 670-676. Int. Assoc. Hydrogen Energy, Beijing.

Radecka, M. (2004). $TiO_2$ for photoelectrolytic decomposition of water. *Thin Solid Films* **451/2**, 98-104.

Rasmussen, N. (1975). Project leader, Reactor Safety Study. Report WASH-1400 NUREG 75/014. US Nuclear Regulatory Commission, Washington, DC.

Rasten, E., Hagen, G., Tunold, R. (2003). Electrocatalysis in water electrolysis with solid polymer electrolyte. *Electronica Acta* **48**, 3945-3952.

Reed, T. (ed.) (1981). "Biomass gasification". Noyes Data Corp., Park Ridge, NJ.

Reuter, K., Frenkel, D., Scheffler, M. (2004). The steady state of heterogeneous catalysis, studied by first-principle statistical mechanics. *Phys. Rev. Lett.* **93**, 116105.

RIT (1997). Accelerator driven systems. 3 pp., Royal Institute of Technology, Stockholm, Website http://www.neutron.kth.se/introduction/.

Robinson, J. (ed.) (1980). "Fuels from Biomass". Noyes Data Corp., Park Ridge, NJ.

Roddy, D. (2004). Making a viable fuel cell industry happen in the Tees Valley. *Fuel Cells Bulletin* Jan. 10-12.

Roh, H., Jun, K., Dong, W., Chang, J., Park, S., Joe, Y. (2002). Highly active and stable $Ni/Ce-ZrO_2$ catalyst for $H_2$ production from methane. *J. Molec. Catalysis A: Chemical* **181**, 137-142.

Roos, M., Batawi, E., Harnisch, U., Hocker, T. (2003). Efficient simulation of fuel cell stacks with the volume averaging method. *J. Power Sources* **118**, 86-95.

Rosa, M. de la, Navarro, J., Díaz-Quintana, A., Cerda, B. de la, Molina-Heredia, F., Balme, A., Murdoch, P., Díaz-Moreno, I., Durán, R., Hervás, M. (2002). An evolutionary analysis of the reaction mechanisms of photosystem I reduction by cytochrome $c_6$ and plastocyanin. *Bioelectrochemistry* **55**, 41-45.

Rosi, N., Eckert, J., Eddaoudi, M., Vodak, D., Kim, J., O'Keeffe, M., Yaghi, O. (2003). Hydrogen storage in microporous metal-organic frameworks. *Science* **300**, 1127-1129.

Rostrup-Nielsen, J. (2000). New aspects of syngas production and use. *Catalysis Today* **63**, 159-164.

Rostrup-Nielsen, J., Aasberg-Petersen, K. (2003). Steam reforming, ATR, partial oxidation: catalysts and reaction engineering. In "Handbook of Fuel Cells" (Vielstich, W., Lamm, A., Gasteiger, H., eds.), Ch. 14. Wiley, Chichester.

Rubbia, C. (1994). The energy amplifier, In "Proc. 8th Journées Saturne, Saclay", pp. 115-123. http://db.nea.fr/html/trw/docs/saturne8/ (last accessed 1999).

Rubbia, C., Rubio, J., Buono, S., Carminati, F., Fiétier, N., Galvez, J., Gelès, J., Kadi, Y., Klapisch, R., Mandrillon, P., Revol, J., Roche, C. (1995). Conceptual design of a fast neutron operated high power energy amplifier. European Organization for Nuclear Research, preprint collection CERN/AT-95-44.

Rubbia, C., Rubio, J. (1996). A tentative programme towards a full scale energy amplifier. European Organization for Nuclear Research, preprint CERN/LHC/96-11(ET). 36 pp. Website http://sundarssrv2.cern.ch/search.html

Runge, E., Gross, E. (1984). Density-functional theory for time-dependent systems. *Phys. Rev. Lett.* **52**, 997-1000.

Saiki, Y., Amao, Y. (2003). Bio-mimetic hydrogen production from polysaccharide using the visible light sensitation of zinc porphyrin. *Biotechnology Bioeng.*, **82**, 710-714.

Saito, M., Takeuchi, M., Watanabe, T., Toyir, J., Luo, S., Wu, J. (1997). Methanol synthesis from $CO_2$ and $H_2$ over a Cu/ZnO-based multicomponent catalyst, *Energy Conversion Management* **38**, S403-S408.

Sandrock, G., Thomas, G. (2001). Database administrators for an online hydride database of the International Energy Agency and the US Department of Energy at http://hydpark.ca.sandia.gov.

Satija, R., Jacobsen, D., Arif, M., Werner, S. (2004). In situ neutron imaging techniques for evaluation of water management systems in operating PEM fuel cells. *J. Power Sources* **129**, 238-245.

Sattler, G. (2000). Fuel cells going on-board. *J. Power Sources* **86**, 61-67.

Sauk, J., Byun, J., Kim, H. (2004). Grafting of styrene to Nafion membranes using supercritical $CO_2$ impregnation for direct methanol fule cells. *J. Power Sources* **132**, 59-63.

Schaefer, A., Horn, H., Ahlrichs, R. (1992). Fully optimized contracted Gaussian basis sets for atoms Li to Kr. *J. Chem. Phys.* **97**, 339.

Schaefer, A., Huber, C., Ahlrichs, R. (1994). *J. Chem. Phys.* **100**, 5829.

Schaeffer, G., Alsema, E., Seebregts, A., Buerskens, L., Moor, H. de, Durstewitz, M., Perrin, M., Boulanger, P., Laukamp, H., Zuccarro, C. (2004). Synthesis report Photex-project ECN Report, Petten.

Scharff, M. (1969). "Elementary quantum mechanics". John Wiley & Sons, London.

Schlamadinger, B., Marland, G. (1996). Full fuel cycle carbon balances of bioenergy and forestry options. *Energy Conversion Management* **37**, 813-818.

Schlapbach, L., Züttel, A. (2001). Hydrogen-storage materials for mobile applications. *Nature* **414** 353-358.

Schlesinger, H., Brown, H. (1940). Metallo borohydrides, III: Lithium borohydride. *J. Am. Chem. Soc.* **62**, 3429-3435.

Schober, T. (2001). Tubular high-termperature proton conductors: transport numbers and hydrogen injection. *Solid State Ionics* **139**, 95-104.

Schultz, M., Diehl, T., Brasseur, G., Zittel, W. (2003). Air pollution and climate-forcing impacts of a global hydrogen economy. *Science* **302**, 624-527.

Schültz, M., Werner, H-J., Lindh, R., Manby, F. (2004). Analytical energy gradients for local second-order Møller-Plesset perturbation theory using density fitting approximations. *J. Chem. Phys.* **121**, 737-750.

Schulze, M., Knöri, T., Schneider, A., Gülzow, E. (2004). Degradation of sealings for PEFC test cells during fuel cell operation. *J. Power Sources* **127**, 222-229.

Schuster, M., Meyer, W., Schuster, M., Kreuer, K. (2004). Towards a new type of anhydrous organic proton conductor based on immobilized imidazole. *Chem. Meterials* **16**, 329-337.

Scrosati, B. (1995). Challenge of portable power. *Nature* **373**, 557-558.

Semelsberger, T., Brown, L., Borup, R., Inbody, M. (2004). Equilibrium products from autothermal processes for generating hydrogen-rich fuel-cell feeds. *Int. J. Hydrogen Energy* **29**, 1047-1064.

Shang, C., Bououdina, M., Song, Y., Guo, Z. (2004). Mechanical alloying and electronic simulations of ($MgH_2$ + M) systems (M = Al, Ti, Fe, Ni, Cu and Nb) for hydrogen storage. *Int. J. Hydrogen Energy* **29**, 73-80.

Shayegan, S., Hart, D., Pearson, P., Bauen, A., Joffe, D. (2004). Hydrogen infrastructure costs: what are the important variables? In "Hydrogen Power – Theoretical and Engineering Solutions, Proc. Hypothesis V, Porto Conte 2003" (Marini, M., Spazzafumo, G., eds.), pp. 499-508. Servizi Grafici Editoriali, Padova.

Shen, P., Shi, Q., Hua, Z., Kong, F., Wang, Z., Zhuang, S., Chen, D. (2003). Analysis of microsystins in cyanobacteria blooms and surface water samples from Meilang Bay, Taihu Lake, China. *Environment Int.* **29**, 641-647.

Shimizu, K., Fukagawa, M., Sakanishi, A. (2004). Development of PEM water electrolysis type hydrogen production system. In "15th World Hydrogen Energy Conference, Yokohama 2004". Hydrogen Energy Systems Soc. of Japan (CDROM).

Shoiji, M., Houki, Y., Ishiyama, T. (2001). Feasibility of the high-speed hydrogen engine. In "Hydrogen Energy Progress XIII, Proc. 13th World Energy Conf., Beijing 2000" (Mao, Z., Veziroglu, T., eds.), pp. 641-647. Int. Assoc. Hydrogen Energy, Beijing.

Shore, L., Farrauto, R. (2003). PROX catalysts. In "Handbook of Fuel Cells" (Vielstich, W., Lamm, A., Gasteiger, H., eds.), Ch. 18. Wiley, Chichester.

Siegel, N., Ellis, M., Nelson, D., Spakovsky, M. von (2003). Single domain PEMFC model based on agglomerate catalyst geometry. *J. Power Sources* **115**, 81-89.

Simbeck, D., Chang, E. (2002). Hydrogen supply: cost estimate for hydrogen pathways – scoping analysis. National Renewable Energy Lab., Report NREL/SR-540-32525, Golden, CO.

Simon, G. de, Parodi, F., Fermeglia, M., Taccani, R. (2003). Simulation of process for electrical energy production based on molten carbonate fuel cells. *J. Power Sources* **115**, 210-218.

Sistiaga, M., Pierna, A. (2003). Application of amorphous materials for fuel cells. *J. Non-Crystalline Solids* **329**, 184-187.

Slater, J. (1928). The self consistent field and the structure of atoms, *Phys. Rev.* **32**, 339-348.

Slater, J. (1951). *Phys. Rev.* **81**, 385.

Sljivancanin, Z., Hammer, B. (2002). Oxygen dissociation at close-packed Pt terraces, Pt steps, and Ag-covered Pt steps studied with density functional theory. *Surface Sci.* **515**, 235-244.

Sluiter, M., Belosludov, R., Jain, A., Belosludov, R., Adachi, H., Kawazoe, Y., Higuchi, K., Otani, T. (2003). *Ab initio* study of hydrogen hydrate clathrates for hydrogen storage within the ITBL environment. In "Proc. ISHPC Conference 2003" (Veidenbaum, A. *et al.*, eds.) , pp. 330-341. Springer-Verlag, Berlin.

SMAB (1978). "Metanol sam drivmedel," Annual Report, Svensk Metanol-utveckling AB, Stockholm.

Smith, B. (2002). Nitrogenase reveals its inner secrets. *Science* **297**, 1654-1655.

Sørensen, B. (1975). Energy and resources. *Science* **189**, 255-260, and in "Energy: Use, Conservation and Supply" (Abelson, P., and Hammond, A., eds.), Vol. II, pp. 23-28. Am. Ass. Advancement of Science, Washington, DC (1978).

Sørensen, B. (1979a). "Renewable Energy". Academic Press, London.

Sørensen, B. (1979b). Nuclear power: the answer that became a question. An assessment of accident risks. *Ambio* **8**, 10-17.

Sørensen, B. (1981). A combined wind and hydro power system. *Energy Policy*, March, pp. 51–55.

Sørensen, B. (1982). Comparative risk assessment of total energy systems. In "Health Impacts of Different Sources of Energy", pp. 455-471. Report IAEA-SM-254/105, Int. Atomic Energy Agerncy, Vienna.

Sørensen, B. (1983). Stationary applications of fuel cells. In "Solid State Protonic Conductors II – for Fuel Cells and Sensors", pp. 97-108 (Goodenough, J., Jensen, J., Kleitz, M., eds.), Odense University Press, Odense.

Sørensen, B. (1984). Energy storage. *Ann. Rev. Energy* **9**, 1-29.

Sørensen, B. (1987). Chernobyl accident: assessing the data. *Nuclear Safety* **28**, 443-447.

Sørensen, B. (1991). Energy conservation and efficiency measures in other countries. Greenhouse Studies No. 8. Commonwealth of Australia, Dept. Arts, Sport, Environment, Tourism and Territories, Canberra.

Sørensen, B. (1993). What is life-cycle analysis? In "Life-cycle Analysis of Energy Systems", pp. 21-53. Workshop Proceedings, OECD Publications, Paris.

Sørensen, B. (1995). History of, and recent progress in, wind–energy utilization. *Ann. Rev. Energy & Environment* **20**, 387–424.

Sørensen, B. (1996a). Life-cycle approach to assessing environmental and social externality costs. In "Comparing Energy Technologies", Ch. 5, pp. 297-331. International Energy Agency, IEA/OECD, Paris.

Sørensen, B. (1996b). Scenarios for greenhouse warming mitigation. *Energy Conversion & Management* **37**, 693-698.

Sørensen, B. (1996c). Does wind energy utilization have regional or global climate impacts? "Proc. 1996 European Union Wind Energy Conference", pp. 191-194. H. Stephens & Ass., Bedford UK.

Sørensen, B. (1997a). Impacts of energy use. In "Human Ecology, Human Economy" (Diesendorf and Hamilton, eds.), pp. 243-266. Allen and Unwin, New South Wales.

Sørensen, B. (1997b). Externality estimation of greenhouse warming impacts, *Energy Conversion Management* **38**, S643–S648.

Sørensen, B. (1998). Brint (Strategy note from Danish Hydrogen Committee). Danish Energy Agency, Copenhagen.

Sørensen, B. (1999). Long-term scenarios for global energy demand and supply: four global greenhouse mitigation scenarios. Final Report from a project performed for the Danish Energy Agency, IMFUFA Texts 359, Roskilde University, pp. 1-166.

Sørensen, B. (2000). Role of hydrogen and fuel cells in renewable energy systems. In "Renewable Energy: the Energy for the 21st Century", Proc. World Renewable Energy Conference VI, Reading, Vol. 3, pp. 1469-1474. Pergamon, Amsterdam.

Sørensen, B. (2002a). Handling fluctuating renewable energy production by hydrogen scenarios. In "14th World Hydrogen Energy Conference", Montreal, 2002. File B101c, 9 pp. CD published by CogniScience Publ. for l'Association Canadienne de l'Hydrogène, revised CD issued 2003.

Sørensen, B. (2002b). Understanding photoelectrochemical solar cells. In "PV in Europe, from PV Technology to Energy Solutions", Roma Int. Conf. ( Bal, J., *et al.*, eds.), pp. 3-8. WIP Munich and ETA Florence.

Sørensen, B. (2003a). Scenarios for future use of hydrogen and fuel cells. In "Hydrogen and Fuel Cells Conference. Towards a Greener World", Vancouver June, CDROM, 12 pp. Published by Canadian Hydrogen Association and Fuel Cells Canada, Vancouver.

Sørensen, B. (2003b). Time-simulations of renewable energy plus hydrogen systems. In "Hydrogen Power – Theoretical and Engineering Solutions. Proc. Hypothesis V, Porto Conte 2003" (Marini, M., Spazzafumo, G., eds.), pp. 35-42. Servizi Grafici Editoriali, Padova.

Sørensen, B. (2003c). Hydrogen scenarios using fossil, nuclear or renewable energy. In "Proc. 1ˢᵗ European Hydrogen Energy Conference, Grenoble 2003", paper CO5/78, CDROM, 14 pp. Published by Association Francaise de l'Hydrogène, Paris.

Sørensen, B. (2004a). "Renewable Energy". 3ʳᵈ ed. Elsevier Academic Press, Burlington. MA. Previous editions 1979; 2000.

Sørensen, B. (2004b). Surface reactions at fuel cell catalysts (in preparation).

Sørensen, B. (2004c). Biological hydrogen production (in preparation).

Sørensen, B. (2004d). Total life-cycle analysis of PEM fuel cell car. In "Proc. 15ᵗʰ World Hydrogen Energy Conf., Yokohama". 29G-09, CD Rom, Hydrogen Energy Soc. Japan.

Sørensen, B. (2004e). Quantum chemical exploration of PEM fuel cell processes. In "Proc. 15ᵗʰ World Hydrogen Energy Conf., Yokohama". 28K-02, CD Rom, Hydrogen Energy Soc. Japan.

Sørensen, B. (2004f). Absorption of hydrogen in Mg lattice (submitted to *Science*).

Sørensen, B. (2004g). Readiness of hydrogen technologies. Presentation at "Hydrogen and Fuel Cell Futures Conference", September, Perth, Australia.

Sørensen, B. (2004h). Fuel cell and electric vehicle comparison (in preparation).

Sørensen, B. (2004i). The last oil? (submitted to *Science*).

Sørensen, B., Petersen, A., Juhl, C., Ravn, H., Søndergren, C., Simonsen, P., Jørgensen, K., Nielsen, L., Larsen, H., Morthorst, P., Schleisner, L., Sørensen, F., Petersen, T. (2001). Project report to Danish Energy Agency (in Danish): Scenarier for samlet udnyttelse af brint som energibærer i Danmarks fremtidige energisystem, *IMFUFA Texts* No. 390, 226 pp., Roskilde University; report download site: http://mmf.ruc.dk/energy.

Sørensen, B., Petersen, A., Juhl, C., Ravn, H., Søndergren, C., Simonsen, P., Jørgensen, K., Nielsen, L., Larsen, H., Morthorst, P., Schleisner, L., Sørensen, F., Petersen, T. (2004). Hydrogen as an energy carrier: scenarios for future use of hydrogen in the Danish energy system. *Int. J. Hydrogen Energy* **29**, 23-32.

Sørensen, B., Sørensen, F. (2000). A hydrogen future for Denmark. In "Hydrogen Energy Progress XIII, Proc. 13ᵗʰ World Energy Conf., Beijing 2000" (Mao, Z., Veziroglu, T., eds.), pp. 35-40. Int. Assoc. Hydrogen Energy, Beijing.

Spakovsky, M., Olsommer, B. (2002). Fuel cell systems and system modeling and analysis perspectives for fuel cell development. *Energy Conversion Management* **43**, 1249-1257.

Spath, P., Mann, M. (2001). Life cycle assesment of hydrogen production via natural gas steam reforming. Revised USDoE contract report NREL/TP-570-27637, National Renewable Energy Lab., Golden, CO.

Spohr, E. (1999). Molecular simulation of the electrochemical double layer. *Electro-*

*chimica Acta* **44**, 1697-1705.

Springer, T., Zawodzinski, T., Gottesfeld, S. (1991). Polymer electrolyte fuel cell model. *J. Electrochem. Soc.* **138**, 2334-2342.

Staroverov, V., Scuseria, G., Tao, J., Perdew, J. (2004). Tests of a ladder of density functionals for bulk solids and surfaces. *Phys. Rev. B* **69**, 075102.

Starz, K., Auer, E., Lehmann, T., Zuber, R. (1999). Characteristics of platinum-based electrocatalysts for mobile PEMFC applications. *J. Power Sources* **84**, 167-172.

Steele, B., Heinzel, A. (2002). Materials for fuel-cell technologies. *Nature* **414**, 345-352.

Steinberg, M., Takahashi, H., Ludewig, H., Powell, J. (1977). Linear accelerator fission product transmuter, Paper for American Nuclear Society meeting, New York, 17 pp.

Stipe, B., Rezaei, M., Ho, W., Gao, S., Persson, M., Lundqvist, B. (1997). Single-molecule dissociation by tunneling electrons. *Phys. Rev. Lett.* **78**, 4410-4413.

Stockie, J. (2003). Modeling hydrophobicity in a porous fuel cell electrode. Paper for "Computational Fuel Cell Dynamics II", Banff Int. Res. Center, USA.

Stockie, J., Promislow, K., Wetton, B. (2003). A finite volume method for multicomponent gas transport in a porous fuel cell electrode. *Int. J. Numerical Methods Fluids* **41**, 577-599.

Stroebel, D., Choquet, Y., Popot, J., Picot, D. (2003). An atypical haem in the cytochrome $b_6f$ complex. *Nature* **426**, 413-418.

Sudan, P., Wenger, P., Mauron, P., Gremaud, R., Züttel, A. (2004). Reversible properties of $LiBH_4$. In "Hydrogen Power – Theoretical and Engineering Solutions, Proc. Hypothesis V, Porto Conte 2003" (Marini, M., Spazzafumo, G., eds.), pp. 433-440. Servizi Grafici Editoriali, Padova.

Sullivan, M., Waterbury, J., Chisholm, S. (2003). Cyanophages infecting the oceanic cyanobacterium *Prochlorococcus*. *Nature*, **242**, 1047-1052.

Summers, W., Gorensek, M., Danko, E., Schultz, K., Richards, M., Brown, L. (2004) Analysis of Economic and Infrastructure Issues Associated with Hydrogen Production from Nuclear Energy. In Proc. 15th World Hydrogen Energy Conf., Yokohama. CD Rom, Hydrogen Energy Soc. Japan.

Superfoss (1981). En dansk industri. Danmarks Radio Skole-TV, Glostrup.

Suppes, G., Lopes, S., Chiu, C. (2004). Plug-in fuel cell hybrids as transition technology to hydrogen infrastructure. *Int. J. Hydrogen Energy* **29**, 369-374.

Svensson, F., Hasselrot, A., Moldanova, J. (2004). Reduced environmental impact by lowered cruise altitude for liquid hydrogen-fuelled aircraft. *Aerospace Sci. Tech.* **8**, 307-320.

Sveshnikov, D., Sveshnikova, N., Rao, K., Hall, D. (1997). Hydrogen metabolism of mutant forms of *Anabaena variabilis* in continuous cultures and under nutritional stress. *FEBS Lett.* **147**, 297-301.

Szczygiel, J., Szyja, B. (2004). Diffusion of hydrocarbons in the reforming catalyst: molecular modelling. *J. Molecular Graphics Modelling* **22**, 231-239.

Takahashi, K. (1998). Development of fuel cell electric vehicles. Paper presented at "Fuel Cell Technology Conference, London, September", IQPC Ltd., London.

Takeuchi, K., Fujioka, Y., Kawasaki, Y., Shirayama, Y. (1997). Impacts of high concentrations of $CO_2$ on marine organisms: a modification of $CO_2$ ocean sequestration, *Energy Conversion Management* **38**, S337-S341.

Takimoto, M. (2004). Development of fuel cell hybrid vehicles in Toyota. In Proc. 15th

World Hydrogen Energy Conf., Yokohama. 01PL-08, CD Rom, Hydrogen Energy Soc. Japan.

Tamagnini, P., Axelsson, R., Lindberg, P., Oxelfelt, F., Wünchiers, R., Lindblad, P. (2002). Hydrogenases and hydrogen metabolism of cyanobacteria. *Microbiol. Mol. Biol. Rev.* **66**, 1-20.

Tanimoto, K., Kojima, T., Yanagida, M., Nomura, K., Miyazaki, Y. (2004). Optimization of the electrolyte composition in a $(Li_{0.52}Na_{0.48})_{2-2x}AE_xCO_3$ (AE = Ca and Ba) molten carbonate fuel cell. *J. Power Souirces* **131**, 256-260.

Tatsumi, K., Tanaka, I., Inui, H., Tanaka, K., Yamaguchi, M., Adachi, H. (2001). Atomic structures and energetics of $LaNi_5$-H solid solution and hydrides. *Phys. Rev.* **B64**, #184105 (10 pp).

Tayhas, G., Palmore, R. (2004). Bioelectric power generation. *Trends Biotechnology* **22**, 99-100.

Taylor, J., Alderson, J., Kalyanam, K., Lyle, A., Phillips, L. (1986). Technical and economic assessment of methods for the storage of large quantities of hydrogen. *Int. J. Hydrogen Energy* **11**, 5-22.

Tchouvelev, A., Howard, G., Agranat, V. (2004). Comparison of standards requirements with CFD simulations for determining sizes of hazardous locations in hydrogen energy station. In "Proc. 15th World Hydrogen Energy Conf., Yokohama". CD Rom, Hydrogen Energy Soc. Japan.

Thomas, L. (1927). *Proc. Cambridge Phil. Soc.* **23**, 542.

Toevs, J., Bowman, C., Arthur, E., Heighway, E. (1994). Progress in accelerator driven trans-mutation technologies, In "Proc. 8th Journées Saturne, Saclay", pp. 22-28. Website: http://db.nea.fr/html/trw/docs/saturne8/ (accessed 1999).

Tokyo Gas Co. (2003). Press Release 16/10/03, Corporate Communications Dept.

Tokyo Gas Co. (2004). Completion of the plans by Tokyo Gas and the Railway Technical Res. Inst. for railway hydrogen station. http://www-tokyo-gas.co.jp

Tomonou, Y., Amao, Y. (2004). Effect of micellar species on photoinduced hydrogen production with Mg chlorophyll-*a* from *spirulina* and colloidal platinum. *Int. J. Hydrogen Energy* **29**, 159-162.

Toyota (1996). High-performance hydrogen–absorbing alloy. Website http://www.toyota.co.jp/e/november_96/electric_island/press.html.

Trebst, A. (1974). *Ann. Rev. Plant Physiol.* **25**, 423-447.

Tromp, T., Shia, R-L., Allen, M., Eiler, J., Young, Y. (2003). Potential environmental impact of a hydrogen economy on the stratosphere. *Science* **300**, 1740-1742.

Troshina, O., Serebryakova, L., Sheremetieva, M., Lindblad, P. (2002). Production of H2 by the unicellular cyanobacterium *Gloeocapsa alpicola* CALU 743 during fermentation. *Int. J. Hydrogen Energy* **27**, 1283-1289.

Tsubomura, H., Matsumura, M., Nomura, Y., Amamiya, T. (1976). Dye sensitised zinc oxide: aqueous electrolyte: platinum photocell. *Nature* **261**, 402-403

Tsuchiya, H., Inui, M., Fukuda, K. (2004). Penetration of fuel cell vehicles and hydrogen infrastructure. In Proc. 15th World Hydrogen Energy Conf., Yokohama. 30A-01, CD Rom, Hydrogen Energy Soc. Japan.

Tsuchiya, H., Kobayashi, O. (2004). Mass production cost of PEM fuel cell by learning curve. *Int. J. Hydrogen Energy* **29**, 985-990.

Tsygankov, A., Fedorov, A., Kosourov, S., Rao, K. (2002a). Hydrogen production by cyanobacteria in an automated outdoor photobioreactor under aerobic conditions. *Biotechnology Bioeng.* **80**, 777-783.

Tsygankov, A., Kosourov, S., Seibert, M., Ghirardi, M. (2002b). Hydrogen photoproduction under continuous illumination by sulphus-deprived, synchronous *Chlamydomonas reinhardtii* cultures. *Int. J. Hydrogen Energy* **27**, 1239-1244.

Tu, H., Stimming, U. (2004). Advances, aging mechanisms and lifetime in solid-oxide fuel cells. *J. Power Sources* **127**, 284-293.

Tzeng, G-H., Lin, C-W., Opricovic, S. (2004). Multi-criteria analysis of alternative-fuel buses for public transportation. *Energy Policy* (in print).

Ueno, Y., Otauka, S., Morimoto, M. (1996). Hydrogen production from industrial wastewater by anaerobic microflora in chemostat culture. *J. Ferment. Bioeng.* **82**, 194-197.

UK Treasury (2004). GDP deflators at market prices.In "Economic data & tools", http://www.hm-treasury.gov.uk/economic_data_and_tools/gdp_deflators.

UKDTI (2003). A fuel cell vision for the UK – the first steps. Taking the White Paper forward. UK Department of Trade and Industry, The Carbon Trust and EPSRC.

Um, S., Wang, C. (2004). Three-dimensional analysis of transport and electrochemical reactions in polymer electrolyte fuel cells. *J. Power Sources* **125**, 40-51.

UN (1996). "Populations 1996, 2015, 2050". United Nations Population Division and UNDP: available at the website http://www.undp.org/popin/wdtrends/pop/fpop.htm.

UN (1997). "UN urban and rural population estimates and projections as revised in 1994". United Nations Population Division and UNDP, Washington. Website: http://www.undp.org/popin/wdtrends/urban.html.

USDoE (2002a). National hydrogen energy roadmap. Towards a more secure and cleaner energy future for America. United States Department of Energy, Washington, DC.

USDoE (2002b). A technology roadmap for Generation IV Nuclear Energy Systems. US Department of Energy NERAC/GIF report, Washington, DC, weblocation: http://gif.inel.gov/roadmap/pdfs/gen_iv_roadmap.pdf.

USDoE (2003). International Energy Annual 2001. Energy Information Administration report DOE/EIA-0219(2001), US Department of Energy, Washington, DC.

USDoE (2004). Hydrogen posture plan. An integrated research, development and demonstration plan. United States Department of Energy, Washington, DC.

Vaillant (2004). Zukunft Brennstoffzellen. http://www.valiant.de.

Vankelecom, I. (2002). Polymer membranes in catalytic reactors. *Chem. Rev.* **102**, 3779-3810.

Venetsanos, A., Huld, T., Adams, P., Bartzis, J. (2003). Source, dispersion and combustion modelling of an accidental release of hydrogen in an urban environment. *J. Hazardous Mat.* **A105**, 1-25.

Verhelst, S., Sierens, R. (2003). Simulation of hydrogen combustion in spark-ignition engines. In "La planète hydrogène", Proc. 14th World Hydrogen Conf., Montréal 2002, CDROM published by CogniScience Publ., Montréal.

Veyo, S., Fukuda, S., Shockling, L., Lundberg, W. (2003). SOFC fuel cell systems. In "Handbook of Fuel Cells", Vol. 4 (Vielstich, W., Lamm, A., Gasteiger, H., eds.), Ch. 93. Wiley, Chichester.

Volbeda, A., Charon, M., Piras, C., Hastchikian, E., Frey, M., Fonticilla-Camps, J. (1995). Crystal structure of the nickel-iron hydrogenase from *Desulfovibrio gigas*.

*Nature* **373**, 580-585.

Vosko, S., Wilk, L., Nusair, M. (1980). Accurate spin-dependent electron liquid correlation energies for local spin density calculations: a critical analysis. *Canadian J. Phys.* **58**, 1200.

VW (2002). Environmental Report 2001/2002: Mobility and sustainability; Schweimer, G., Levin, M. (2001). Life cycle inventory for the Golf A4 (internal report), Volkswagen AG.

VW (2003). Lupo 3 litre TDI, Technical Data, Volkswagen AG, Wolfsburg.

Wagner, U., Geiger, B., Schaefer, H. (1998). Energy life cycle analysis of hydrogen systems. *Int. J. Hydrogen Energy* **23**, 1-6.

Wang, C-Y. (2003). Two-phase flow and transport. In "Handbook of Fuel Cells", Vol. 3 (Vielstich, W., Lamm, A., Gasteiger, H., eds.), Ch. 29. Wiley, Chichester.

Wang, L., Murta, K., Inaba, M. (2004). Development of novel highly active and sulphur-tolerant catalysts for steam reforming of liquid hydrocarbons to produce hydrogen. *Appl. Catalysis A: General* **257**, 443-47.

Wang, M. (2002). Fuel choices for fuel-cell vehicles: well-to-wheels energy and emission impacts. *J. Power Sources* **112**, 307-321.

Wang, X., Gorte, R. (2002). A study of steam reforming of hydrocarbon fuels on Pd/ceria. *Appl. Catalysis A: General* **224**, 209-218.

Wang, Y., Balbuena, P. (2004). Roles of proton and electric field in the electroreduction of $O_2$ on Pt(111) surfaces: results of an ab-initio molecular dynamics study. *J. Chem. Phys.* **B108**, 4376-4384.

Wang, Z., Wang, C., Chen, K. (2001). Two-phase flow and transport in the air cathode of proton exchange membrane fuel cells. *J. Power Sources* **94**, 40-50.

Weber, A., Ivers-Tiffée, E. (2004). Materials and concepts for solid oxide fuel cells (SOFCs) in stationary and mobile applications. *J. Power Sources* **127**, 273-283.

Weht, R., Kohanoff, J., Estrin, D., Chakravarty, C. (1998). An *ab initio* path integral Monte Carlo simulation method for molecules and clusters: application to $Li_4$ and $Li_5^+$. *J. Chem. Phys.* **108**, 8848-8858.

Weiss, M., Heywood, J., Drake, E., Schafer, A., AuYeung, F. (2000). On the road 2020. Report MIT EL 00-003, Laboratory for Energy and Environment, Massachussetts Institute of Technology, Cambridge, MA.

Weiss, M., Heywood, J., Schafer, A., Natarajan, V. (2003). Comparative assessment of fuel cell cars. Report LFEE 2003-001 RP, Massachussetts Institute of Technology, Cambridge, MA.

Wenk, S., Qian, D., Wakayame, T., Nakamura, C., Zorin, N., Rögner, M., Miyake, J. (2002). Biomolecular device for photoinduced hydrogen production. *Int. J. Hydrogen Energy* **27**, 1489-1493.

Weydahl, T., Gruber, A., Gran, I., Ertesvåg, I. (2003). Mathematical modelling and numerical simulations of different diffusion effects in hydrogen-rich turbulent combustion. In "La planète hydrogène", Proc. 14th World Hydrogen Conf., Montréal 2002, CDROM published by CogniScience Publ., Montréal, 9 pp.

Wieland, S., Melin, T., Lamm, A. (2002). Membrane reactors for hydrogen production. *Chem. Eng. Sci.* **57**, 1571-1576.

Wilkinson, D., Vanderleeden, O. (2003). Serpentine flow design. In "Handbook of Fuel Cells", Vol. 3 (Vielstich, W., Lamm, A., Gasteiger, H., eds.), Ch. 27. Wiley, Chichester.

Williams, K., Eklund, P. (2000). Monte Carlo simulations of $H_2$ psysisorption in finite-diameter carbon nanotube ropes. *Chem. Phys. Lett.* **320**, 352-358.

Wise, D. (1981). *Solar Energy* **27**, 159-178.

Wolfbauer, G. (1999). The electrochemistry of dye sensitized solar cells, their sensitizers and their redox shuttles. Ph. D. Thesis, Monash University, Melbourne.

Woo, Y., Oh, S., Kang, Y., Jung, B. (2003). Synthesis and characterization of sulphonated polyimide membranes for direct methanol fuel cell. *J. Membrane Science* **220**, 31.45.

Woodward, J., Orr, M., Cordray, K., Greenbaum, E. (2000). Enzymatic production of biohydrogen. *Nature* **405**, 1014-1015.

World Energy Council (1995). "Survey of Energy Resources" 17th ed. World Energy Conference, London.

Wurster, R. (1997a). PEM fuel cells in stationary and mobile applications. Paper for Biel Conference (http://www.hyweb.de/knowledge).

Wurster, R. (1997b). Wasserstoff-Forschungs- und Demonstrations-Projekte. Kryotechnik 26. Feb. 1997, VDI-Tagung (http://www.hyweb.de/knowledge).

Wurster, R. (1998). Paper presented at Deutsche Kälte-Klima-Tagung, Würzburg.

Wurster, R. (2003). GM well-to-wheel-Studie – Ergebnisse and Schlüsse. LB Systemtechnik website http://www.HyWeb.de.

Wurster, R. (2004). Daily use of hydrogen in road vehicles and their refueling infrastructure: safety, codes and regulation. In "Proc. HYFORUM: Clean Energies for the 21st Century", Beijing. Available at http://www.lbst.de.

Yagishita, T., Sawayama, S., Tsukahara, K., Ogi, T. (1996). Photosynthetic bio-fuel cells using cyanobacteria, In "Renewable Energy, Energy Efficiency and the Environment: World Renewable Energy Congress", vol. II, pp. 958-961. Pergamon, Elmsford, NJ.

Yamada, Y., Matsuki, N., Ohmori, T., Mametsuka, H., Kondo, M., Matsuda, A., Suzuki, E. (2003). One chip photovoltaic water electrolysis device. *Int. J. Hydrogen Energy* **28**, 1167-1169.

Yamaguchi, M., Horiguchi, M., Nakanori, T., Shinohara, T., Nagayama, K., Yasuda, J. (2001). Development of large-scale water electrolyzer using solid polymer electrolyte in WE-NET, In "Hydrogen Energy Progress XIII", Vol. 1 ("Proc. 13th World Hydrogen Energy Conf., Beijing 2000"; Mao and Veziroglu, eds.), pp. 274-281. Int. Assoc. Hydrogen Energy & China Int. Conf. Center for Science and Technology, Beijing.

Yang, C., Srinivasan, S., Bocarsly, A., Tulyani, S., Benziger, J. (2004). A comparison of physical properties and fuel cell performance of Nafion and zirconium phosphate/Nafion composite membranes. *J. Membrane Sci.* **237**, 145-161.

Yasuda, I., Shirasaki, Y., Tsuneki, T., Asakura, T., Kataoka, A., Shinkai, H., Yamaguchi, R. (2004). Development of membrane reformer for high-efficient hydrogen production from natural gas. In "15th World Hydrogen Energy Conference, Yokohama 2004". Hydrogen Energy Systems Soc. of Japan (CDROM).

Yinxing, S. (1637). *High skills in materials production* (Tian gong kai wu). China.

Yokata, O., Oku, Y., Sano, T., Hasegawa, N., Matsunami, J., Tsuji, M., Tamura, Y. (2000). Stoichiometric consideration of steam reforming of methane on Ni/Al2O3 catalyst at 650°C by using a solar furnace simulator. *Int. J. Hydrogen Energy* **25**, 81-86.

Yu, L., Liu, H. (2002). A two-phase flow and transport model for the cathode of PEM fuel cells. *Int. J. Heat Mass Transfer* **45**, 2277-2287.

Yu, L., Wuye, D., Xianchen, C., Bin, M. (2001). Aerospike engine and single-stage-to-orbit return transportation. In "Hydrogen Energy Progress XIII, Proc. 13th World Energy Conf., Beijing 2000" (Mao, Z., Veziroglu, T., eds.), pp. 654-663. Int. Assoc. Hydrogen Energy, Beijing.

Zaluska, A., Zaluski, L., Ström-Olsen, J. (2001). Structure, catalysis and atomic reactions on the nano-scale: a systematic approach to metal hydrodes for hydrogen storage. *Appl. Phys.* **A72**, 157-165.

Zapp, P. (1996). Environmental analysis of solid oxide fuel cells. *J. Power Sources* **61**, 259-262.

Zehr, J., Waterbury, J., Turner, P., Montoya, P., Omoregle, E., Steward, G., Hansen, A., Karl, D. (2001). Unicellular cyanobacteria fix $N_2$ in the subtropical North Pacific Ocean. *Nature* **412**, 635-638.

ZFK (1999). Hydrogen research at Forschungszentrum Karlsruhe. In "Proc. EIHP Workshop on Dissemination of Goals, Preliminary Results and Validation of Methodology", European Commission DG XII, acsessed 2003 at the website http://www.eihp.org/eihp1/workshop/intro.html.

Zhang, Q., He, D., Li, J., Xu, B., Liang, Y., Zhu, Q. (2002). Comparatively high yield methanol production from gas phase partial oxidation of methane. *Appl. Catalysis A: General* **224**, 201-207.

Zhang, T., Chu, W., Gao, K., Qiao, L. (2003). Study of correlation between hydrogen-induced stress and hydrogen embrittlement. *Materials Sci. Eng.* **A147**, 291-299.

Zhang, T., Liu, H., Fang, H. (2003). Biohydrogen production from starch in wastewater under thermophilic condision. *J. Environm. Managem.* **69**, 149-156.

Zhang, Y., Suenaga, K., Colliex, C., Iijima, S. (1998). Coaxial nanocables: silicon carbide and silicon oxide sheathed with boron nitride and carbon. *Science* **281**, 973–975.

Zhou, L., Zhou, Y., Sun, Y. (2004). A comparative study of hydrogen adsorption on superactivated carbon versus carbon nanotubes. *Int. J. Hydrogen Energy* **29**, 475-479.

Zhu, Q., Li, J., Wei, J. (2001). Production and utilization of hydrogen in China. In "Hydrogen Energy Progress XIII, Proc. 13th World Energy Conf., Beijing 2000" (Mao, Z., Veziroglu, T., eds.), pp. 105-109. Int. Assoc. Hydrogen Energy, Beijing.

Zhu, Y., Ha, S., Masel, R. (2004). High power density direct formic acid fuel cells. *J. Power Sources* **130**, 8-14.

Zittel, W., Wurster, R. (1996). "Hydrogen in the Energy Sector". Ludwig–Bölkow–ST Report: http://www.hyweb.de/knowledge/w–i–energiew–eng.

Zouni, A., Witt, H., Kern, J., Fromme, P., Krauss, N., Saenger, W., Orth, P. (2001). Crystal structure of photosystem II from *Synechoccocus elongatus* at 3.8Å resolution. *Nature* **409**, 739-743.

Züttel, A. (2004). Hydrogen storage methods. *Naturwissenschaften* **91**, 157-172.

Züttel, A., Rentsch, S., Fischer, P., Wenger, P., Sudan, P., Mauron, P., Emmenegger, C. (2003). *J. Alloys Compounds* **356**, 515.

Züttel, A., Wenger, P., Sudan, P., Mauron, P., Orimo, S. (2004). Hydrogen density in nanostructured carbon, matals and complex materials. *Materials Sci. Eng.* **B108**, 9-18.

# SUBJECT INDEX